Innovative Research in Life Sciences

Innovative Research in Life Sciences

Pathways to Scientific Impact, Public Health Improvement, and Economic Progress

E. Andrew Balas, MD, PhD

Augusta University
Augusta, GA, USA

Lessons from award winning scientists and leading research universities

Registered Office
John Wiley & Sons, Inc., 111 River Street, Hoboken, NJ 07030, USA

Editorial Office
111 River Street, Hoboken, NJ 07030, USA

For details of our global editorial offices, customer services, and more information about Wiley products visit us at www.wiley.com.

Wiley also publishes its books in a variety of electronic formats and by print-on-demand. Some content that appears in standard print versions of this book may not be available in other formats.

Library of Congress Cataloging-in-Publication Data

Names: Balas, E. A. (E. Andrew), author.
Title: Innovative research in life sciences : pathways to scientific impact, public health improvement, and economic progress / E. Andrew Balas.
Description: First edition. | Hoboken, NJ : Wiley, 2019. | Includes bibliographical references and index. |
Identifiers: LCCN 2018025831 (print) | LCCN 2018026695 (ebook) | ISBN 9781119225874 (Adobe PDF) | ISBN 9781119225881 (ePub) | ISBN 9781119225867 (hbk)
Subjects: | MESH: Public Health | Research | Entrepreneurship | Economic Development | Inventions
Classification: LCC RA425 (ebook) | LCC RA425 (print) | NLM WA 20.5 | DDC 362.1–dc23
LC record available at https://lccn.loc.gov/2018025831

Cover Design: Wiley
Cover Images: © procurator/Getty Images; © Vijay Patel/Getty Images

Set in 10/12pt Warnock by SPi Global, Pondicherry, India

Printed in the United States of America

V10005671_110118

Contents

Preface

It is right to call things after the ends they realize.

Aristotle (350 BCE)[1]

To succeed with focus, speed, and efficiency, researchers need a good understanding of how science works, the big picture of outcomes, and how results can be produced. There is a pressing need to look beyond publishing scientific papers, understand what comes afterward, and see how science can change people's lives. Often, the devil is in the big picture, not in the details. With a broader view, the fundamental concept of innovation can add value to science and the work of research laboratories.

Like in most other endeavors, research can be most effective when the targets are in clear view. This book should help the beginning researcher in choosing areas of exploration and boosting productivity. It should also help the midcareer scientists to get a broader understanding of what is going well and what needs to be added to enhance chances of success. Senior scientists should also get a valuable resource to enhance mentoring of junior researchers.

In this scholarly undertaking, three-dimensional life sciences innovation is defined as the creation of new value-producing resources or endowing existing resources with enhanced potential for creating societal benefits: scientific contributions, improvement in public health, and economic development. This approach is essentially an adaptation and generalization of the definition of innovation by Peter Drucker in research. Not every life sciences innovation produces equal value in all listed dimensions, but the most successful research innovations have an impressive presence in the above-described three-dimensional space.

Basic research and curiosity-driven studies have always been and will always remain at the center of scientific progress. On the other hand, there is also a great opportunity for more synergism between basic research and innovation. For obvious reasons, applied and clinical research have a closer, more obvious connection with innovation.

---

1 Aristotle *De Anima* 2.4. Written 350 BCE. Translated by J.A. Smith. *The Internet Classics Archive.*

Distorted trivializations of the concept of research innovation should never deter anyone from discussing the full range of human creativity. In other words, the focus on research innovation should not be interpreted as an attempt to turn every researcher into a business entrepreneur and foolishly measure success by the number of patents. It should also not be portrayed as an effort to put direct health-care application as the sole yardstick of meaningful research.

This book is about breakthrough ideas, serial innovators, and award-winning scientists: the intellectually rich and the scientifically famous. By studying hundreds of award-winning scientists, serial innovators, and also research universities, it is focused on how meritorious ideas are born and become landmark scientific discoveries. The primary focus of this book is what happens before the publication is submitted or the innovation becomes protected. This phase is the most neglected but most exciting in the process of scientific creativity and innovation. Particularly, twelve competencies of innovative biomedical researchers have been identified for detailed analysis and description.

The methodology of this book is focused on a large variety of statistical databases and a vast number of stories about individual discoveries. The focus is on the birth of great ideas and their evolution to become practically significant scientific discovery and in many cases valuable intellectual property. Each chapter includes generous bibliographic lists to substantiate conclusions and recommendations.

After highlighting overarching concepts and significant challenges, this compilation is about some overarching competencies and driving principles of successful researchers in life sciences. Particularly, stories have been carefully compared to see when the unique becomes similar and the similarity becomes a noticeable trend.

Based on extensive research, a large variety of methodologies have been applied, and their results were integrated.

Research on Research In many ways, this book adopts the historical analysis method often used in diplomacy and military science with particular attention to the more recent scientific discoveries. The focus is not on the timeline of events but either on typical situations and how they have been handled efficiently or on exemplary skills that turned out to be very useful in a variety of situations. The case vignettes of stories should add a further sense of reality to otherwise abstract and principled discussions. The life and work of award-winning scientists should illustrate the determinants of research and inspire to achieve what appeared to be beyond reach in the past.

Synthesis of Discovery Processes This retrospective study drew upon the rapidly expanding and excellent archival sources related to Nobel Prizes and other major scientific awards (Lasker Award, Japan Prize, Queen Elizabeth Medal, and others). Information about prize motivations, winners, and published interviews with winners helped to explore the commonalities of the process behind significant scientific and public health accomplishments. The analysis also covered the scientific, public health, and economic impact after publication of the discovery.

Review of Serial Innovators   In this book, serial innovators have been identified as those who have developed five or more widely used new products or healthcare services. The literature was reviewed, and also nominations were solicited from colleagues. The resulting pool of serial inventors collectively provides a range of perspectives on the process of innovation. The work of these innovators is analyzed with the help of their publications, newspaper interviews, and the growing number of video-recorded statements.

Scientific Root Analysis   Landmark life sciences discoveries have been explored by starting from the final public health impact and backtracking to the initial identification of the problem through the scientific process (e.g. fluoridation of water as a public health achievement in the prevention of dental caries). The methodology was applied to a broad range of public health achievements (e.g. improvement in transportation safety, dramatically increased life expectancy of cystic fibrosis patients, improvement in lymphoma survival, and others).

Innovation Performance Analysis   Among the primary sources, a large number of pertinent national databases have been used to connect university research with innovation, examine characteristics of performing and nonperforming intellectual property, and correlate these features with available literature. The research and innovation data sources include NLM PubMed, NSF Science and Engineering Indicators, AUTM Statistics, Public Access to Court Electronic Records (PACER), United States Patent and Trademark Office, and many others. To the extent possible, international statistical data are also provided for comparison.

Litigation Review   When innovators and institutions go to their lawyers and the court, it not only signals major conflict but also defines the boundaries of innovation and inventor recognition. Correspondingly, a review of court cases focused on the undesirable conflicts between innovators and universities/ research institutions to explore the opportunities for prevention and development of more productive innovation culture. In this study, a significant number of court cases have been collected from recent decades for analysis.

Expert Networks   This project has immensely benefited from the bouncing of ideas and debating various interpretations with academic colleagues nationwide and internationally. Ad hoc discussions with leaders of research laboratories helped to identify pointers to innovative work. Hallways of national conferences, Skype and telephone discussions, interactions with industry leaders, and copious email correspondence have all been very useful. Every major observation had to be discussed with others to hear different views and understand possible interpretations.

Development of the reference lists extensively relied on original sources, including research studies, databases, digital research libraries, published archives, and expert knowledge:

- In processing and synthesizing, information from primary sources served as the backbone of this study and its conclusions. These sources include the peer-reviewed *original research reports* and *national statistical reports*.
- Information from secondary sources has been used to further elaborate and interpret ideas and themes of the primary sources. These sources include recorded interviews with researchers of major discoveries, peer-reviewed research syntheses, and philosophy of science publications.
- Tertiary sources – like web searches, newspaper articles, or conversations with scientists – were helpful by serving as pointers and extensively used to locate primary sources information.

The only criterion for our selection of research studies was the potential for protecting and improving individual and community health. Therefore, we did not exclude scientifically sound transportation safety innovations or organizational effectiveness studies that showed their value in protecting health and improving health outcomes. We also did not exclude relevant physics theories, commercialization principles, military history, or opioid addiction studies just because they might be politically controversial. In reviewing scientific studies, we were ready to consider anything and everything that has relevance to understanding life, inner workings of science, practical impact, and human health outcomes.

To support the interpretation of facts and concepts, this book sends the most important messages of each chapter in several ways, including narrative discussion, conceptual graphic models, quotations from renowned scientists, case vignettes of award-winning researchers, and others. This book is designed to send the consistent messages through several channels.

The development of this book received immense support from eminent researchers and thought leaders. Particularly, the following chapter reviewers should be recognized for their outstanding intellectual contributions: Elena Andresen (Oregon Health & Science University), Howard Bleich (Harvard University), Fran Butterfoss (Old Dominion University), Scott Evans (University of Utah), David Fleming (University of Missouri), Jean-Paul Gagnon (University of North Carolina), Steve Gnatz (Loyola University Chicago), Miklos Gratzl (Case Western Reserve University), Susan Fagan (University of Georgia), Joe Kornegay (Texas A&M University), Laura Magaña (Association of Schools and

Programs of Public Health), Dan Masys (University of Washington), Farah Magrabi (University of New South Wales), Zoltan Néda (Babes-Bolyai University), Gerry Pepe (Eastern Virginia Medical School), and Benny Zeevi (DFJ Tel Aviv Venture Partners). Special recognition should go to the PhD students of my biomedical research innovation laboratory, particularly Marlo Vernon and Nadine Mansour. Finally, I want to thank our artist, Elvira Bojadzic in Sarajevo (BiH), for the fine illustrations of scientific concepts.

The biggest thank you goes to my family. My parents gave us lasting values and inspired my scientific endeavors. My predecessors fought against oppression of independent thinking and for the recognition of minority thought. Special appreciation goes to my wife for the loving support in the midst of many challenges and to my children and grandchildren who represent the inspiring future.

In many ways, innovation is the key when looking at the long-term impact of research as opposed to seeing only the short-term results of publications or impact factors. Reading this book should give insight how the most innovative scientists launch projects, focus on promising research opportunities, and elevate research to unprecedented heights. Effective improvement of the research process calls for a total focus on the outcomes and long-term vision of science.

Albatross can be a remarkable analogy in going for the ultimate outcomes of research and charting the most efficient path to achieving them. In the game of golf, albatross is a very rare but far-reaching shot that is three under par. According to National Geographic, the bird albatross may efficiently glide hundreds of miles without flapping and resolutely fly more than 10 000 miles to deliver a meal to its youngster. In developing your research agenda, think about the long-term outcomes.

E. Andrew Balas

Part One

Outcomes of Research

1

Pathways of the Research Innovator

Scientists must have the courage to attack the great unsolved problems of their time.

Otto Warburg (1964)[1]

In recent decades, there has been remarkable progress in advancing life sciences. Discoveries are pouring out of laboratories and research universities all over the world. Science is bringing us closer to realizing the dream of understanding, treating, and preventing major diseases and opening up new, unprecedented economic development opportunities. We live in the exponential times of life sciences: not just the number of discoveries is growing, but also the benefits to people and society are multiplying.

In general terms, research has not done a great job in defining its end product. Better understanding how scientific ideas, life-changing practices, or technologies are generated should help to see the trends of success and learn from the inspiring stories. By choosing areas of interest, researchers make decisions that shape futures and change lives. Ultimately, research should become better targeted, more accomplished, and effective more rapidly.

Scientific research is known for leading to peer-reviewed, replicable, and generalizable knowledge. The dopamine neurotransmission model in the brain discovered many years ago will also work next year. It can be used to treat many patients with comparable effects anywhere in the world. The new model of physiologic function can be confirmed by other researchers. Peer review means disclosing methodology and findings to be evaluated by experts not affiliated with the study.

Eugene Wigner's (1960) article on the unreasonable effectiveness of mathematics in the natural sciences elegantly describes the essence of reproducibility

---

1 Warburg, O. (1964). Prefatory chapter. *Annual Review of Biochemistry*, 33 (1): 1–15.

Innovative Research in Life Sciences: Pathways to Scientific Impact, Public Health Improvement, and Economic Progress, First Edition. E. Andrew Balas.
© 2019 John Wiley & Sons, Inc. Published 2019 by John Wiley & Sons, Inc.

and generalizability in science. As Erwin Schrodinger (1932) noted earlier, certain regularities in the events could be discovered in spite of the perplexing complexity of the world. Wigner pointed out that the laws of nature are concerned with such regularities. There is also a "succession of layers of 'laws of nature', each layer containing more general and more encompassing laws than the previous one and its discovery represents a deeper penetration into the structure of the universe than the layers recognized before." Wigner also highlighted the generalizability of the laws of nature: "it is true not only in Pisa, and in Galileo's time, it is true everywhere on the Earth, was always true, and will always be true."

A better understanding of long-term outcomes should make research more streamlined and dissemination of discoveries more effective. When producing peer-reviewed, replicable, and generalizable results, researchers always make important disclosure decisions, either knowingly or not. Examining various choices and their practical implications should improve understanding of consequential scientific discoveries, support researchers striving to innovate, and facilitate the development of more useful research infrastructures. This chapter clarifies concepts, defines terminologies, and introduces a model framework for biomedical research innovation.

Diverse Outcomes of Science

Among the many possible outcomes, models of important relationships, laws of nature, represent a crucial stepping stone in the progress of science. Some of them are complex, while others are simple relationships. When widely published, greater understanding and new models can not only change the usual course of health care but also serve as launch pads for further successful research.

The research concepts of better understanding, new knowledge, and penetrating insight can often be captured by scientific models: verbal, graphic, physical, or mathematical representations of an important feature of the world. The double helix of the DNA, the causative role of *Helicobacter pylori* in gastric ulcer, and rituximab-mediated immune destruction of lymphomas are all examples of models abundantly validated by subsequent studies and patient experiences.

The creative process in academia is called research or scholarly activity. When productive, the creative process leads to results that have great theoretical significance and practical value. Innovative biomedical research has repeatedly proved its value by finding cures for major diseases, improving public health, and generating economic prosperity.

In most academic institutions, the *peer-reviewed* research article and competitive extramural research funding have become the gold standard in expectations and most common pathways of delivering scholarly productivity results (Anderson et al. 2013). Most academic institutions require a certain quantity and impact factor of peer-reviewed research articles. It is noteworthy that the health sciences area is unique in its singular focus on peer-reviewed articles (Anderson et al. 2013; Gelmon et al. 2013; Smesny et al. 2007).

With advances in applied life sciences over many decades, there have also been growing numbers of biomedical innovations – not only to improve human life but also contribute to economic development. Major categories of results generated by biomedical research innovation include (i) products, (ii) services, or (iii) practice recommendations (i.e. guidelines, processes, systems, and organizational structures).

Of the top 10 Achievements in Public Health from 2001 to 2010 identified by the CDC, a decline in vaccine preventable diseases was among the most spectacular scientific achievements (Centers for Disease Control and Prevention 2011). Two vaccine products, in particular, were singled out: the pneumococcal conjugate vaccine and the rotavirus vaccine. An estimated 211 000 serious pneumococcal infections and 13 000 deaths were prevented during 2000–2008 after the pneumococcal conjugate vaccine was introduced (Pilishvili et al. 2010). Similarly, vaccinations for the rotavirus now prevent an estimated 40 000–60 000 hospitalizations each year according to 2011 statistics (Centers for Disease Control and Prevention 2009; Tate and Parashar 2011; Yen et al. 2011). Rotavirus and pneumococcal vaccines also resulted in practice recommendations by the CDC to include these products in the regular schedule of vaccinations for infants and children.

The top 10 Achievements in Public Health also include successful breast, cervical, and colorectal cancer screening services. Particularly, colorectal cancer death rates decreased from 25.6 per 100 000 population to 20.0 for men and from 18.0 per 100 000 to 14.2 for women between 1998 and 2007 (Kohler et al. 2011).

Working in Switzerland, Andreas Grüntzig developed the first balloon angioplasty and successfully used it in patient care in 1977. This product and practice recommendation have saved numerous lives and made them more comfortable. Further refinement included the addition of a heart stent product, left behind after the procedure. The resulting nonsurgical service is used in multiple ways, allowing for devices and drugs to be utilized directly (Gruentzig 1982; Holmes et al. 1984).

According to the classic definition, innovation is the design, invention, development, and implementation of new or altered products, services, processes, systems, or organizational models to create new value for customers and financial returns (Schramm et al. 2008). Removing barriers to the development of innovative biomedical research has the potential to affect millions of people by finding solutions to major global public health concerns.

Successful biomedical research innovation cannot be equated with business success. Many new initiatives highlight the need for much more innovation in areas where business success is limited or nonexistent. For example, there is a great need for innovation in the treatment of rare and esoteric diseases as highlighted by the NIH Office of Rare Diseases Research Bench-to-Bedside (B2B) Awards and the FDA Office of Orphan Products Development. These programs seek to advance the evaluation and development of products for the diagnosis and treatment of often overlooked rare diseases. Increasingly, public–private partnerships are recommended for the development of noncommercial innovations (Nwaka and Ridley 2003).

There have been many commercial successes that later turned out to be health outcome failures. For example, a major maker of pomegranate juice made sweeping claims, citing university studies and researchers, that its juice reduced the rate of heart disease, prostate cancer, and erectile dysfunction. In 2012, the company received a cease-and-desist order after FTC determination that there was insufficient evidence to support claims. This order will remain in effect for 20 years unless they present at least two well-controlled randomized clinical trials substantiating their claims.

Best of Both Worlds: Scientific and Innovative

Innovation is often defined by the common criteria of being novel, non-obvious, and useful. Unlike naturally occurring DNA, practically valuable synthesized sequences can meet innovation criteria and can be protected as intellectual property accordingly (Golden and Sage 2013).

There is an apparent synergism between biomedical research and beneficial innovation. Society has no apparent benefit from research results that are not novel, not obvious, or useless by failing to benefit further research or the practice of health care. The criteria for innovation represent a more subjective or judgmental assessment. Nevertheless, they capture what is needed to benefit society.

The best discoveries of applied sciences are not only *reproducible* and *generalizable* but also *novel*, *non-obvious*, and *useful*. Scientific research leads to replicable and generalizable knowledge, but it is also expected to be novel, non-obvious, and useful.

In other words, the best applied scientific results not only meet but also significantly exceed innovation requirements by offering broadly usable and trustworthy solutions for a new product or service design (Balas and Elkin 2013). In the infrequent case of commercialization, a third set of sustainability considerations is added, including market demand, business model, and environmental impact.

Meanwhile, research interest in reviews of patented innovations has also increased in the scientific community. There are a growing number of articles that review new technologies based on published patents, among other sources (Freschi et al. 2012; Horstkotte and Odoerfer 2012; Talevi et al. 2014; Telang et al. 2012). Patent reviews assist researchers interested in innovation because they identify available and unexplored technologies and highlight opportunities for new directions.

Opportunities That Are Not Just Timely But Also Timeless

The usual assumption that biomedical research accidentally bumps into meaningful discovery or disclosable IP may be intermittently true but is probably more often misinforming. Particularly, the enormous publication and patenting productivity of serial innovators challenge this usual assumption. Well-planned studies have always been viewed as best chances of good results. Therefore, a researcher needs to recognize not only when to start a scientific investigation in a particular area but also when to stop it and switch to a more promising field.

In recent decades, the complex and often controversial relationship between university research and innovation has been gradually highlighted. Better and earlier understanding of the kind of health sciences research that leads to impactful evolution in future research and public health is essential for effective research innovation. It is well known that the overwhelming majority of patents are nonperforming, never licensed, or utilized (Ledford 2013). Therefore, it is a vital interest to identify factors that lead to well-performing IP.

Like any other organized human endeavor, research needs to set targets to guide activities. Target selection is typically influenced by results from other diverse scientific areas. Traditionally, the targets are expressed as research objectives, hypotheses, and questions. It is reasonable to assume that the outcomes and products of research are going to play an increasingly important role in targeting research.

Target selection often starts with the development of a model based on already available data. For example, the discovery of the role of papillomavirus in cervical cancer by Harald zur Hausen was largely triggered by the epidemiologic studies showing the relationship between viral infections and cancers. In his declaration of war on cervical cancer, he wrote that "The condyloma (genital wart) agent has been entirely neglected thus far in all epidemiological and serological studies relating not only to cervical and penile, but also to vulvar and perianal carcinoma. This is particularly unusual in view of the localization of genital warts, their mode of venereal transmission, the number of reports on malignant transition, and the presence of an agent belonging to a well characterized group of oncogenic DNA viruses" (zur Hausen 1976). In other words, the hypothetical model of infectious origin became the target locator and ultimately the Nobel Prize-winning result of his research.

Concepts of forceful *research targeting* harmoniously coexist with accidental discoveries. The most frequently cited accidental classic is the dirty dish with staphylococci in Alexander Fleming's laboratory that led to the discovery of penicillin. "I certainly didn't plan to revolutionize all medicine by discovering the world's first antibiotic" – he stated later. The most newsworthy, contemporary example of serendipity is Pfizer's failed angina drug study that led to the discovery of Viagra.

One of the most practical discoveries in injury prevention also did not come from problem-oriented bioengineering research based on targeted technical specifications of the societal need, but from an accidental discovery. While working as a research associate for DuPont in 1964, Stephanie Kwolek was looking for a lightweight but also strong fiber to be used in tires. The original target was never fully achieved, but during the research, she instead discovered Kevlar, which is five times stronger than steel by weight. Today, Kevlar is widely used in combat helmets, ballistic vests, protective gloves, tennis rackets, racing boats, and many other areas.

The innumerable lessons of targeted and accidental discoveries tell us that we need to develop a better understanding of selecting and deselecting research targets based on good models and the chances of successful innovation benefiting society. When accidental discoveries come up, as they often do, the primary responsibility of researchers is recognizing them and fully developing their potential.

Balancing Research and Innovation

Most appropriately, discussions about scientific innovation should refer to the full range of scholarly creativity, including new models, research methodologies, peer-reviewed publications, IP disclosures, and tech transfer products. Congruently, the terms productivity, efficiency, and quality improvement should consider the full range of scientific innovation without overemphasizing one particular line of activity. The prevalent single-line evaluations, for example, counting only publications or patents, tend to misguide scholarly creativity and appear to be negligibly useful in promoting actual scientific progress.

The NIH Roadmap for Medical Research was launched in September 2004 to address basic steps of translation: basic science research translated to humans (T1 translation) and secondarily translated into clinical practice (T2 translation). The Clinical and Translational Science Awards (CTSA) program was designed to support diverse research teams working in collaboration toward a common goal (Blumberg et al. 2012). However, a review by the Institute of Medicine concluded that the lack of transparency in reporting and also lack of high-level common metrics are significant barriers to overall program accountability (Leshner et al. 2013).

For obvious reasons, the actual progress of science cannot be measured by the number of peer-reviewed research publications or successfully filed patents in any particular field. For example, there were 26 273 human subject studies on low back pain indexed in the PubMed database according to recent searches (February 2015); among them 2779 human subject publication type randomized clinical trials mention low back pain; there were 990 studies found for low back pain in http://clinicaltrials.gov; and there were 2480 patents in the USPTO US Patent Collection Database (www.uspto.gov). In spite of this vast amount of research and development, the treatment of low back pain is far from being fully resolved and continues to need creative prevention and new interventions.

Nevertheless, the number of peer-reviewed scientific publications does give some level of information about the scientific productivity of academic institutions. In the biomedical field, counting PubMed indexed publications may be a reasonable approximation if applied in a much larger set of indicators. Citations of research publications may provide further insight and indeed are used in the evaluation of individual researchers.

National statistics also highlight that innovation success is not a simple correlate of research expenditures: greater spending on research does not equal greater innovative results. The Association of University Technology Managers (AUTM) data suggest that at some universities $20 million research spending leads to a new startup company while at other universities it may take $200 million of research funding to launch a startup company (The Science Coalition 2013). Defining and harnessing the differences between these efforts is of great importance to funders and institutions alike.

The Carnegie Foundation for the Advancement of Teaching classified 207 universities as "very high research activity" or as "high research activity" in the United States (Carnegie Foundation for the Advancement of Teaching 2010). Out of this group, 187 institutions respond to the Association of University Technology Managers Licensing Activity Survey.

According to the AUTM survey, about half of all cumulative active licenses come from 18 universities (Balas and Elkin, 2013). Each of these universities produced an average of 1007 active licenses, creating a "Monument Valley" of high-performing institutions towering over less productive efforts. The remaining 134 universities produce an average of 140 cumulative active licenses (15 universities did not provide data).

An analysis of the 2013 AUTM Licensing Survey, a review of World of Science indexed publications from 2013, and the Integrated Postsecondary Education Data System (IPEDS) 2012–2013 report found that per institution averages (±SEM) were as follows: instructional and research faculty, 2099 ± 164; research expenditure, $362M ± $45M; publications, 3239 ± 368; IP disclosures, 133.2 ± 14; patent applications, 83.3 ± 10.6; patent awards, 33.5 ± 3.9; startup companies initiated, 4.8 ± 0.5; licenses 29.4 ± 3.1; and gross income, from licenses \$13M ± \$27M. The top 10% institutions averaged were as follows: research expenditure, \$848M; publications, 7882; IP disclosures, 33; patent applications, 176; patent awards, 80; startup companies initiated, 11; total licenses, 83; and gross licensing income, \$34M.

A recently published review of university innovation successes further underscored the particular challenges of the biomedical research (The Science Coalition 2013). The vast majority of revenue-producing early successes of university startup companies come from the information technology field, while biomedical startup companies tend to be cash burners for a prolonged period.

The experience of an institution's technology transfer office may also affect their innovation productivity. Years of technology transfer office program existence significantly correlates with greater research expenditures, more licenses and options, greater number of startups, greater adjusted gross income, and greater royalty income in 2013 (all $P < 0.05$). For an individual researcher, the experience and length of existence of an institution's technology transfer office may be a significant factor in the success of promoting one's research discoveries.

An insightful analysis of the licensing results of six universities found that 56% of successful licensing contacts came from faculty inventors, 19% from marketing by TTO staff, 10% from the company (licensee), 7% from the research sponsor, and the rest from miscellaneous unknown sources (Jansen and Dillon 2000). Frequently, professors not just produce great innovations but also build valuable personal networks in the industrial sector (e.g. business card received at conferences, graduate students who have taken positions in industry, companies seeking expertise in academia, and others).

The number of patents, licensing revenues, and job creation of university startup companies are often used as innovation indicators of economic significance. A study of the Massachusetts Institute of Technology (MIT) research concluded that 25 800 active companies founded by MIT alumni employ nearly 3.3 million people and generate annual world sales of \$2 trillion (Roberts 2009).

Essential Concepts of Research Innovation

A practical, useful model of innovation needs to integrate terminology to improve the visibility of common challenges and also consequences of variations, in both regulatory and institutional policies. It should support evaluation of public health and economic impact, assess the role of organizational culture and inventor recognition, highlight opportunities for better functioning policies, and show ways to increase biomedical innovation that benefits society and improves public health.

Innovation is the creation of new wealth-producing resources or endowing existing resources with enhanced potential for creating wealth, according to Peter Drucker (1985). In focusing on life sciences research, three-dimensional or triple innovation can be defined as the creation of new knowledge, health, and wealth resources (e.g. new scientific models, improvements in public health, and economic development).

In other words, *research innovation* is the production of replicable, generalizable scientific discoveries that lead to new models, products, services, or practices benefiting research, people, and society. Again, according to Peter Drucker, innovation is the change that creates a new dimension of performance (Drucker 1985).

The person responsible for the creative result is often called *researcher, author,* inventor, discoverer, scientist, scholar, designer, creator, assignee, investigator, or analyst. Research laboratories are identified as teams of researchers focusing on a significant area of scientific investigations and having specialized methodological capacities and competencies.

The process of research reaches a conclusion when *disclosure* is made (i.e. decision to disseminate the results). Synonyms for results include discovery, practice recommendation, invention, prototype, source program, information system, and others.

Nondisclosure remains a frequently exercised but undesirable research outcome. While exercising such option is currently entirely at author/researcher discretion, it is a major and growing concern of research integrity. Based on self-reported clinical trial outcomes, over 25% of trial reports never published, mainly due to "negative" results and lack of interest (Dickersin et al. 1992). An estimated 50% of innovations with commercial potential are never disclosed, and the negative impact on public health is potentially huge (Thursby et al. 2009).

When applied research results are disclosed, they are supposed to lead to valuable outcomes. In biomedical research, the major outcome categories include further productive research, public health impact, and economic impact:

- *Scientific results* are systematic descriptions of difficult-to-observe objects or phenomena to explain and predict behavior under varying circumstances. The scientific model can be material, graphical, narrative, mathematical, or computational approximation of a real system that leaves out all but the most essential variables. By referencing to existing and commonly accepted knowledge, scientific models are used in the construction and demonstration of scientific theories.
- *Public health outcome,* in the context of the public's health, refers to the general health of a population and the desired distribution of health. Public health includes prevention of diseases, promotion of health, cure of diseases,

prolonged life expectancy, and conditions in which people can be healthy. It can be concerned with the population as a whole or geographic populations such as nations or groups like employees, ethnic groups, disabled persons, prisoners, or others.

- *Economic outcome* is a general improvement of living standards and economic health of a specific locality. Economic development involves advancement of human capital, improvement of infrastructure, improvement of health and safety, and other advances of the general welfare of citizens. Economic development outcomes of innovation can be realized through institutional revenue generated, startup company initiation and success, commercialization of new products, cost savings, and jobs created.

Research Innovation Pathways to Effects

The origins of innovation recognition are simple, clear, and compelling in Article One of the US Constitution: "Congress shall have Power...To promote the Progress of Science and useful Arts, by securing for limited Times to Authors and Inventors the exclusive Right to their respective Writings and Discoveries" (Art. I, Section 8, cl. 8.).

Today, there are numerous complex, divergent categories of writings and discoveries that lead to different practical implications, levels of legal protection, and rewards to authors and inventors. A myriad of laws, regulations, and business expectations has emerged that contribute to making the innovation field much broader but difficult for academic researchers to access successfully. This variety of terminology also provides a pretext to variations in processing and recognition of innovation.

To highlight connections between biomedical research and discovery outcomes, the Research Innovation Pathways to Effects (*RIPE*) *model* conceptualizes the transfer of innovative ideas to future research, public health practice, and the general economy (Figure 1.1). Five research disclosure pathways have the greatest significance to biomedical innovation. They are governed by dissimilar laws, offer variable rewards to inventors, and produce divergent practical results. When research discoveries are made, the results can be made available to the public (general or limited) on one of the following pathways of disclosure:

1) *Direct (PRP) disclosure.* Most frequently, *peer-reviewed publication (PRP)* is the chosen or default pathway for dissemination. It also indicates that the work was reviewed and deemed acceptable by other scientists with relevant expertise in the field. Through this line of disclosure, knowledge becomes reliably, publicly, and essentially freely or for a nominal fee available to anyone who might be interested. On the other hand, limited readership and practical

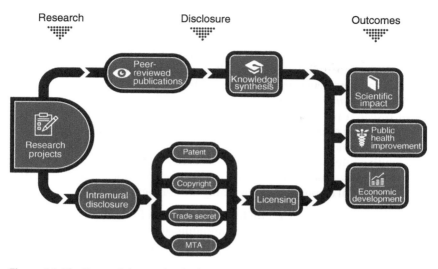

Figure 1.1 The Research Innovation Pathways to Effect (RIPE) model of research discovery disclosure.

impact are frequent concerns due to the large volume and variable quality of scientific articles. Recently, Jeremy Grimshaw and others suggested that most PRP reported research becomes actionable through scientific reviews that synthesize knowledge for practical implementation (Grimshaw et al. 2012).

2) *Staged (IP) disclosure* leads through *intellectual property (IP)* protection of results that appear to be not only novel, useful, and non-obvious but also have commercial potential. It occurs in three subsequent steps: intramural disclosure, legal protection, and extramural disclosure. Typically, the process starts with confidential intramural disclosure to the technology transfer office of the research institute to assess the potential for IP protection and commercialization. Based on the type of legal protection, four IP disclosure pathways are particularly relevant to biomedical research innovation (Figure 1.1):

a) A *patent* creates intellectual property right granted by the government to an inventor to exclude others from making, offering for sale, selling, using, or importing the invention for a limited time, usually for a period that begins when the patent issues and ends 20 years after the date that the application for the patent was filed.

b) *Copyright* is an exclusive right to the use and distribution granted to the author of an original work. Copyright protects the expressive aspect of the innovation. The default time a copyright is enforceable is the life of the author plus 70 years in most countries.

c) A *trade secret* has three parts: information, reasonable measures taken to protect the information, and the economic value it derives from not being publicly known. It is essentially limited extramural disclosure only to those who are intended users. It is protected at the state level so that requirements may vary from state to state.

d) *Material transfer agreement* (MTA) is a contract regarding the transfer of research materials to a recipient that intends to use it for research purposes (e.g. chemical compounds, biological material, reagents, cell lines, plasmids, vectors, and software). Typically, MTAs cover rights to resulting intellectual property, rights to data and use of results generated by the work, publication rights, indemnification and liability, jurisdiction for legal disputes, and governing law for legal disputes.

Frequently, the practically useful innovative results fit multiple intertwining categories. Many innovations can be communicated through any of the listed pathways. For example, medical natural language processing software can be disclosed through any of the above pathways (i.e. peer-reviewed publication, patent, copyrighted documents, trade secret, or material transfer agreement).

While the technicalities of the disclosure may suggest otherwise, researchers and their employing institutions have a large degree of freedom in choosing the disclosure pathway. Obviously, different pathways represent different positioning for practical impact and author rewards. For example, disclosing a new medication through peer-reviewed publication without IP protection would undermine commercialization, manufacturing, and ultimately broad public access. Patenting is the well-functioning disclosure pathway for new drugs.

An additional special type of intellectual property, the use of trademarks also has great potential in scientific communications. Of course, trademarks cannot be viewed as substantial channels of communicating the details of scientific discoveries. On the other hand, they can be very helpful in making discoveries better recognized and easier to remember. When commercialization is at stake, registering a trademark can provide the much-needed stronger protection of the brand by identifying and distinguishing the original researcher/creator from others and indicating the genuine source.

Branding of a new clinical intervention or research method helps to stand out, be remembered, and become the preferred choice. Successful examples include memorable names of landmark systems, studies, and methodologies (e.g. BLAST, Framingham study, zero defect data, radioimmunoassay). In the world of science, many more names and acronyms are created but seldom used by other than their creators. When the intervention or research method is more widely used, the brand becomes valuable and deserves protection. In academia, the strict deterrents of plagiarism alone provide sufficient protection for the brand in most cases. Interestingly, the usefulness of many identifiers is not limited by the fact that official registration and protection as a trademark is often not pursued by the researchers or their institution.

Preferentially choosing peer-reviewed publications for disclosure does not make the channels of intellectual property protection and subsequent commercialization irrelevant. Researchers striving to be in touch with reality need to learn about the societal need for their results, the full range of disclosure channels, and also basic tactics of negotiation with business interests (see Chapter 18).

Learning from Award-winning Scientists and Serial Innovators

Successes of serial research innovators suggest that scientific discovery and innovation cannot be considered just a matter of luck. More frequently good planning and hard work are in the background. The most accomplished researchers and their laboratories have a unique sense of innovation opportunities and also the skills to make them broadly successful and accessible. The successes of innovative researchers indicate skills that are likely to surpass one time or incidental inventors. Successful serial innovation requires meeting great challenges by applied and translational health research.

Historical exploration and reverse engineering of biomedical innovations with the greatest public health benefits could offer many lessons for future research projects. Beyond inspirational value, historical references are also helpful not only to illuminate the research and innovation culture but also to pinpoint the dichotomy of intellectual property protection and dissemination through scientific commons in transferring research results to practice.

Studying the performance of serial research innovators from academia offers unique insight. The list below summarizes the performance of several university researchers who have built a track record of a large number of peer-reviewed publications, numerous patents, notable commercialization successes, and major contributions to better health care:

- Tillman Gerngross, PhD, Dartmouth College (PRP: 23, IP: 24). Most notable achievements: Humanizing the glycosylation machinery in yeast to produce human therapeutic proteins, including antibodies, with fully human carbohydrate structures. Public health impact: Discovered novel and efficient ways to produce new drug proteins through yeast. Economic impact: Cofounded GlycoFi, which is sold to Merck for $400 million. Adimab, cofounder, biotech startup valued at $500 million, privately held. Venture partner with SV Life Sciences.
- Michael Merzenich, PhD, University of California, San Francisco (PRP: 200+, IP: 50+). Most notable achievements: Cochlear implant and sensory cortex mapping. Public health impact: Improved quality of life for the deaf. Understanding of brain function and training informs further brain research. Economic impact: Global hearing implants market is projected to exceed $2 billion in 2017.

- Andrew Schally, MD, Tulane University, Baylor College of Medicine (PRP: 2200+, IP: 29). Most notable achievements: Structure of LH-RH and Nobel Prize in Physiology, 1977. Public health impact: Luteinizing hormone-releasing hormone, which inhibits the growth of prostate cancer. Most widely used prostate cancer treatment.
- Mark Skolnick, PhD, University of Utah (PRP: 139, IP: 9). Most notable achievements: Skolnick directed the group that discovered the breast cancer susceptibility gene BRCA1; found the full-length sequence of BRCA2. His group developed restriction fragment length polymorphism (RFLP) method for genetic mapping. Public health impact: Women with harmful BRCA1 mutation or BRCA2 mutation have nearly 50% chance of developing breast cancer, and genetic testing provides early detection of the risk. Economic impact: He launched three companies, and among them is the biotechnology company Myriad Genetics, Inc., in Salt Lake City.
- Edward Taylor, PhD, Princeton University (PRP: 450, IP: 52). Most notable achievements: Alimta, cancer drug for mesothelioma. Public health impact: Most common drug in use for mesothelioma treatment. Economic impact: Princeton received $524 million from 2005 to 2012 in license income, mostly from Lilly. Alimta earned $1.2 billion in the United States and $2.7 billion globally in 2013.
- Elias Zerhouni, MD, PhD, Johns Hopkins University (PRP: 212, IP: 8). Most notable achievements: High-resolution CT development for heart and lung study and cancer diagnosis; computed tomographic densitometry for lung cancer detection. Economic impact: Founded five startup companies based on inventions and research from Johns Hopkins University.
- Jackie Yi-Ru Ying, PhD (PRP: 350, IP: 180), is a nanotechnology pioneer, former MIT professor, and currently director of the Institute of Bioengineering and Nanotechnology in Singapore. Most notable achievements: Nanomedicine applications, drug delivery, cell and tissue engineering, medical implants, and biosensors, among others. Economic impact: Her work has been instrumental in launching 11 spin-offs. Public health impact: One of her inventions led to the founding of SmartCells, Inc., a spin-off that developed a technology capable of autoregulating the release of insulin, depending on the blood glucose levels for diabetes management. It was acquired by pharmaceutical giant Merck for more than $500 million to further develop this technology for clinical trials.

Road to Meaningful Research Disclosure

According to recent reviews, science policies have developed a hypercompetitive culture where learning about grant writing and competing for grants appear to be more important than spotting health needs, recognizing scientific opportunities, and choosing research targets (Alberts et al. 2014). More appropriately, researchers should look into what makes good sense in choosing biomedical research targets, as opposed to being completely driven by requests for proposals of various funding agencies.

The steps after disclosure are widely discussed and relatively well defined in the business and law literature. Usual university technology transfer discussions tend to center on commercialization and business development of intellectual property (IP) disclosures already made, without much emphasis on how you get to meaningful disclosures. Commercialization after disclosure has been the focus of many reviews and books. The fundamental research process leading to well-performing research disclosures has been largely neglected, but it is the focal point of this discussion.

Contrary to this trend, there is a need to study more intensely the fundamental research process that can produce meaningful invention disclosures. In spite of huge public interest in the process of innovation, there has not been enough research on the road leading to meaningful innovation disclosure. Comprehensive reviews have not focused on how individual researchers and research laboratories become better sources of discoveries. More should be known about how research can reliably lead to practically valuable disclosures.

In response to pressing societal need to increase the productivity of research innovation and in light of the above-described model of research innovation pathways, researchers should become better educated about pathways of research disclosure and the societal outcomes of research. Many students of biomedical PhD programs graduate, and many biomedical researchers spend years in the laboratory without good understanding of public health needs, without reading a patent or copyright registration, without knowing what happens with scientific discoveries after the research is completed, and without understanding technology transfer or commercialization.

There is an emerging need to increase awareness of the roads to practical impact and innovation. Researchers should be knowledgeable about public health needs, able to protect their ideas, and get a better chance to share the benefits of their research. To promote meaningful disclosures, students and practitioners of biomedical research need to learn about the ultimate outcomes of biomedical research; recognition of public health needs; process of choosing promising research targets; basic steps of research disclosure and technology transfer; experiences of innovative research laboratories and serial innovators; methods of successful collaboration with communities and industries; use of the IP literature on patents, copyrights, and case studies of trade secrets; legal and regulatory environment of intellectual property protection; basics of launching new products, services, and companies; and institutional environment and resources of supporting innovation.

There is particular need to learn from research on research and biomedical innovation. We need to understand consequential discoveries, support researchers striving to innovate, and facilitate the development of more useful institutional policies and legislation. Defining and harnessing the differences between institutions in technology transfer is of great importance to sponsors and research institutions alike to ensure successful use of research funding.

Research leadership of institutions can do more in promoting the full range of scholarly creativity and should ultimately be judged based on their ability to do so. To assess performance and inform researchers, institutional effectiveness should measure the major impact of scholarly creativity including peer-reviewed publication, intellectual properties, and successes of practical application. Generation of new ideas is largely dependent on an organizational culture that promotes and protects research innovation, which is likely to have significant further research and public health impact. Academic institutions should bring innovation to the center of scholarly discussions.

Life sciences are at a remarkable moment of opportunity. Research is leading to understanding, treating, and preventing a growing number of diseases. Investment in biomedical research has done many wonders. To realize the exciting opportunities, life sciences research needs talented researchers who can build studies and also societal support, including government and private funding, to achieve ambitious goals for the future.

Beyond learning about pathways and creating institutional infrastructures, a variety of important measures are emerging to promote high-impact and innovative research. We call them boosters of research innovation and chapters of a major section provide further insight. After all, every difficulty, every unknown, and every crisis in health care is a need and opportunity for innovation. Academic institutions do not innovate, only creative individuals and laboratories do. At the epicenter of great discoveries is the talented, innovative, and well-prepared researcher working in the laboratory.

References

Alberts, B., Kirschner, M.W., Tilghman, S., and Varmus, H. (2014). Rescuing US biomedical research from its systemic flaws. *Proceedings of the National Academy of Sciences of the United States of America* 111 (16): 5773–5777. doi: 10.1073/pnas.1404402111.

Anderson, M.G., D'Alessandro, D., Quelle, D. et al. (2013). Recognizing diverse forms of scholarship in the modern medical college. *International Journal of Medical Education* 4: 120–125.

Balas, E.A. and Elkin, P.L. (2013). Technology transfer from biomedical research to clinical practice measuring innovation performance. *Evaluation & the Health Professions* 36 (4): 505–517.

Blumberg, R.S., Dittel, B., Hafler, D. et al. (2012). Unraveling the autoimmune translational research process layer by layer. *Nature Medicine* 18 (1): 35–41. doi: 10.1038/nm.2632.

Carnegie Foundation for the Advancement of Teaching (2010). *Classification Description*. http://carnegieclassifications.iu.edu/ (accessed 31 January 2014).

Centers for Disease Control and Prevention (2009). Rotavirus. In: *Epidemiology and Prevention of Vaccine-Preventable Diseases (The Pink Book)*, 11e, 245–256. Washington, DC: Public Health Foundation.

Centers for Disease Control and Prevention (2011). Ten great public health achievements – United States, 2001–2010. *MMWR. Morbidity and Mortality Weekly Report* 60 (19): 619–623.

Dickersin, K., Min, Y.I., and Meinert, C.L. (1992). Factors influencing publication of research results. Follow-up of applications submitted to two institutional review boards. *JAMA* 267 (3): 374–378.

Drucker, P.F. (1985). The discipline of innovation. *Harvard Business Review* 63 (3): 67–72.

Freschi, C., Ferrari, V., Melfi, F. et al. (2012). Technical review of the da Vinci surgical telemanipulator. *The International Journal of Medical Robotics and Computer Assisted Surgery* 9 (4): 396–406.

Gelmon, S.B., Jordan, C., and Seifer, S.D. (2013). Community-engaged scholarship in the academy: an action agenda. *Change: The Magazine of Higher Learning* 45 (4): 58–66.

Golden, J.M. and Sage, W.M. (2013). Are human genes patentable? The supreme court says yes and no. *Health Affairs (Millwood)* 32 (8): 1343–1345. doi: 10.1377/hlthaff.2013.0707.

Grimshaw, J.M., Eccles, M.P., Lavis, J.N. et al. (2012). Knowledge translation of research findings. *Implementation Science* 7 (1): 50.

Gruentzig, A. (1982). Results from coronary angioplasty and implications for the future. *American Heart Journal* 103 (4): 779–783.

Holmes, D.R. Jr., Vlietstra, R.E., Smith, H.C. et al. (1984). Restenosis after percutaneous transluminal coronary angioplasty (PTCA): a report from the PTCA registry of the National Heart, Lung, and Blood Institute. *The American Journal of Cardiology* 53 (12): C77–C81.

Horstkotte, E. and Odoerfer, K. (2012). Towards improved therapies using nanopharmaceuticals: recent patents on pharmaceutical nanoformulations. *Recent Patents on Food, Nutrition & Agriculture* 4 (3): 220–244.

Jansen, C. and Dillon, H.F. (2000). Where do the leads for licenses come from? Source data from six US institutions. *Industry and Higher Education* 14 (3): 150–156.

Kohler, B.A., Ward, E., McCarthy, B.J. et al. (2011). Annual report to the nation on the status of cancer, 1975–2007, featuring tumors of the brain and other nervous system. *Journal of the National Cancer Institute* 103: 714–736.

Ledford, H. (2013). Universities struggle to make patents pay. *Nature* 501 (7468): 471–472.

Leshner, A.I., Terry, S.F., Schultz, A.M., and Liverman, C.T. (2013). *The CTSA Program at NIH: Opportunities for Advancing Clinical and Translational Research*. Washington, DC: National Academies Press.

Nwaka, S. and Ridley, R.G. (2003). Virtual drug discovery and development for neglected diseases through public–private partnerships. *Nature Reviews Drug Discovery* 2 (11): 919–928.

Pilishvili, T., Lexau, C., Farley, M.M. et al. (2010). Sustained reductions in invasive pneumococcal disease in the era of conjugate vaccine. *The Journal of Infectious Diseases* 201 (1): 32–41. doi: 10.1086/648593.

Roberts, J.D. (2009). A perspective distilled from seventy years of research. *The Journal of Organic Chemistry* 74 (14): 4897–4917. doi: 10.1021/jo900641t.

Schramm, C., Arora, A., Chandy, R. et al. (2008). *Innovation Measurement: Tracking the State of Innovation in the American Economy.* http://users.nber.org/~sewp/SEWPdigestFeb08/InnovationMeasurement2001_08.pdf (accessed 9 June 2018).

Schrodinger, E. (1932). *Uber Indeterlninismus in der Physik.* Leipzig: J. A. Barth.

Smesny, A.L., Williams, J.S., Brazeau, G.A. et al. (2007). Barriers to scholarship in dentistry, medicine, nursing, and pharmacy practice faculty. *American Journal of Pharmaceutical Education* 71 (5): 91.

Talevi, A., Gantner, M.E., and Ruiz, M.E. (2014). Applications of nanosystems to anticancer drug therapy (part I: nanogels, nanospheres, nanocapsules). *Recent Patents on Anti-Cancer Drug Discovery* 9 (1): 83–98.

Tate, J.E. and Parashar, U.D. (2011). Monitoring impact and effectiveness of rotavirus vaccination. *Expert Review of Vaccines* 10 (8): 1123–1125. doi: 10.1586/erv.11.94.

Telang, M., Bhutkar, S., and Hirwani, R. (2012). Analysis of patents on preeclampsia detection and diagnosis: a perspective. *Placenta* 34 (1): 2–8. doi: 10.1016/j.placenta.2012.10.017.

The Science Coalition (2013). *Sparking Economic Growth 2.0: Companies Created from Federally Funded University Research, Fueling American Innovation and Economic Growth.* Washington, DC: The Science Coalition. http://www.sciencecoalition.org/reports/Sparking%20Economic%20Growth%20FINAL%20 10-21-13.pdf (accessed 21 June 2018).

Thursby, J., Fuller, A.W., and Thursby, M. (2009). US faculty patenting: inside and outside the university. *Research Policy* 38 (1): 14–25.

Wigner, E.P. (1960). The unreasonable effectiveness of mathematics in the natural sciences. Richard Courant lecture in mathematical sciences delivered at New York University, May 11, 1959. *Communications on Pure and Applied Mathematics* 13 (1): 1–14.

Yen, C., Armero Guardado, J.A., Alberto, P. et al. (2011). Decline in rotavirus hospitalizations and health care visits for childhood diarrhea following rotavirus vaccination in El Salvador. *The Pediatric Infectious Disease Journal* 30 (1 Suppl): S6–S10. doi: 10.1097/INF.0b013e3181fefa05.

Zur Hausen, H. (1976). Condylomata acuminata and human genital cancer. *Cancer Research* 36 (2 Part 2): 794–794.

2

First Dimension

Scientific Impact

When we say that we understand a group of natural phenomena, we mean that we have found a constructive theory which embraces them.
Albert Einstein (1919)[1]

As a result of huge societal investments, many impressive benefits, and growing enthusiasm, the future of life sciences research has never been brighter. Nothing compares with the magic of smart questions and amazing scientific discoveries that represent an entirely new understanding of previously hidden, misunderstood, or incomprehensible processes of nature.

Science, human progress, and excitement go together. Understanding the role of dendritic cells, identification of the gene editing system, exploration of regulators of vesicle transport in cells, and discovery that mature cells can be reprogrammed to become pluripotent are just a few recent shining discoveries from biomedical research. Many landmark discoveries shed new lights from previously unimaginable angles.

In recent decades, there have been unprecedented scientific advances spurred by dramatic advances in technology. As a result of better understanding of important natural phenomena, a myriad of new prevention strategies and novel cures can be expected (Collins et al. 2016). In the years to come, creativity and imagination of life sciences are going to have a rapidly growing impact on our lives. None of this is science fiction anymore.

To meet the escalating expectations, better articulation of scientific outcomes is needed to support the planning of more effective research projects (Macilwain 2010). Growing recognition of inefficiencies in scientific production, particularly flattening of innovation in several areas and unacceptably high rate of non-repeatable research results, urge better understanding of the scientific process and its ultimate impact. Fortunately, talented theorists of

---

1 Einstein, A. (1919). Time, space, and gravitation. *London Times* (28 November).

Innovative Research in Life Sciences: Pathways to Scientific Impact, Public Health Improvement, and Economic Progress, First Edition. E. Andrew Balas.
© 2019 John Wiley & Sons, Inc. Published 2019 by John Wiley & Sons, Inc.

science help understand of what constitutes significant scientific accomplishment and valuable intellectual contribution.

Intellectual Impact and Scientific Pluralism

In his will, Alfred Nobel established awards to recognize scientific discoveries "for the greatest benefit to mankind." In physiology or medicine, the particular emphasis is on discoveries that have changed the scientific paradigm. Interestingly but not surprisingly, many Nobel Prizes in physics and chemistry have also been awarded for discoveries that directly benefit life sciences and serve as the foundation for better health care. Very appropriately, the Nobel Prize selection excludes consideration of lifetime achievements or scientific leadership.

Paraphrasing expectations of the Polya Prize, great researchers can be recognized for their outstanding creativity, imaginative exposition, and distinguished contribution to the progress of science. Thanks to the great progress in science, we can ask questions that people never imagined possible to answer or even ask before. At the center of great science and recognition for scientific discoveries is the appreciation for intellectual contribution.

Attacking great mysteries has always been the motivation of many scientists. Not surprisingly, cancer is one of the most frequent motivators of ambitious researchers. Many of them fail to reach the grand objective, but the progress they make already generates many benefits along the way. For example, Nobel Laureate Otto Warburg spent many years of researching the causes and evolution of cancer but never got even close to it. However, he made many landmark discoveries in his ambitious investigations.

Scientific pluralism states that the multiplicity of representation of everything worth knowing is an advantage in scholarly investigations and practical applications (Kellert et al. 2006). Pluralism is a commitment to maintain multiple systems of knowledge in each field of inquiry (Chang 2012). In other words, the lack of single overarching theory that presently characterizes many scientific areas does not necessarily constitute a deficiency. Science has multiple aims, not just one overriding aim.

Scientific pluralism offers many benefits by greatly facilitating applications of scientific principles to practical problem-solving (Chang 2012). Applying the most fitting theoretical frameworks to the solution of a practical problem can greatly facilitate the design process and support the smooth functioning of complex technologies. For example, the GPS technology applies to (i) satellites that are kept in place by Newtonian physics, (ii) atomic clock that is ruled by quantum mechanics, (iii) corrections by general and special relativity, (iv) interpretations in a geospatial network, and (v) communications to people from a flat Earth point of view.

Scientific pluralism also offers benefits by opening minds to alternative explanations. Allowing multiple territories or systems simultaneously provides insurance against unpredictability, compensation for the limitations of each theory, and multiple satisfaction of any given aim. Scientific pluralism is a different but related concept to critical thinking, the concept increasingly highlighted in modern educational theory. Critical thinking ensures that we have evidence and good reasons to accept or reject what is suggested by other people (Bowell and Kemp 2014).

Tolerance of alternative explanations can also be beneficial when many researchers are drawn to the same fashionable but ultimately unproductive idea. Such scientific fashions can become a tremendous waste of resources especially when other approaches could be explored more productively. Nobel laureate Harald zur Hausen spoke about the early years of his studies on the role of papillomavirus in the development of cervical cancer. At a conference in Florida, the dominant view was search for the causative role of herpes virus, and zur Hausen's comments about the papillomavirus were unwelcome (Smith 2008). Today, one may wonder how much taxpayer and foundation money was spent on researching the wrong virus at that time.

In developing new projects, researchers need a sense of purpose and clarity to achieve great outcomes. There has to be a passion for the end product, not just for the curiosity-triggering research. Philosopher of mathematics and science, Imre Lakatos pointed out that truly scientific programs successfully predict novel facts. Conversely, pseudoscience can be recognized by repeatedly failing to predict novel facts (Lakatos 1978).

As described in more details below, novel scientific models, well-substantiated observations, unprecedented methodologies, and theories well grounded in evidence need to be considered as meaningful intellectual contributions to further research.

Howard Temin was a US-born geneticist, virologist, and social activist. He majored in biology at Swarthmore College and received a doctorate in animal virology working in Professor Dulbecco's laboratory at CalTech. In his doctoral dissertation in 1959, Temin noted that the Rous sarcoma virus must have "some kind of close relationship with the genome of the infected cell." After completing training, he worked at the University of Wisconsin in Madison. He started talking about a "provirus" that was genetic material introduced by the virus.

Temin's uniquely strong skills and also supremely self-confident personality made him an ideal challenger of the central dogma of molecular biology that transcription happens only from the DNA to the RNA. For 10 years, he was saying that RNA tumor viruses have to be able to copy themselves into DNA to make sense of their role in tumor genesis. Nobody believed him for years. To say the least, his hypothesis was not welcomed by the scientific community, and, in some cases, the rejection was quite public. For example, one of the leading professors of the time had a major conference presentation, and afterward someone asked why he did not mention Temin's work at all. He responded that he "gave the attention to Temin's work it deserves."

In 1969, Temin began searching for the particular enzyme that was responsible for the transfer of viral RNA into proviral DNA. The discovery of the reverse transcriptase is one the most consequential accomplishments of modern life sciences. "For their discoveries concerning the interaction between tumor viruses and the genetic material of the cell," Howard Martin Temin, Renato Dulbecco, and David Baltimore received the Nobel Prize in 1975. Afterward, Temin not only remained hardworking in research but also repeatedly traveled to the Soviet Union to support persecuted scientists. Throughout his career, Temin had a profound effect on teaching students and colleagues by being a terror and a mentor simultaneously.

You Cannot Make a Good Landing Out of a Bad Approach

The wisdom of seasoned aviators also has a message for life sciences researchers interested in increasing the intellectual impact of their projects and results. It is impossible to plan projects successfully and perform research effectively if the ultimate products and goals of science are either not considered or misidentified. Researchers need to have a passion for the end product, not just being passionate about the curiosity-driven process of research.

Certainly, planning for successful research cannot use the peer-reviewed publication as the ultimate outcome when, in fact, it is only an intermediary on the road to scientific progress and societal benefits. You do not plan a ride only to the next village but to the ultimate destination of the trip. You do not plan a house to build only the foundation but to have a complete, valuable home ready to be used by future owners. It is hard to imagine effective production in any enterprise that ignores long-term outcomes or practical value considerations. In science, planning beyond isolated publications and achieving broadly reproducible results are essential.

Not infrequently, researchers and research funding agencies talk about the process of research without much consideration of desired outcomes. In applied translational sciences, many overseers focus on the information system support for research, the structure of administrative support for the research, and the availability of various knowledge resources without spending much time on thinking about the expected outcomes and the measurement of societal benefits. It is like talking about car manufacturing by focusing on information system support, the structure of administration, and availability of knowledge resources without being interested in the number and quality of cars produced.

Intellectual impact of research should never be confused with the so-called impact factor of scientific journals, largely a marketing tool based on the cumulative number of citations of articles published. Citation of articles is heavily influenced by factors other than intellectual impact (e.g. variability based on the field of study, methodology papers tend to be more cited, and others). According to some studies, journals of physics and chemistry have the highest impact factors, and arts and humanities and social sciences have the lowest impact factors (Hamilton 1991).

The most frequently cited papers are major methodologies that become indispensable resources for many researchers (Van Noorden et al. 2014). At the time of writing this book, the record holder with over 300 000 citations is the article titled "Protein measurement with the folin phenol reagent" (Lowry et al. 1951). This paper describes a landmark methodology for quantifying protein.

Meanwhile, the groundbreaking Nobel Prize-winning Watson and Crick article on the DNA double helix structure attracted less than 5% of citations in comparison with the Lowry methodology papers in the *Proceedings of the National Academy of Sciences*. For very good reasons, journal impact factor is not a factor in awarding the Nobel Prize.

In the research community, high-impact journals are considered influential in their particular field. In recent years, the prestige of high-impact journals has been questioned by studies showing the high rate of non-repeatable research published in these journals and also the alleged role of building up impact factor numbers by luxury brand marketing techniques of their publishers (Schekman 2013).

It was pointed out that these journals are building an image, scarcity, and brand not proportional to the scientific value. The so-called impact factor is primarily a marketing tool and not an accurate representation of scientific value. Schekman also called universities and promotion committees not to judge papers by where they are published. Instead, the focus should be on the quality of science, not on the journal's brand.

Another commonly misguided approach is the overzealous focus on writing grant applications. At least once a week, most researchers get an email invitation to one or more uniquely effective, life-changing grant writing course – a close scientific relative of the get-rich-quick schemes.

Checklist research cannot lead very far. Some researchers are aimlessly drifting because they are just trying to respond to calls for proposals that come from funding agencies. Checklist science: let us check what they want line by line and make sure it is promised in the grant application. Promising a project in one direction Monday and in an entirely different direction on Wednesday makes very little sense. Such approach is unlikely to create any value scientifically or produce meaningful revenues. The reviewers will

Harold Varmus is a native of New York who graduated from Amherst College and Columbia University College of Physicians and Surgeons. His first serious exposure to the excitement of experimental success was as a clinical associate at NIH. It also sparked his interest in molecular biology and tumor virology. After working for NIH, he joined the laboratory of Michael Bishop at the University of California, San Francisco. At UCSF, he gradually rose from postdoctoral fellow to faculty member and full professor.

The productive collaboration of Harold Varmus and Michael Bishop progressed for many years, and they jointly discovered the role of retroviral oncogenes. This achievement was recognized by their shared Nobel Prize in 1989. The Nobel lecture of Varmus was a brilliant sequence of model presentations, which he called hypotheses, showing the evolution of understanding based on gradually unfolding experimental evidence (Frängsmyr 1990).

After receiving his Nobel Prize, Varmus became very active in scientific leadership and advocacy for more support. He served as director of the National Institutes of Health and later as director of the National Cancer Institute. In between, he was president and CEO of the Memorial Sloan Kettering Cancer Center. More recently, he became cofounder of Public Library of Science (PLOS), an important nonprofit publisher. An eminent scientist, outdoor enthusiast, and family man, Varmus became one of the most impressive forces, driving of reshaping biomedical research.

quickly recognize that the applicant just mirrors expectations of the call for proposals with an empty promise on paper but without any relevant competence or commitment.

First of all, researchers have to be interested in reality, an important but enigmatic part of nature, and only afterward they should look for grant funding. The appropriate approach starts with ambitious and engaged research in a focused area of interest. There is no reason for jumping into diverse grant applications without having a defined research interest. After doing homework about societal needs and research opportunities, the subsequent grant applications make sense if they are in line with the focused research agenda and help to understand nature's enigma.

Intellectual Products of Scientific Endeavors

To understand how scientific discoveries evolve, it is worthwhile to look into the major types of research products that come out of successful projects. Importance of exploring types of research outcomes is further underscored by the need to look beyond peer-reviewed publications.

For obvious reasons, data are always about the past. Theories and models lead into the future. Not only the role of science but also the specific task of scientific theories and models can be illuminated the following way: "The idea of growth and the concept of empirical character are soldered into one" (Lakatos 1978). Progressive scientific theories have their novel facts confirmed, and degenerated scientific theories are those whose predictions of novel facts are refuted. When theory lags behind the facts, we are dealing with degeneration of research programs.

A better understanding of intellectual outcomes should make projects leading to them more streamlined and efficient. In reviewing the principal products of research studies, Table 2.1 provides a side-by-side comparison based on the area of applicability, scientific foundation, method of expression, and predictive value.

In many ways, *scientific theories* represent the broadest intellectual products of scientific investigation, called research. Based on a wide-ranging synthesis of accumulated research evidence, the scientific theory is a well-substantiated,

Table 2.1 Comparison of principal research products.

Major products of research	Scientific theories (frameworks)	Scientific models	Scientific observations	Novel methodologies
Area of applicability	General understanding of a wide range of phenomena	Explanation of a specific phenomenon	Specific and actual	Specific and actual
Scientific foundation	Grounded in a wide range of past observations	Observational and experimental data collection	Observational and experimental data collection	Measurement of new phenomena or disruptive technology
Method of expression	Declared principles or mathematical formulas	Narrative, pictorial, physical, or mathematical representations	Recorded facts and data	Description of methodology with initial results, prototyping, or production
Predictive value	Many diverse but accurate predictions	Repeatable in a variety of situations	Leaves interpretation to the reader	Useful in many more research and production projects
Example	Germ theory of infectious diseases	Causative role of *Helicobacter pylori* in gastric ulcer	Estimated prevalence of obesity in the United States	Cohen–Boyer method of splicing genes to make recombinant proteins

logical framework of reference describing an important aspect of the natural world that not just explains past observations but also effectively predicts future behavior. Scientific theories or frameworks are repeatedly tested and confirmed through multiple observations and extensive experimentation.

Albert Einstein eloquently described the concept of principle-based modeling (Einstein 1919): "Theories of principle...employ the analytic...method. Their starting-point and foundation are not hypothetical constituents, but empirically observed general properties of phenomena, principles from which mathematical formula are deduced of such a kind that they apply to every case which presents itself."

In the field of life sciences, probably the most famous scientific framework of reference is Charles Darwin's theory of biological evolution. It became widely known and increasingly accepted after the publication of his book titled *On the Origin of Species by Means of Natural Selection* (Darwin 1859). In the center of the theory is the gradual transformation of one species into another and differential survival or reproduction of individuals based on their ability to adapt to variable circumstances. The long series of differential survival and gradual transformation has been leading to the evolution of species.

Darwin's theory of natural selection and evolution became a profoundly significant reference point in life sciences discussions and had an unparalleled impact on the progress of science. It has positively influenced the understanding of the origins of life, paleontology and extinct species, crossbreeding and hybridization, genetic inheritance, mutations, and ultimately genome sequencing. Interestingly, the nearly universal acceptance of Darwin's theory also made it a political litmus test for accepting the role of evidence in modern natural sciences.

Accumulation of independent models and observations often calls for an overarching theoretical framework. In response to the desire to have an overarching theory for the earlier hodgepodge of ideas about probability and statistics, the renowned mathematician Andrey Kolmogorov developed an overarching theoretical framework. In his landmark book, he noted: "the theory of probability as a mathematical discipline can and should be developed from axioms in exactly the same way as geometry and algebra" (Kolmogorov 1950). He is considered the founder of modern and rigorous probability theory, one of the most powerful theoretical foundations of modern biomedical research. He also made many landmark contributions to information theory, harmonic analysis, and set theory.

Scientific models are greatly simplified but useful representations of reality, actual objects, mechanisms, phenomena, or processes. The development of the model needs to be rooted in observations and validated through repeated scientific experiments. Regarding expression, models can be narrative, graphic, physical, or mathematical. Development of responsive models is an important success factor that will be discussed later among the essential skills of award-winning scientists.

By definition, understanding nature or particular phenomena means a validated scientific model of expository excellence. It would be a minimalist expectation from a scientist to measure something and present results without an ability or attempt to interpret them. Understandably, conclusions from theoretical or practical perspectives are routine parts of peer-reviewed scientific publications.

Scientific theories provide general framework, while models support understanding of specific phenomena in nature. Obviously, development of useful scientific theories is a much greater challenge than modeling a specific function. Due to the largely unexplored, enormous complexities of living systems, modeling is much more frequent in life sciences. The major difference between scientific theories or frameworks and scientific models is the extensiveness of natural phenomena explained by them.

Model-based research studies are built on a broad consideration of past scientific observations and on new experiments targeted toward confirming the model. As Nobel laureate John Gurdon pointed out, when you develop a well-substantiated model and subsequently conduct an experiment to validate it, then the results will always be interesting, regardless of the rejection or confirmation of the model (Smith 2012).

Scientific observation is recording of facts about a natural phenomenon (e.g. observations of nature, experimental data collections). Many original research studies fall into this category. In many ways, evidence-based scientific model building or clinical applications rely on observational studies. The accuracy of scientific observation is an essential requirement for usefulness in furthering research and practice.

Promoting and supporting vast amounts of scientific observations is one of the most controversial parts of modern science. The universe is infinitely large and complex, and many things can be measured, but not all of them make good sense. "Not everything that can be counted counts" was posted on the desk of Albert Einstein (Cameron 1957). When measurements ignore past observations, or launched without grounding in scientific theory or model, the results can be easily misleading and ultimately uninterpretable.

Recently, more and more researchers started outsourcing observational or experimental studies (e.g. therapy analytics, first-in-human studies, rat colonoscopy, contract fermentation and subcultures). One example is Science Exchange, an online service that processes outsourced research of scientific laboratories at major universities and commercial contract research organizations (Wadman 2013).

There are several major forces pushing toward the production of scientific observational studies. First, peer-reviewed journals strongly prefer original research studies that present new data. Second, the scientific productivity of researchers is often measured in the number of publications, particularly original research studies (Pfirman et al. 2007). Third, pursuing observational studies does not require extensive theoretical work and is often based on the shaky assumption that some surprising results should always come out if large volumes of data are produced. Obviously, such approach is very risky due to flawed statistical testing and misleadingly significant results (Ioannidis 2005).

In spite of many benefits, the huge volume of observational studies and projects that are built on a large number of hypotheses generate tremendous risks regarding statistical analysis. Shooting the sky approach to large databases and observational data is unlikely to be very effective. In fact, the misleading statistical analyses often lead to non-repeatable research.

The final and special contribution to the progress of science is the development of *new scientific methodologies* that are useful in a large variety of research projects. Quite understandably, publications describing some of the most useful methodologies are all times winners when it comes to citation by other researchers. There are two kinds of novel research methodologies: (i) leading-edge measurement of a previously non-measurable new characteristic of living organisms (e.g. Willem Einthoven's practical electrocardiogram that received the Nobel Prize in 1924) and (ii) better, faster, and cheaper measurement methodologies that hugely improve and expand the measurement of an important characteristic (i.e. so-called disruptive innovation).

Peer-reviewed scientific articles about the novel and widely applicable research methodologies can become enormously popular and highly cited (see the article of Lowry et al. (1951) as described above). The major advances of life sciences gave rise to many transformational technologies and novel methodologies.

Being a great methodologist is one important way to become a highly accomplished scientist. A new measurement is more important than an old theory. Being on the cutting edge of science often requires not just well-informed application of novel methodologies but also further development in support of innovative explorations. Due to the significance of this challenge, a separate chapter focuses on the competencies of technical and methodological developments.

Ronald Fisher was a mathematician who made many landmark contributions to genetics and more importantly to biostatistics. He was born in London, educated in Cambridge, and worked first for the city of London and later as a statistician at the Rothamsted Agricultural Experiment Station, before becoming university faculty member.

Among others, he developed the analysis of variance (ANOVA), Fisher's z distribution (later called F distribution), linear discriminant analysis, principles of randomization in experimental design, and the Fisher exact test, a great tool not just for research in life sciences but also for teaching the basic principles of statistical hypothesis testing (Bliss 1964). His book titled *Statistical Methods for*

Research Workers is one of the most influential biostatistics textbooks of all times (Fisher 1925). It popularized the concept of *P* value and suggested the use of 0.05 or 1 in 20 as the threshold of significance. His book also introduced the concept of meta-analysis.

His statistical studies also led to many landmark accomplishments in population genetics. Among them Fisher–Kolmogorov equation, a nonlinear reaction–diffusion equation (Thompson 1990). In the early part of his academic career, Ronald Fisher was, quite controversially, the professor of eugenics at University College London until the department was abolished. In the later part of his life, he grew into a formidable promoter of biostatistics studies and also became the founder of the International Biometric Society.

Reality Means Reproducibility

Reproducibility is the quintessential quality threshold of good science. Invariably, successful practical applications depend on accurate forecasting of outcomes. When a reproducible scientific discovery becomes the basis of a functioning prototype and, subsequently, commercialized product, it is supposed to lead to predictable and consistently beneficial outcomes. The Salk vaccine prevents polio with a high degree of accuracy based on the research discovery of Jonas Salk and his team. The Kutta–Zhukovsky theorem defines aerodynamic lift, and, based on such design, thousands of airplanes take off every day.

Reproducibility of research results can be defined as the requisite foundation to predict novel facts in life sciences. This is in line with expectations identified by Lakatos: "A given fact is explained scientifically only if a new fact is predicted with it" (1978). More specifically, there are three levels of reproducibility of published scientific results and underlying models:

- *Transparency* is the availability of complete information about the methodology that is sufficient for replication of the published study. Intermittently, journal space limitations or unreliable authorship can limit transparency for replication. A significant number of computational studies and commercially driven projects fail to provide full disclosure, effectively obstructing replication (Drummond 2009).
- *Replication* is repeating the research study and reaching the same primary outcome conclusions. This first replication can be self-replication that is conducted by the originating laboratory or replicative study performed by another laboratory based on the transparent methodology of the original study. Due to sampling variation and negligible deviations in methodology, exact duplication of every single result would not only be highly unlikely but also in effect unnecessary.

- *Triangulation* is confirmation of the study results and underlying models through entirely different methodologies and experiments. It relies on the convergence of measurements taken from different points. In his studies leading to the discovery of in vitro fertilization, Robert Edwards studied the successful in vitro maturation of eggs in rats, mice, and hamster (Johnson 2011). Triangulation confirmed reproducibility of his preclinical research results and accurately projected success in human trials.

Based on the many oral history recordings available, award-winning scientists frequently make references to evidence from multiple and diverse sources that are coming from unrelated laboratories, represent profoundly different methodologies, but point to the same direction of understanding a natural phenomenon. Conversely, the same scientists hardly ever highlight the level of statistical significance as a reason for trusting certain research results.

Triangulation is the integration of multiple observations and research results in the study of the same natural phenomenon. As in geometry and survey, the concept originally assumed two different points of observation, but generalization expands to measurements from multiple different points. Certainly, triangulation is a powerful technique that facilitates validation of data by cross-checking and profoundly deepens understanding of the investigated natural phenomena. Indeed, triangulation is a very potent logical tool for the development of new models and theories.

In his classic sourcebook of sociological methods, Norman Denzin (1978) identified four basic types of triangulation: (i) data triangulation means collecting the observations is multiple sampling strategies, (ii) investigator triangulation represents reliance on multiple researchers, (iii) methodological triangulation uses more than one method to gather data, and (iv) theoretical triangulation is the use of several different perspectives in data analysis.

One May Not Know in the Beginning How the End Will Look Like

To understand what concepts of research mean, it is useful to look at the *boundaries of science* and see what is outside and unacceptable. As pointed out earlier, the demarcation criterion of pseudoscience is the failure to make any novel predictions of previously unknown phenomena. It is eye opening to observe the many attempts of selling irreproducible assumptions to the public.

As Darwin's theory of evolution became universally accepted, the opposing view of intelligent design is widely questioned. In essence, it states that living organisms are best explained as being created by an intelligent designer, not an undirected process such as natural selection. There are various explanations provided for such assumptions, mostly referring to the complexity of living systems. In recent years, federal courts, school board deliberations, and the American Academy of Religion took the position that intelligent design cannot withstand scientific scrutiny and should not be taught in science classes (Kitzmiller v. Dover 2005; Goldenberg 2007; Moore and The AAR Religions in the Public-School Task Force 2010).

List of popular but in fact unscientific assumptions and prejudices is very long. Their common characteristic is the exaggerated impact statements without proportional substantiating evidence. One of the most illustrative among them is the politically motivated hate campaign against genetically modified food (GMO) in spite of the lack of demonstrated human harm that would be caused by such food. The belief that childhood immunization causes autism, a clear impediment to lifesaving immunizations, has also been investigated multiple times and proved to be unsubstantiated. The health impacts of the Chernobyl nuclear meltdown accident also appear to be exaggerated when in fact they were more in line with other, much less publicized industrial and mining accidents.

When scientific-sounding but in fact unscientific assumptions drive the development and sale of various products, the consequences can be not only much less beneficial than expected but in some cases directly harmful. Many traditional herbal remedies enjoy a high level of trust in spite of never being substantiated by sound scientific theories or observations.

For example, a weight-loss regimen that includes *Aristolochia fangchi*, an herb used in traditional Chinese medicine, turned out to be carcinogenic and trigger of rapidly progressing chronic kidney disease. The relationship between the herbal remedy and fatal consequences has not been recognized for a long time, partly due to prolonged latency and partly due to uncommon genetic susceptibility (Grollman and Marcus 2016).

Interestingly, most clinical interventions used in practice have never been tested in clinical trials. They do not meet the rigorous scientific criteria of evidence-based medicine. However, many of them are obviously very effective. Removal of the inflamed appendix is the nearly universal cure, and there is no need for clinical trials. Common experience shows that skipping surgery for open fractures can cause major illness and lead to death. Repositioning and healing the fractured limb is the right treatment without clinical trials.

The scientific experiment is not the only source of information that corroborates replicable observations. As a satirical article pointed out, jumping out of an airplane with or without an open parachute does not need randomized controlled clinical trial (Smith and Pell 2003). However, not requiring controlled scientific evidence has its major risks as several interventions widely believed to be beneficial turned out to be rather useless in controlled clinical trials. The Cochrane Collaboration on evidence-based medicine has several stunning systematic reviews of popular but ultimately useless clinical procedures (Garner et al. 2013).

Table 2.2 Side-by-side comparison of basic and applied research.

	Basic science	Applied research
Driver	Curiosity	Problem
Fundamental question	How this stuff works	How this problem could be solved
Funding	Largely government	Largely private
Industry relations	Rarely	Often recommended
Scientific result	New model	Practical solution
Understanding nature	Definitely	Some extent
Economic impact	Consequential application	Intellectual property
Public health impact	Through applied research	Direct
Overall impact	Broader	Particular
Chance for science prize	Fair	Minimal

Curiosity-driven science has been and undoubtedly will remain the essential ingredient of human progress (Table 2.2). Advancing basic science is going to take energy and thoughtfulness and the participation of a lot of dedicated researchers. The progress of life sciences faces many challenges, but they are all addressable.

The landmark report of the National Academies of Science (Committee on Prospering in the Global Economy of the 21st Century 2007) advocated strengthening the nation's traditional commitment to long-term basic research that has the potential to be transformational and generate the flow of new ideas that fuel the economy and enhance the quality of life.

Scientific discoveries succeed in a multidimensional space of intellectual contributions, public health improvement, and economic development. Among these dimensions, the intellectual and scientific contribution has always been and will remain at the center. Good science is the foundation for economic and public health progress. With improved management of the research process and evidence-based planning, the perspective of scientific and human progress has never been more exciting.

References

Bliss, C.I. (1964). RA Fisher's contributions to medicine and bioassay. *Biometrics* 20 (2): 273–285.

Bowell, T. and Kemp, G. (2014). *Critical Thinking: A Concise Guide*. New York: Routledge.

Cameron, W.B. (1957). The elements of statistical confusion: or: what does the mean mean? *AAUP Bulletin* 43 (1): 33–39.

Chang, H. (2012). Scientific pluralism and the mission of history and philosophy of science. Inaugural Lecture, Department of History and Philosophy of Science, University of Cambridge (11 October).

Collins, F.S., Anderson, J.M., Austin, C.P. et al. (2016). Basic science: bedrock of progress. *Science* 351 (6280): 1405–1405.

Committee on Prospering in the Global Economy of the 21st Century (2007). *Rising Above the Gathering Storm: Energizing and Employing America for a Brighter Economic Future*. Washington, DC: National Academies Press.

Darwin, C. (1859). *On the Origins of Species by Means of Natural Selection*, 247. London: Murray.

Denzin, N.K. (1978). *Sociological Methods: A Sourcebook*. Chicago: Aldine Publishing Company.

Drummond, C. (2009). Replicability is not reproducibility: nor is it good science. *Proceedings of the Evaluation Methods for Machine Learning Workshop at the 26th ICML*, Montreal, Canada.

Einstein, A. (1919). Time, space, and gravitation. *London Times* (28 November).

Fisher, R.A. (1925). *Statistical Methods for Research Workers*. Guildford: Genesis Publishing Pvt Ltd.

Frängsmyr, T. (ed.) (1990). *Les Prix Nobel. The Nobel Prizes 1989*. Stockholm: Nobel Foundation.

Garner, S., Docherty, M., Somner, J. et al. (2013). Reducing ineffective practice: challenges in identifying low-value health care using Cochrane systematic reviews. *Journal of Health Services Research & Policy* 18 (1): 6–12.

Goldenberg, S. (2007). Creationists defeated in Kansas school vote on science teaching. *The Guardian* (15 February).

Grollman, A.P. and Marcus, D.M. (2016). Global hazards of herbal remedies: lessons from Aristolochia. *EMBO Reports* 17 (5): 619–625.

Hamilton, D.P. (1991). Who's uncited now? *Science* 251 (4989): 25.

Ioannidis, J.P. (2005). Why most published research findings are false. *PLoS Medicine* 2 (8): e124.

Johnson, M.H. (2011). Robert Edwards: the path to IVF. *Reproductive Biomedicine Online* 23 (2): 245–262.

Kellert, S.H., Longino, H.E., and Waters, C.K. (eds.) (2006). *Scientific Pluralism [electronic resource]*, vol. 19. Minneapolis: University of Minnesota Press.

Kitzmiller v. Dover: Intelligent Design on Trial (20 December 2005). Berkeley, CA: National Center for Science Education.

Kolmogorov, A.N. (1950). *Foundations of the Theory of Probability*. New York: Chelsea Publishing.

Lakatos, I. (1978). *The Methodology of Scientific Research Programs*. Cambridge: Cambridge University Press.

Lowry, O.H., Rosebrough, N.J., Farr, A.L., and Randall, R.J. (1951). Protein measurement with the Folin phenol reagent. *Journal of Biological Chemistry* 193 (1): 265–275.

Macilwain, C. (2010). Science economics: what science is really worth. *Nature* 465 (7299): 682–684.

Moore, D.L. and The AAR Religions in the Public-School Task Force (2010). *Guidelines for Teaching About Religion in K-12 Public Schools in the United States* (April). The American Academy of Religions

Pfirman, S., Balsam, P., Bell, R. et al. (2007). Maximizing productivity and recognition, part 1: publication, citation, and impact. *Science* doi: 10.1126/science.caredit.a0700155.

Schekman, R. (2013). How journals like *Nature, Cell,* and *Science* are damaging science. *The Guardian* 9: 12–23.

Smith, A. (2008). Transcript of the telephone interview with Harald zur Hausen immediately following the announcement of the 2008 Nobel Prize in Physiology or Medicine. Nobelprize.org. Nobel Media AB 2014.

Smith, A. (2012). Interview with the 2012 Nobel Laureates in Physiology or Medicine Sir John B. Gurdon and Shinya Yamanaka (6 December). *Nobelprize.org.*

Smith, G.C. and Pell, J.P. (2003). Parachute use to prevent death and major trauma related to gravitational challenge: a systematic review of randomised controlled trials. *British Medical Journal* 327 (7429): 1459.

Thompson, E.A. (1990). RA Fisher's contributions to genetical statistics. *Biometrics* 46: 905–914.

Van Noorden, R., Maher, B., and Nuzzo, R. (2014). The top 100 papers. *Nature,* 514(7524), 550–553.

Wadman, M. (2013). NIH mulls rules for validating key results. *Nature* 500 (7460): 14–16.

3

Second Dimension

Public Health Value

The exciting thing is to see somebody who is doomed to die, live and be happy.

Willem J. Kolff (1991)[1]

Historically, basic science research has tremendously improved lives and benefited public health (e.g. development of new antibiotics, a better understanding of cancer-causing agents, compatibility in blood transfusion, and others). In many ways, improvement in public health, individual wellness, and community health status is the hallmark of ultimate success in biomedical research.

As illustrated most strikingly over the past 100 years, science allowed to nearly double the human lifespan and formidably expand understanding of health and illness. Since 1950, the mortality rates per 1000 population have fallen dramatically to less than half what it was in the beginning of this period (Figure 3.1). The spectacular drop in overall mortality rates and the corresponding increase in life expectancy can be directly linked to the progress in life sciences and subsequent public health actions (e.g. immunizations, seat belts, control of infectious diseases).

The concept of public health highlights the ultimate societal benefits, enhanced health, longer life, and reduced illnesses and disabilities in the community. The modern concept of public health emerged in the nineteenth century. It was established by four factors: (i) decision-making based on data and evidence, (ii) a focus on populations rather than individuals, (iii) a goal of social justice and equity, and (iv) emphasis on prevention rather than curative care (Koplan et al. 2009).

---

1 Kolff, W. (1991). Interview: Willem Kolff, pioneer of artificial organs (15 November). American Academy of Achievement, Salt Lake City, UT. http://www.achievement.org/achiever/willem-j-kolff/ (accessed 3 September 2016).

Innovative Research in Life Sciences: Pathways to Scientific Impact, Public Health Improvement, and Economic Progress, First Edition. E. Andrew Balas.
© 2019 John Wiley & Sons, Inc. Published 2019 by John Wiley & Sons, Inc.

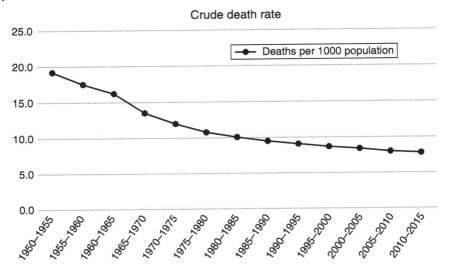

Figure 3.1 Crude death rate between 1950 and 2015. *Source:* Data from United Nations (2015).

According to the landmark Institute of Medicine report, "public health is what we, as a society, do collectively to assure the conditions for people to be healthy" (IOM 1988). The World Health Organization (WHO) defines its mission exclusively in societal and public health terms to publish and disseminate scientifically rigorous public health information of international significance that enables policy-makers, researchers and practitioners to be more effective; it aims to improve health, particularly among disadvantaged populations.

Improving Health in Mission Statements of Science

Scientific knowledge represents the fundamental reference point by showing how disease alters physiology, chemistry, or cell function; how the living system fights to protect function; and how an understanding of important functions may promise cure or prevention of diseases.

Basic biomedical research has no explicit disease link, but it is helpful to have a working knowledge of the primary physiologic systems and pathophysiologic reactions that appear to manifest in various diseases and respond to environmental stressors. This is true not just for research on the human body but also on the effects and vulnerabilities of infectious agents, like viruses, bacteria, parasites, and others.

Since 1901, the Nobel Prize in Physiology or Medicine has been awarded for outstanding discoveries in the fields of life sciences once a year. In his will, Alfred Nobel stated: "The said interest shall be divided into five equal parts, which shall be apportioned as follows: /- - -/ one part to the person who shall have made the most important discovery within the domain of physiology or medicine." After some wavering in the first few decades, the Nobel Prize in Physiology or Medicine decidedly turned toward rewarding fundamental discoveries about the nature and behavior of living systems.

Together with Berkeley professor Jennifer Doudna, French microbiologist Emmanuelle Charpentier, director of the Max Planck Institute for Infection Biology in Berlin, is the codiscoverer of the bacterial CRISPR/Cas9 immune system, a promising tool for genome editing. In a recent interview, she made the point about the motivation of her basic science pursuits: "I have started with bacterial genetics very early on but with always hope that one day I may find something, a mechanism that could be useful either for therapeutic purpose or technology purposes" (L'Oréal Foundation 2016).

Innovation is in the mind of the MIT team also working on the CRISPR/Cas9 immune system. When a member of his team proposes a project, leader of the laboratory Feng Zhang, PhD, asks: "Will it be a 'hack,' clever but inconsequential, or innovation?" (Begley 2015).

Defining the ultimate *mission of science* is not only a task that has great practical significance in directing the attention of the scientific community and researchers nationwide but also a tremendous challenge in itself. As one may suspect, there are many ways to express how research aims to benefit society.

Reviewing various expressions, mission statements of the major scientific organizations is very revealing about the directions and priorities:

- In its 2015 Annual Report, the National Academy of Medicine sets the foremost goal of shaping policy, advancing science, cultivating leadership, and, ultimately, improving human health.
- The prominent science philanthropy and stellar supporter of basic science research, the Howard Hughes Medical Institute states the following objectives: "promotion of human knowledge within the field of the basic sciences…and the effective application thereof for the benefit of mankind." In the fiscal year 2015, the Howard Hughes Medical Institute (HHMI) invested $666 million in biomedical research and $85 million in science education.
- The National Institutes of Health has the mission to "seek fundamental knowledge about the nature and behavior of living systems and the application of that knowledge to enhance health, lengthen life, and reduce illness and disability." Currently, NIH spends about $33 billion on supporting biomedical research annually.
- The Centers for Disease Control and Prevention (CDC) has the mission of developing and applying disease prevention and control, environmental health, and health promotion and health education activities designed to

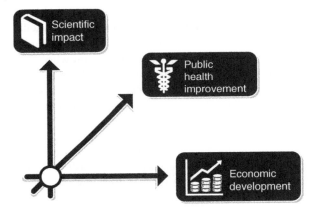

Figure 3.2 Three dimensions of success in life sciences innovation.

improve the health of the people of the United States. Obviously, this mission is almost entirely societal and applied to public health. The CDC budget was $11.5 billion in the fiscal year 2016.

- In comparison, the Stanford University's Office of Technology Licensing has the mission to promote the transfer of university technology for society's use and benefit. The Harvard Office of Technology Development promotes the public good by advancing science, fostering innovation, and translating new inventions into useful products that are available and beneficial to society.

The above list of mission statements underscores the general desire to make science responsive to public needs. As described earlier, the benefits of life sciences innovation are realized in a three-dimensional space of success (Figure 3.2).

Obviously, the road from basic science research to societal benefits like public health improvement is long and challenging. Improvement efforts need to address not only health care but also social determinants of health as pointed out by the World Health Organization (i.e. social gradient, stress, early life, social exclusion, work, unemployment, social support, addiction, food, transport) (Wilkinson and Marmot 2003). The transition often takes several decades but can be effective and successful. Correspondingly, no one should expect immediate practical benefits from basic science research.

The Road from Basic Research to Better Health

Detecting and curing disease at the individual level is one of the ultimate rewards of science; it also remains the most forward-looking opportunities to improve the life of large groups of people and communities in general.

NIH Director Francis Collins pointed to two of the three major outcome expectations of biomedical research: "the nation that invests in biomedical research will reap untold rewards in its economy and the health of its people" (Collins 2015).

For obvious reasons, landmark public health improvements tend to arise from fundamental scientific discoveries. The history of smallpox variolation and vaccination illustrates how a series of scientifically sound observations also become public health achievement leading to complete victory by eradicating a disease.

It was a common observation that survivors of smallpox became immune to the disease and they were called to nurse the afflicted patients in the Roman times. Variolation with the subcutaneous injection of smallpox virus into non-immune individuals was practiced in Africa, India, and China during the Middle Ages. In 1777, George Washington made sure that all his soldiers were variolated before new military operations.

In the eighteenth century, British physician Edward Jenner made tremendous progress by advocating the cowpox virus vaccination. Subsequently, smallpox vaccination programs have been developed in many countries worldwide. In 1800, Thomas Jefferson launched the National Vaccine Institute (Riedel 2005). As a result, the last naturally occurring smallpox case was reported in the United States in 1949. Ultimately, the World Health Organization declared eradication of smallpox worldwide in 1979.

Emil von Behring, a German scientist, is rightly considered being one of the founding fathers of modern immunology. He was born in Prussia and studied at the Military Medicine Academy of Berlin. After a few years of practicing medicine, he became a professor at the University of Marburg in spite of considerable opposition by the faculty.

While in the military, he already made some important observations regarding immunity and developed the humoral theory of immunity against infectious agents. Later in the Marburg years, his attention turned to the study of tetanus toxin and subsequently identification of the diphtheria toxin based on the pioneering work of Friedrich Loeffler, an associate of Robert Koch in Berlin (MacNalty 1954).

Behring's diphtheria research resulted in the discovery of serum therapies, an important method of passive immunization. His diphtheria serum therapy resulted in a dramatic drop in mortality. He later discovered that immunity could be produced by doses of diphtheria toxin neutralized by antitoxin, leading to the principle of active immunization. For his discoveries, von Behring received many recognitions, among them the first Nobel Prize ever awarded in Physiology or Medicine in 1901.

Public health has also been effective in inspiring and informing *basic research*. There are many excellent examples of basic researchers motivated by major public health needs and speaking passionately about the potential clinical and public health impact in their studies. In the history of life sciences, many important discoveries came out of research designed to find treatment for cancer but never had any significant impact on cancer. Several major research projects conducted by Otto Warburg or Albert Szent-Györgyi can illustrate the difference between motivation and actual results. The chapter on humanism and innovation provides further information about the complex interaction.

The pioneering cancer immunotherapy rituximab illustrates the winding road of ideas after initial discoveries to actual public health improvement. In this particular example, the evolution of immunology spans well over a century, starting with many key observations long before the discoveries of von Behring.

Immunotherapy has been a long-standing aspiration of research on cancer treatment. However, the series of failures appeared endless. Meanwhile, the discovery and characterization of B cells occurred in the mid-1960s. Max Cooper and Robert Good were the first to suggest a functional division of labor between antibody-producing B cells from the chicken bursa of Fabricius and T cells from the intact thymus for delayed-type hypersensitivity (Cooper et al. 1965). The discovery integrated information from experimental animal models, clinical studies of patients with immune deficiency diseases, and cell surface molecule characterization technologies.

The first promising discovery was the production of antibodies that target one specific portion, or epitope, of protein (Köhler and Milstein 1975). In 1984, Milstein and Köhler received the Nobel Prize in Physiology or Medicine "for theories concerning the specificity in development and control of the immune system and the discovery of the principle for production of monoclonal antibodies." In 1980, Stashenko and others identified a protein known as CD20 on mature B cells and almost all B-cell lymphomas and chronic lymphocytic leukemias.

The next decisive step was *in vitro* discovery of the functional monoclonal antibody, called rituximab, against the CD20 protein at IDEC Pharmaceuticals' laboratories in 1991 (Grillo-López 2000). As a result of an IDEC Pharmaceuticals and Genentech, Inc. collaboration, application in the treatment of non-Hodgkin's lymphoma was expedited. The first patient entered in phase I trials in March 1993, and the last patient entered in the phase III study in March 1996. Market approval for the treatment of relapsed or refractory, CD20-positive B-cell, low-grade or follicular non-Hodgkin's lymphoma was granted by the US Food and Drug Administration in 1997 and by the European Union in 1998.

Between 1985 and 1997, non-Hodgkin's lymphoma death rates had been continuously increasing in the United States. After the FDA approval in 1997, the trend reversed, and the death rates have been continuously decreasing (Howlader et al. 2016). The fast-tracked FDA approval made the rapid introduction into clinical practice possible. The elapsed time between successful completion of the clinical trial to widespread use and public health impact was far shorter than the estimated average time of 17 years for typical preventive care interventions (Balas and Boren 2000).

Life science is progressing at a pace unimaginable before. Today, there are more than 6000 diseases with known molecular basis. An entirely new treatment paradigm is emerging with the concept of taking cells out of the body, reengineer them, and putting them back to treat diseases and restore health. The first such treatment was just approved by FDA.

Breaking Down Silos: Basic Science and Public Health

Occasionally, the concept of *public health* is confused with the very limited practices of public health departments (e.g. water sanitation, immunization programs, health education, and others). Some basic scientists might become concerned with the risks of becoming subservient to another, occasionally very narrowly defined and essentially technical field.

In life sciences, the problem-driven and the *curiosity-driven research* often define themselves as separate and nonoverlapping. Once a high-ranking researcher summed it up: "HHMI is not particularly involved in public health – our primary mission is the support of basic biomedical research." Many basic scientists have mixed sentiments when it comes to recognizing public health as one of the ultimate, long-term outcomes of basic sciences.

In the light of the above much broader definitions of public health, such confusion is unwarranted. To prevent misperception, sometimes people talk about the public's health or occasionally population health or community health. This book uses the term public health based on the broadest definitions to describe the concept of societal benefits of life sciences.

Curiosity and desire to understand nature have always been and remain powerful drivers of basic research but with the ultimate purpose in mind. There is no reason to distress basic science researchers with unrealistic expectations like immediate societal or public health benefits.

The CDC has a much-cited list of the 10 great *public health achievements* in the twentieth century (Centers for Disease Control and Prevention (CDC) 1999). Cross-referencing this list with the list of Nobel Prize winners provides some insight into the distinction between basic science and public health research:

1) Control of infectious diseases (24 Nobel Prizes, including the first one in 1901).
2) Declines in deaths from heart disease and stroke (about a dozen Nobel Prizes).

3) Safer and healthier foods (about a dozen Nobel Prizes).
4) Immunizations (five Nobel Prizes).
5) Family planning (Edwards, several Peace Prize winners).
6) Healthier mothers and babies.
7) Fluoridation of drinking water.
8) Tobacco as a health hazard.
9) Workplace safety.
10) Motor vehicle safety.

The list also illustrates that basic science research has benefited public health repeatedly and in major ways. On the other hand, it is noteworthy that about half of the cited public health achievements that saved or helped millions of lives had no chance to be recognized as fundamental discoveries about the nature and behavior of living systems.

Wide-ranging recognition of public health achievements has been a challenge not just for award selection in medicine and physiology but also in the recognition systems of the public health community. For example, Bela Barenyi, one of the most prolific inventors in the history of automotive industry and pioneer of passive safety in car design, has never been fully understood, embraced, and recognized by either the life sciences or the public health community. His work has been one of the most influential contributions to achieving the historic improvement in motor vehicle safety.

The discovery of DDT (dichlorodiphenyltrichloroethane) is perhaps the only classic public health discovery and infection control measure that received the Nobel Prize in Physiology or Medicine. The DDT was synthesized and even patented many years before the ultimate discovery by Paul Hermann Müller, a Swiss chemist who worked for J. R. Geigy AG in Basel. His love for nature, plants, and the science of botany led to the study of plant protection and insect control. His studies on the differences between insect and human metabolism helped to identify chemicals with great pesticide potential. Ultimately, DDT is credited for saving millions of lives through mosquito control and prevention of a large number of diseases including malaria, dengue fever, and typhus.

More than 60 years after the discovery, DDT is still used but less widely due to resistance and environmental concerns. Today, DDT is banned or restricted in many countries because research findings had shown that pesticides can cause cancer as well as damage to the environment. Public health is finding new evidence-based approaches for mosquito control avoiding the use of chemicals.

In spite of controversies surrounding recognition of scientific and public health achievements, there is little doubt about the profound impact of basic science research on improving health in the community. Breaking down silos has great practical and theoretical significance. Basic science can guide prevention of diseases, lead to development of new interventions, and also influence policy actions. In a speech at the Society for Women's Health Research 2016 annual gala, Food and Drug Administration Commissioner Robert Califf summed it up: "no hocus-pocus gets around the importance of good basic science."

Putting Science to Work and Change Lives

Scurvy was a devastating disease and cause of many deaths especially among sailors on long sea voyages. Since the medieval times, repeated observations have been made regarding the value of citrus fruit, fresh vegetable food greens, and lemons or lime in preventing the disease.

Life scientists and leading research institutions are increasingly compelled to make their relevance to society clear in mission statements and other prominent announcements. When major diseases like cancer continue distressing

---

Albert Szent-Györgyi was an eminently productive and successful physiologist. He was descendent of a Hungarian landowner family in Marosvasarhely, Transylvania. After graduating in the medical school in Budapest (today called Semmelweis University), he quickly focused his research on the chemistry of cellular respiration.

At several major European universities, Szent-Györgyi worked extensively on the reactions of energy production in cells. His landmark discovery was the fumaric acid and several other key steps of what later became known as the tricarboxylic acid (TCA) cycle or the Krebs cycle. Among others, he also worked extensively on the identification of the mysterious antiscorbutic factor. At that time, shortage of practical sources of antiscorbutic factor hampered research progress.

It was a fortunate turning point when Szent-Györgyi accepted an appointment at the University of Szeged in Hungary. This area is the home of paprika that is particularly rich in vitamin C and represented the unprecedented source of the antiscorbutic factor for identification. Public health observations over more centuries have been instrumental in zeroing on the cause of scurvy and identification of vitamin C, the Nobel Prize-winning discovery of Albert Szent-Györgyi and his research fellow (Svirbely and Szent-Györgyi 1932). Using paprika, as a source of antiscorbutic factor, Szent-Györgyi successfully identified L-ascorbic acid as vitamin C prevents scurvy.

During World War II, Nobel laureate Szent-Györgyi made several but unsuccessful attempts to save his homeland from Hitler's war. After the war, he escaped communism and immigrated to the United States. In later years, his research focused on muscle function and the role of free radicals in cancer.

many families, and new diseases are spreading rapidly, the public wants to know how we are going to understand what is going on and how we can find cures to save lives.

Past achievements and public benefits of landmark scientific discoveries raise expectations very high. When millions and billions are spent, society also wants to know what returns and benefits can be expected. The pressure to make promises is enormous.

Some level of public health *outcome awareness* has the promise of increasing effectiveness and productivity in life sciences. It is often stated that major federal sponsors of biomedical research tend to fund what is physical and chemical. However, public health considerations could significantly strengthen the foundation of high-impact basic research by making projects better targeted.

Along these lines, NIH started to require grant applicants to expressly state in their application how their research "is relevant to public health." Obviously, this requirement is largely driven by public accountability for results and also by the desire to effectively present the case in congressional appropriations. Most basic scientists would interpret this requirement as "clinical relevance." Societal benefit is often understood as curing diseases as opposed to thinking about the direct impact on populations, including perhaps advancing prevention and wellness.

To secure support, basic scientists increasingly make references to *societal benefits* of their research. In a blog posted on 1 May 2008, NIH staff advised applicants to wow reviewers with public health relevance statements that can also be understood by a general lay audience. Subsequently, these public health relevance statements can be used by NIH for portfolio analyses, to develop reports to Congress, and to communicate the importance of research to the public.

The gradually growing pressure to produce applied research results and demonstrate public health impact lead to some undesirable patterns as well. For example, the National Institute of Neurological Disorders and Stroke (NINDS) noticed a significant decrease in the number of basic grants awarded between 1997 and 2012, mainly due to the declining ratio of basic science applications. The alarming trends led to the joint letter of all NIH directors underscoring that only funding for a broad portfolio of basic, translational, population, and clinical research can lead to a healthier future (Collins et al. 2016).

Several studies show that the preclinical research has a particular responsibility for delivering reliable results for the next steps of applied and clinical testing. An analysis of the success rates of 148 randomized controlled clinical trials found that the primary reasons for failure are lack of efficacy (56%), safety concerns (28%), shifting strategic priorities (7%), commercialization (5%), and operational issues (5%) (Arrowsmith and Miller 2013). Overly optimistic but ultimately unsustainable preclinical and animal research "discoveries" can lead to unnecessary clinical trials, failed commercialization, and dashed hopes of patients and public health.

A review of *preclinical studies* on amyotrophic lateral sclerosis reported that none of the tested compounds showed clinical efficacy in a group of more than 50 peer-reviewed research publications describing promising treatments that extend the lifespan of the mouse model. Apparently, most observed and published effects were only measurements of noise in the distribution of survival and not actual drug effects (Scott et al. 2008).

High-impact Research Driven by Public Health Needs

Ideally, both basic and applied sciences are part of successful scientific initiatives in life sciences. Public health could also better orient, support, and learn from basic research. More communication is needed between public health and basic science researchers.

The history of science is rich in landmark discoveries stemming from overlooked, disowned, and unrecognized but very real and major public health needs. In the 1970s population growth and contraception were recognized issues of public health significance. However, infertility and its treatment were viewed as nonexistent, dangerous, or irrelevant health issues. Moreover, research on *in vitro* fertilization was labeled as undesirable political and moral morass or worse. In spite of all the odds against it, in vitro fertilization was recognized, developed, and successfully tested. It also led to the Noble Prize of Robert Edwards. Today, about 60 000 IVF procedures are performed in the United States annually. Many happy families and babies can celebrate the success.

Unlike basic sciences, many applied biomedical research projects have a more immediate goal of guiding public health action. The selected targets of these projects make them uniquely positioned and effective in responding to societal needs.

Thomas Dawber is considered the founder epidemiologist of the Framingham Heart Study, the widely recognized benchmark for cardiovascular epidemiological studies. At the time of launching it, there was great uncertainty regarding the causes and evolution of cardiovascular diseases. The focus of this study was on factors suspected of causing predisposition to coronary heart disease. It was believed that epidemiological research is a prelude to prevention (Oppenheimer 2005)

This landmark cardiovascular risk factor cohort study started in 1948. The Framingham study was designed as more than a decade-long follow-up of people free initially of atherosclerotic or hypertensive cardiovascular disease. The study started with 5200 adult participants, and after 30 years, only 3% of participants dropped out (Kannel 1988).

The study of that size and that long duration required exceptional talents not just regarding science but also in securing continued funding, politics of the profession, and support by the larger community. As the study unfolded, there have been several changes and adjustments in the direction and methodology. However, the original priorities and the opportunities for major discoveries have been successfully retained and produced great results (Mahmood et al. 2014).

The longest-serving director of the study was Thomas Royle Dawber from 1949 to 1966 (Richmond 2006). With colleagues, he published more than 100 scientific papers. One of them was the landmark study in the *Annals of Internal Medicine* in 1961 that identified major risk factors of coronary heart disease (Kannel et al. 1961).

Among other findings, the Framingham study demonstrated that cigarette smoking, elevated cholesterol level, and elevated blood pressure play significant roles in the development of heart disease. Later it showed that exercise and elevated HDL levels reduce the risk of heart disease. The Framingham Risk Score predicts 10-year risk of future coronary heart disease. More recently, the Framingham study became an important source of genetic risk studies of cardiovascular diseases.

Clinical research that is driven by the needs of patients and public health expectations can have an enormously powerful impact through gradual changes. Life expectancy of patients with many major illnesses illustrates the exceptional power of small, incremental improvements in health-care services. The studies that lead to recognition of such best practices and resulting improvements are called health services and outcomes research.

Cystic fibrosis is an example of impressive results of small changes without magic pill discoveries. Over the last 50 years, life expectancy of patients nearly quadrupled from around 10 to 40 years of age, all without scientific breakthroughs in treatment. Today, some patients successfully reach retirement age.

Experts credit better coordination of care and a multitude of smaller but more skillfully used treatment options in the dramatic improvements in health outcomes: airway clearance techniques, inhaled medications, good nutrition, regular physical activity, and so-called modulator therapies to correct culpable dysfunction of the defective protein.

Cystic fibrosis patient care is best supported by accredited centers that have a full team of physicians, social worker, respiratory therapist, dietitian, and nurse. The capable, prepared, and comprehensive team approach shows its major benefits.

Activities of the Cystic Fibrosis Foundation have been particularly effective in improving the services: registry of patient cases, analyses of best practices, and exploration of therapeutic needs and opportunities.

The discovery of the genetic basis of cystic fibrosis decades ago has not had much impact on the dramatically improving outcomes. A few recently released drugs might turn the important genetic discoveries into future improvements.

Of course, patients want a cure, not just care. If you get hookworm disease, taking one single pill will eliminate the disease. Tetanus (lockjaw) is completely prevented by vaccination. Acute appendicitis will be cured by surgical removal of the appendix. Not always but most of the time, these treatments give the perfect cure.

Unfortunately, such perfect cure is not possible most of the time. Particularly, chronic diseases have no cure – that is why they go on for years. Many of these diseases are deadly without treatment, some of them rather quickly. The only meaningful option is to manage chronic diseases by a multitude of well-selected health-care interventions. Chronic diseases are the perfect example where population-based interventions are key to effectively manage them and reverse the impact of world epidemics.

There are many other diseases also illustrating the power of outcomes research. According to data from the Centers for Disease Control and Prevention, the death rate from coronary heart disease fell by 38% from 2003 to 2013 (Mozaffarian et al. 2016). One of the major reasons for the drop is that hospitals treat at least half their patients within an hour. No single breakthrough can be credited for the progress made; the drop has been the cumulative effect of better prevention, diagnosis, and treatment. Other examples of "no magic pills but much better practices" include childhood cancers, asthma, and many other diseases.

While exciting promises of major scientific discoveries get copious coverage, the accomplishments of meticulous outcomes research are rarely mentioned in the popular press. In spite of the improved general interest in supporting medical research, health services research remains grossly underappreciated.

The publicly funded meticulous, time-consuming search for better prevention, early detection, more effective treatment, and improved outcomes is essential for improving the life expectancy of patients with major and potentially devastating diseases, like cancer, heart disease, diabetes, and many others.

To advance basic biomedical research, the emerging precision medicine initiative of NIH takes into account individual variability in environment, genes, and lifestyle for each person to guide the development of effective prevention and treatment (Collins and Varmus 2015). Precision medicine should advance clinical practice, open up new opportunities for basic biomedical research studies, improve patient outcomes, and therefore advance the general health of the public. Critics of precision medicine highlight social inequalities and factors beyond genes or biology that influence population health and disadvantage citizens (Bayer and Galea 2015).

Biomedical research innovation and improvement in public health have a complex relationship. With the growing recognition of the amount of wasted, non-repeatable research and a simultaneous slowdown in innovation, the progress of basic biomedical research is becoming an immediate and urgent public health concern. In moving forward, more communication and collaboration are needed between public health and basic science researchers.

Ultimately, biomedical research needs to become successful from inception through *proof-of-concept* clinical studies to sustainable large-scale implementation to improve clinical outcomes and public health. By building on past successes, public health professionals could probably better orient, support, and also learn from basic research. It is reasonable expectation that public health outcome considerations will continue making a tremendous difference by positively influencing and accelerating basic science research.

The experience of benefiting from science demands new words in our vocabulary. As Queen Elizabeth noted in her 2017 Christmas speech, "I don't know that anyone had invented the term 'platinum' for a 70th wedding anniversary when I was born. You weren't expected to be around that long."

Science is the land of hope and great triumphs. Life expectancy has been going up worldwide, many major infectious disease epidemics are history, cardiovascular mortality rates have been going down, HIV/AIDS is turned into a chronic disease, and cancer death rates are going down.

Basic biomedical research has profoundly and positively influenced public health. We are in the midst of a remarkable stream of scientific advances. With further improvements and community health outcome considerations, the future of biomedical research has never been brighter. This is going to have a tremendous effect on the individual as well as community health in the years to come. In life sciences, we are at the threshold of enormously exciting times.

References

Mozaffarian, D., Benjamin, E.J., Go, A.S. et al. (2016). Heart disease and stroke statistics – 2016 update: a report from the American Heart Association. *Circulation* 133 (4): e38–e360.

Arrowsmith, J. and Miller, P. (2013). Trial watch: phase II and phase III attrition rates 2011–2012. *Nature Reviews Drug Discovery* 12 (8): 569–569.

Balas, E.A. and Boren, S.A. (2000). Managing clinical knowledge for healthcare improvement. *Yearbook of Medical Informatics* 2000 (2000): 65–70.

Bayer, R. and Galea, S. (2015). Public health in the precision-medicine era. *New England Journal of Medicine* 373 (6): 499–501.

Begley, S. (2015). Meet one of the world's most groundbreaking scientists. *Stat* (6 November).

Centers for Disease Control and Prevention (CDC) (1999). Ten great public health achievements – United States, 1900–1999. *MMWR: Morbidity and Mortality Weekly Report* 48 (12): 241.

Collins, F.S. (2015). Exceptional opportunities in medical science: a view from the National Institutes of Health. *JAMA: the Journal of the American Medical Association* 313: 131–132.

Collins, F.S. and Varmus, H. (2015). A new initiative on precision medicine. *New England Journal of Medicine* 372 (9): 793–795.

Collins, F.S., Anderson, J.M., Austin, C.P. et al. (2016). Basic science: bedrock of progress. *Science* 351 (6280): 1405–1405.

Cooper, M.D., Peterson, R.D., and Good, R.A. (1965). Delineation of the thymic and bursal lymphoid systems in the chicken. *Nature* 205: 143–146.

Grillo-López, A.J. (2000). Rituximab: an insider's historical perspective. *Seminars in Oncology* 27 (6 Suppl 12): 9–16.

Howlader, N., Noone, A.M., Krapcho, M. et al. (eds.) (2016). *SEER Cancer Statistics Review, 1975–2013*. Bethesda, MD: National Cancer Institute http://seer.cancer.gov/csr/1975_2013, based on November 2015 SEER data submission, posted to the SEER web site, April 2016.

Institute of Medicine (US) Committee for the Study of the Future of Public Health (1988). *The Future of Public Health* (vol. 88, no. 2). National Academy Press, Washington, DC.

Kannel, W.B. (1988). Contributions of the Framingham study to the conquest of coronary artery disease. *American Journal of Cardiology* 62: 1109–1112.

Kannel, W.B., Dawber, T.R., Kagan, A. et al. (1961). Factors of risk in the development of coronary heart disease – six-year follow-up experience: the Framingham study. *Annals of Internal Medicine* 55 (1): 33–50.

Köhler, G. and Milstein, C. (1975). Continuous cultures of fused cells secreting antibody of predefined specificity. *Nature* 256: 495–497.

Koplan, J.P., Bond, T.C., Merson, M.H. et al. (2009). Towards a common definition of global health. *The Lancet* 373 (9679): 1993–1995.

L'Oréal Foundation (2016). Pr. Emmanuelle Charpentier – L'Oréal-UNESCO Laureate 2016 – Germany (14 May) [Video file]. https://www.youtube.com/watch?v=xldariJBojY (accessed 21 June 2018).

MacNalty, A.S. (1954). Emil von Behring. *British Medical Journal* 1 (4863): 668.

Mahmood, S.S., Levy, D., Vasan, R.S., and Wang, T.J. (2014). The Framingham heart study and the epidemiology of cardiovascular disease: a historical perspective. *The Lancet* 383 (9921): 999–1008.

Oppenheimer, G.M. (2005). Becoming the Framingham study 1947–1950. *American Journal of Public Health* 95 (4): 602–610.

Richmond, C. (2006). Thomas Royle Dawber: founder epidemiologist of the Framingham heart study. *British Medical Journal* 332 (7533): 122.

Riedel, S. (2005). Edward Jenner and the history of smallpox and vaccination. *Proceedings (Baylor University Medical Center)* 18 (1): 21.

Scott, S., Kranz, J.E., Cole, J. et al. (2008). Design, power, and interpretation of studies in the standard murine model of ALS. *Amyotrophic Lateral Sclerosis* 9 (1): 4–15.

Stashenko, P., Nadler, L.M., Hardy, R.U., and Schlossman, S.F. (1980). Characterization of a human B lymphocyte-specific antigen. *The Journal of Immunology* 125 (4): 1678–1685.

Svirbely, J.L. and Szent-Györgyi, A. (1932). The chemical nature of vitamin C. *Biochemical Journal* 26 (3): 865.

United Nations, Department of Economic and Social Affairs, Population Division (2015). *World Population Prospects: The 2015 Revision*. Custom data acquired via website (28 August). New York: United Nations.

Wilkinson, R.G. and Marmot, M. (eds.) (2003). *Social Determinants of Health: the Solid Facts*. Geneva: World Health Organization.

4

Third Dimension

Economic Development

Good drugs save lives and a major side effect is they can also make you rich.
Raymond Schinazi (2011)[1]

Research in life sciences has been the source of not only exciting discoveries and significant improvements in public health but also economic development. In fact, key life sciences discoveries almost always have some undeniable business impact as they reach the practice of health care or public health. Often, this economic impact happens long after the publication of landmark research results and many other synergistic discoveries pave the way toward success (e.g. recognition of the role of DNA and genetically engineered cells decades later).

Countless forerunners of modern science have been credited with having an impact on businesses. Louis Pasteur, the profoundly influential discoverer of the germ theory of infectious diseases, was also known for the development of pasteurization, a process that kills bacteria in milk, beer, and other liquid beverage and prevents souring. Among others, his biomedical research discoveries opened ways to a large variety of novel food products and services that support good hygiene and prevention of diseases.

The significance of defining the economic dimension of life sciences research is also recognized at the national level. In requests for appropriations and congressional testimonies, congressmen and their legislative assistants expect science agency directors to routinely highlight the economic benefits of publicly funded research. "The impact of scientific research extends far beyond disease. Throughout history, advances in science and technology strengthened our economy, raised our standard of living, enhanced our global leadership" – stated the director of the National Institutes of Health at the Congressional House Appropriations hearings (Zerhouni 2008).

---

1 Berkrot, B. (2011). Gilead could have had Pharmasset cheap: founder. *Reuters* (22 November).

Innovative Research in Life Sciences: Pathways to Scientific Impact, Public Health Improvement, and Economic Progress, First Edition. E. Andrew Balas.
© 2019 John Wiley & Sons, Inc. Published 2019 by John Wiley & Sons, Inc.

Expectations are high for future economic returns on investments in research. Ben Bernanke, former chairman of the Federal Reserve, summed it up: "With full employment in sight, further economic growth will have to come...primarily from increases in productivity...monetary policy can no longer be the only game in town. Fiscal-policy makers in Congress need to step up. As a country, we need to do more to improve worker skills, foster capital investment and support research and development" (Bernanke 2015).

Economic Impact of Life Sciences Research

The economic impact of research in life sciences is a critical third dimension of research success. As it was pointed out earlier, the best of both worlds, science and invention, can often be realized simultaneously in life sciences innovation (see Figure 4.1).

It is a natural and ubiquitously appearing dimension in addition to the scientific intellectual contribution and public health impact. There are very few, if any, major life sciences discoveries that have not led to significant changes in the economic dimension.

A large variety of terms have been proposed to define the societal impact of research: societal benefits, usefulness, public values, practical knowledge, societal relevance, and others (Bornmann 2013). The various terms cover social, cultural, environmental, and economic returns of publicly funded research. At the national level, the impact-of-science studies tend to focus on the economic dimensions of societal benefits.

Pasteur's quadrant is a conceptualization of research introduced by Donald E. Stokes, political science professor and former dean of the Woodrow Wilson School of Public and International Affairs at Princeton University (Stokes 1997). In the quadrant model of scientific research, studies are inspired by the

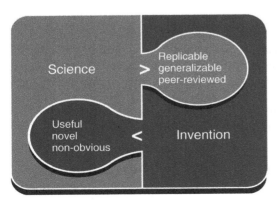

Figure 4.1 The best of both worlds in life sciences innovation.

quest for fundamental understanding and consideration of use (i.e. fundamental understanding + consideration of use).

The pure basic research aims fundamental understanding without consideration of potential use (i.e. yes + no in the quadrant). The pure applied research aims practical use without the desire to gain fundamental understanding (i.e. no + yes). The use-inspired basic research is exemplified by Pasteur's work, hence the name of the quadrant (yes + yes).

For many academic researchers, peer-reviewed publication of study results is the end of the road, and no consideration is given to what happens in the afterlife. It is simply assumed that someone else will read the paper, work on practical implementation, and achieve great societal benefits. In reality, a huge proportion of publications is never cited, not replicable, not feasible for practical application, or otherwise not worthy of use, and there is no one else to pick up the costly tasks of technology transfer to practice.

When the research results have an impact, they change the direction of subsequent projects, get translated into practice, and are used for the development of new products and services. Consequently, the ability to get the necessary support and financial sustainability becomes essential for continued and expanded impact. In economic terms, innovation happens in one of the two broad categories:

Sustaining innovation represents gradual or radical improvements in technologies (Christensen 1997). Development of insulin pumps, the discovery of the effective ingredient of hepatitis C drug Sovaldi, and development of the rotavirus vaccine or bionic prosthesis are examples of such innovation. As they open entirely new avenues of treatment and disease prevention, most celebrated life sciences innovations fall into this category.

Disruptive innovation defines technology that represents a new value proposition by being faster and cheaper than currently used technologies. In other words, disruptive innovation also introduces novel business models. Such innovation not only creates new and massively larger markets but also eventually disrupts existing markets, displacing established companies that fail to adapt.

The dramatic decline in the price of human genome sequencing is an example of such beneficial disruption with growing societal impact. Disruptive methodologies make a quantum leap by developing superior technologies (e.g. 454 DNA sequencing, Illumina, and SOLiD DNA sequencing). For example, the cost of human genome sequencing went down from $100 million in 2001 to around $1000 in 2015, thanks to the new technologies (Wetterstrand 2016).

The Diabetes Control and Complications Trial (DCCT) led to significant practice changes in diabetes management and greatly improved prevention of diabetes complications (DCCT 1993). The new, intensive diabetes management greatly reduced the prevalence of vision loss and cut cardiovascular complications by half. This profoundly influential diabetes study did not directly present any marketable and consumable products like drugs or devices, university patents, or intellectual property generating licensing revenues. However, the consequential practice change led to a wide range of improved products and services amounting to a very significant economic development impact.

After the DCCT publication, a range of new services and companies have been launched, creating jobs, generating economic progress, and also leading to many more productive years in the life of patients with diabetes. A comprehensive study of the Business Case for Diabetes Self-Management showed an excellent return on investment when comparing the cost of these programs with the improved quality-adjusted life-year (Kilpatrick and Brownson 2008).

Notably, results of the DCCT stimulated the development of several new insulin products that provide better support for intensive blood sugar management and more consistent blood sugar levels. The expectations of intensive management also led to the development of new blood sugar meters. Most notably, the DCCT substantially increased the demand for insulin pumps. In 2002, total sales of pumps and related supplies exceeded $1 billion per year (Kanakis et al. 2002). Ten years later, 14% of patients in England and Wales, 41% of patients in Germany and Austria, and 47% of patients in the United States were using insulin pumps (Forlenza et al. 2016). An entire industry of patient management services has emerged to provide reminders, records, and logistical support for continuous diabetes care.

The intertwining relationship between life sciences discoveries and economic progress is not surprising especially in light of significant public health impact of every major business functions. Along these lines, John Quelch, a professor at Harvard T.H. Chan School of Public Health and Harvard Business School pointed out that every company has a public health footprint (Quelch 2015). The four aspects of the public health footprint include consumer health, employee health, community health, and environmental health. Any research that ultimately impacts one or more of these four aspects also influences the success of businesses.

On the other hand, partnering with business at the time of conducting the research has been a matter of intense debates in the scientific community and the general public. It offers significant benefits both ways but also uniquely important risks. The chapter on engaged research provides more information about the role of business partnerships and corporate support in conducting research.

Blockbuster Impact of University Research

The societal impact of countless life sciences innovations is enormous. In addition, many of these products have an economic impact in the millions of dollars or more. Most remarkably, pharmaceutical drugs that generate at least $1 billion for the company in annual sales are called *blockbuster drugs*.

The following examples illustrate the wide-ranging and enormous economic impact of well-targeted and managed biomedical research innovation.

The effects and metabolism of cholesterol were the subjects of numerous breakthrough research projects over several decades. A major part of the credit goes to Nobel Prize winners Michael Brown and Joseph Goldstein for the discoveries related to the regulation of cholesterol metabolism. Subsequently, Akira Endo, a Japanese pharmaceutical researcher, developed the first statin drug to reduce serum cholesterol, a significant factor behind cardiovascular mortality. In the subsequent search for more potent and safe statin drugs, the industry research team led by Bruce D. Roth developed atorvastatin, which would become the drug Lipitor. For several years, Lipitor provided one-quarter of the total revenues of Pfizer. This treatment got an enormous boost when a research study showed that it decreases not only bad cholesterol but more importantly coronary death in comparison with placebo (Scandinavian Simvastatin Survival Study Group 1994). Ultimately and over a period of 15 years, Lipitor generated $130 billion revenues in sales (News in Brief 2011).

Infliximab (Remicade) illustrates not only the value of blockbuster drugs but also the momentous role of university research. At the New York University School of Medicine, Jan Vilček and Junming Le developed a monoclonal antibody against TNF-alpha, a major promoter of inflammation in numerous chronic inflammatory autoimmune diseases in the 1980s. The industry partnership with Centocor Biotech, Inc., was instrumental in making the development successful. Centocor later became a subsidiary of Johnson & Johnson.

Remicade was first approved for use in the treatment of Crohn's disease in 1998. Today, Remicade is effectively used in the treatment of Crohn's disease, rheumatoid arthritis, ankylosing spondylitis, ulcerative colitis, psoriasis, psoriatic arthritis, and many other inflammatory diseases. In 2007, Royalty Pharma purchased a portion of NYU's worldwide royalty interest in Remicade for $650 million in up-front cash plus additional payments should yearly sales of Remicade exceed certain agreed thresholds. Again, a significant portion of the multiyear revenues went back to the research and inventors.

Jan Vilček was born into a physician family in Bratislava, Czechoslovakia (now Slovakia). During the Holocaust, he and his family had to hide for an extended period. After the war, he graduated in the Comenius University medical school, got his PhD, and started working as a researcher in immunology and microbiology. He became particularly interested in the study of cytokines, natural regulators of the immune system.

In 1965, Jan Vilček immigrated to the United States and joined the New York University Medical School. Continuing and expanding the very promising research, he wrote 350 publications and developed 46 patents. Among his many research and development successes, the discovery of infliximab was undoubtedly the most prosperous (Turner 2016). The resulting blockbuster drug generated significant royalty revenues for NYU and also for him. In 2013, Vilček received the National Medal of Technology and Innovation from President Obama.

Together with his wife Marica, Jan Vilček made major philanthropic donations. Most notable among them is $120 million to New York University to fund scholarships, research, and the new medical student residence hall. Furthermore, they also launched the Vilček Foundation to raise awareness of immigrant achievements and foster appreciation of the arts and sciences. Jan Vilček's life exemplifies not only pursuit of good science but also the generous appreciation of arts and diversity of cultures.

Genetic engineering research done by Stanley Cohen of Stanford University and Herbert Boyer of the University of California, San Francisco, led to three landmark patented technologies in the early 1970s. The method of splicing genes to make recombinant proteins proved to be foundational to the biotechnology industry (Cohen 2013). The patented technologies were licensed to 468 companies, applied in an estimated 2442 new products, and generated over US$35 billion in sales. The landmark Cohen–Boyer patents have been cited by over 260 more patents as prior art (Feldman et al. 2007).

The universities and the inventors themselves also benefited from the technological breakthroughs. Over the 17-year lifetime of the patents, the two universities received more than $250 million in licensing revenues before the patent expired in December 1997. Meanwhile, Cohen and Boyer have waived rights to personal royalties (UPI 1984).

In 1987, John J. Kopchick, professor of molecular biology at the Heritage College of Osteopathic Medicine, and his graduate student Wen Chen discovered pegvisomant, the growth hormone receptor antagonist. It was approved by the FDA under the commercial name Somavert. Approximately, 40 000 individuals are diagnosed with acromegaly worldwide. Ohio University and the inventors received more than $80 million from the license to Pfizer (Ohio Research 2012). Kopchick published over 300 manuscripts related to the structure and action of the growth hormone and its relationship to obesity, diabetes, and aging. In 2014, John and Char Kopchick gave a $2 million donation to support three newly established scientific and educational award programs at Ohio University.

Gilead's Sovaldi (sofosbuvir) is a drug for the treatment of chronic hepatitis C. It had the best-selling entry in pharmaceutical history by sales reaching $5.8 billion in the first half of 2014 (Cookson 2014). Sovaldi was designed by Raymond Schinazi at Emory University and Michael Sofia of Pharmasset. Schinazi cofounded and sold several pharmaceutical companies for hundreds of millions of dollars, including Triangle Pharmaceuticals, Idenix Pharmaceuticals, Pharmasset, and RFS Pharma, LLC. Schinazi published over 420 peer-reviewed articles and held more than 80 US patents.

It is also very revealing how one of the most entrepreneurial biomedical researchers of all times, Raymond Schinazi, summed up the history of Pharmasset, the company that was sold to Gilead for $11 billion: "I coined that name. It's actually 'pharmaceutical assets' and the idea was to create assets that would be sold to companies. That was the initial business plan" (Berkrot 2011). Obviously, knowing what companies desire and value can tremendously influence such research and development. It can greatly expedite the commercialization of all results, including research targets and also accidental by-products. At the time of the Gilead sale, Schinazi owned 4% of Pharmasset.

Blockbuster drugs should not be confused with the most expensive drugs. In 2010, Alexion Pharmaceuticals' Soliris, the only therapy approved for the treatment of patients with paroxysmal nocturnal hemoglobinuria to reduce hemolysis, was priced at $409 500 a year. It was the world's single most expensive drug (Herper 2010). In general terms, the fewer patients a drug helps, the more it costs. Conversely, the lack of financial return potential severely limits pharmaceutical research that otherwise could pay more attention to diseases of the developing world.

Good Science Must Come First in Commercialization

While the rewards of success can be enormous, life sciences researchers need to fully recognize the many hazards of technology transfer on the road to societal impact. Early recognition of long-term implications can keep staying focused. Well-targeted, rigorous research projects can go a long way toward impact.

In general terms, the impact of scientific research may take a long time to come through. The lengthy transfer time represents a particular challenge of outcome evaluations as links to the university can be forgotten.

To maximize the chances of societal benefits and minimize expenses along the way, researchers need to know what happens to their research results after they get out of the laboratory either through publication in a peer-reviewed scientific journal or intellectual property disclosure and subsequent commercialization.

From a technology development and transfer perspective, the scientific process of biomedical research can be divided into the following phases:

1) *Basic research* exploring the fundamental function without direct disease relevance.
2) *Preclinical research* investigates compounds and technologies in the laboratory using biochemistry, cell lines, animal models, computational simulation, and other experimentation without human subjects.

3) *Clinical research* that studies, based on the reassuring safety and efficacy projections of preclinical research, the effect of new compounds or technologies on human subjects. Appropriate controlled clinical trials can be:
   a) Phase I trial: In a small group of human subjects, safety is evaluated, tolerance is tested, and a safe dosage range is specified.
   b) Phase II trial: In a larger group of people, the effectiveness of the intervention is evaluated, and information is collected about safety and side effects.
   c) Phase III trial: In large groups of people and often multicenter study, effectiveness is confirmed, the frequency of side effects is estimated, and a comparison is made with commonly used treatments.

As described earlier, choices of the communication channel have a profound influence on the success of bench-to-bedside technology transfer. Evaluating the commercial potential of a new technology is not a trivial matter, and it is not for someone lacking business skills.

Louis Pasteur was a French chemist and microbiologist, father of modern hygiene, and one of the most important pioneers of microbiology and immunology. As a child, he was influenced by the screams of villagers having their wolf bites treated with a hot iron by the local blacksmith to prevent rabies. He became interested in science at the urging of his father.

In his PhD work, he discovered different stereospecific forms of the same molecule, tartaric acid from plants. As professor of chemistry, Pasteur visited local distilleries with his students and offered his help. Indeed, the father of one of his students asked him to investigate souring in alcohol. This request led to the discovery of fermentation and also the demonstration of organisms causing it. Subsequently, Pasteur was asked by the Department of Agriculture to investigate devastating silkworms that threatened the silk industry. His landmark studies lead to better understanding of infections and infectious diseases. He also pointed out factors, such as temperature, humidity, ventilation, and sanitation, that influence susceptibility to disease. His further studies on chicken cholera, anthrax, and rabies advanced understanding of infections and vaccine development. Controversies started when some of his studies could not be replicated, and others turned out to be replications of previous discoveries (Ligon 2002).

As a biotechnology pioneer, Pasteur worked with distilleries, development of sheep and cattle vaccines, and "pasteurization," heating to a below boiling point for a brief period for food preservation and prevention of infection transmission. When his fame started generating donations and subscriptions, he was able to build the Pasteur Institute in Paris. Ultimately, his studies led to enormous cost savings in agriculture and developed food industry technologies still used today (Smith 2012). His work also laid the scientific foundation of modern asepsis in health care.

Without Good Science, There Is No Positive Technology Transfer

Commercialization of new technologies should never be separated from science or attempt to go against scientific principles. The celebration of successful commercialization of life sciences research results should never be equated with lucrative adventures without scientific basis.

Sometimes, overzealous business interests try to get ahead of replicable research by promising and selling more than what is supported by sound science. Statements that outside organizations will evaluate the technology or data will be presented at some point in the future ring hollow. Publicly disclosed evidence on the replicability and real-world value of new technologies is essential to gain public trust and warrant continued reimbursement. Refusal to publish data in peer-reviewed journals is a hazardous defiance of scientific convention that also undermines long-term business viability.

When people spend on useless products or harmful services, sooner or later the complaints will start, and victims demand refund and compensation. In other cases, share values tumble, and investors flee. The business is halted and regulators crack down (e.g. consumer protection agencies, Federal Trade Commission, Food and Drug Administration, Centers for Medicare and Medicaid Services, and many others). Careless infatuations with the business potential of scientific sounding ideas tend to come to an ugly end.

The following examples illustrate how success limited to the economic dimension and profit making can be undermined and invalidated by the deficiencies in the scientific contribution and public health value dimensions. One-dimensional financial success, without repeatable research results and public health benefits, does not work for the long run in life sciences.

The blockbuster Merck drug Vioxx illustrates how defects of the scientific process and also lack of attention to public health concerns can undermine the very existence of one of the financially most successful drugs ever developed (Topol 2004). Vioxx was an effective blockbuster prescription drug to relieve signs and symptoms of arthritis, acute pain, and painful menstrual cycles.

Several years after the introduction and widespread use of Vioxx, clinical data started to emerge that Vioxx might cause an increased risk of cardiovascular diseases such as heart attack and strokes during chronic use. In 2004, the FDA issued a public health advisory concerning the use of Vioxx, and Merck decided withdrawal of this product. Around the time of withdrawal, several more controversies erupted including the alleged withholding of important scientific information about various risks and lawsuits by families of deceased patients to get compensation from the manufacturer.

Launched in 2003, Theranos, a Palo Alto, California-based company, made its mission to revolutionize laboratory diagnostics. It started with the correct observation that contemporary diagnostics use large amounts of blood while the microfluidics technology is maturing. According to various estimates, much smaller amounts should be sufficient for accurate measurements. Based on this promise and with significant venture capital support, the privately owned company achieved spectacular growth, making the owner one of the youngest billionaires.

Later, question marks started to accumulate around the company and its blood testing device. Apparently, the Theranos testing device, designed to use only a few drops of finger-prick blood, had not been the subject of open peer-reviewed research studies (Paradis 2016). Peer review is a time-honored and valuable exercise not just in science but also in the commercialization of life sciences technologies. Stanford Professor Ioannidis was among the first to raise concerns about the lack of Theranos publications: "How can the validity of the claims be assessed if the evidence is not within reach of other scientists to evaluate and scrutinize?" (Khan 2017).

The Wall Street Journal started raising concerns about whether the company's technology worked (Carreyrou 2015). In 2015, an investigation by the Food and Drug Administration found that the company's small blood collection device, called nanotainer, was an "uncleared" medical device. Apparently, Theranos was built up by very attractive marketing but without the necessary scientific foundation.

In 1984, the inventor of interleukin-2 fusion protein toxin, a genetically engineered compound with cancer-fighting potential, joined a university. Five years later, human trials begin on interleukin-2. The scientific idea and the partnering company Seragen were enthusiastically embraced by the president of the university, a former philosophy professor. He started investing millions of dollars of university money into the company. Meanwhile, the company went public in April 1992. At some point, the university's stake reached $84 million.

When the FDA approval was slower than expected and the great promises of the new drug turned out to be much narrower than expected, the company lost its funding (Barboza 1998). In 1998, Seragen stock, which was $12 a share at its height, closed at 45¢ a share. Subsequently, the company was delisted by NASDAQ. Ultimately, the university lost 90% of its investment. Apparently, a dysfunctional, high-risk arrangement existed among an enthusiastic biomedical researcher, entrepreneur marketer, and an angel investor all working within the same institution. It is especially dangerous when a grant funded and subsidized university is misled by hopes and conflicts of interest.

Brain training is a billion-dollar industry that is expected to further grow in the future. Clearly, activities of the marketers of brain training technologies have attracted the attention of the Federal Trade Commission. Recently, a University of California researcher experienced trouble by claiming that certain technologies improve the vision of users among people of all ages, genders, and visual abilities (FTC 2016). In the absence of randomized, double-blind, and adequately controlled evidence substantiating the health benefits, the FTC ordered discontinuation of such claims and imposed a significant fine.

Impact of Life Sciences on the National Economy

The aggregate economic impact of life sciences research goes far beyond major gains for universities and companies. Through its budgetary need for support and more importantly through high-impact commercialized results, life sciences research has had a significant and rapidly growing impact on economies at national and international levels. A recent landmark analysis of biomedical research in the United States focused on total public and private investment and personnel (economic inputs) and on resulting patents, publications, drug and device approvals, and value created (economic outputs) (Moses et al. 2015). The analysis urged new investments to realize the clinical value of past scientific discoveries and opportunities to improve care (e.g. repatriation of foreign capital, new innovation bonds, administrative savings, and public–private risk sharing collaborations).

There have been many more very insightful studies on the economic impact of university innovation and scientific research. The short-term economic impact of life sciences research can include new jobs, revenues, taxes, better turnover, increased profit, new startup companies, and expansion of established companies into new areas, innovative products, and services. The long-term economic impact may include more productive citizen working years as a result of better treatments, reduction in health-care costs, and improved environmental footprint. Particularly the following categories of economic impact may occur.

Direct Budget Impact. Spending on research and subsequent development is a significant component of the budget of most organizations and national economies. Currently, the budget of the National Institutes of Health's is about $37 billion (NIH 2016). NIH supports about 490 000 jobs each year. Unfortunately, due to lack of increases combined with inflation, the NIH budget has lost nearly 25% of its purchasing power over the recent decade (Collins 2015).

Adding the research and development budgets of the Department of Agriculture, Department of Defense, Department of Energy, Department of Health and Human Services, National Aeronautics and Space Administration, and National Science Foundation, the total federal spending on research is approximate $63 billion annually. More than half of the annual spending from these funds goes toward life sciences research, particularly biomedical research and health sciences (Yamaner 2016).

Using direct budget or grant funding as a measure of economic impact is a double-edged sword. Obviously, spending more taxpayer money on anything, not just research, will pay for more jobs. However, using already received grant funding as a reason to ask for more grant funding is probably not the circular outcome expected by taxpayers. Therefore, such statements should be avoided in project proposals and reports of accomplishments.

One of the most common but least impressive conclusions in scientific abstracts is that "our research demonstrates the need for more research in this area." The result of research cannot be simply a request for more grant funding. Productive scientists are expected to generate more compelling, impactful results for use and benefit of society.

Consequential Economic Impact. The research results coming out of laboratories funded by NIH and other major sponsors of life sciences provide the foundation for the US biomedical industry, which generated $69 billion for the national GDP and supported 7 million jobs in 2011 (NIH 2016).

Among the most insightful reports of the economic impact of life sciences is the study by the Battelle group on the human genome project (Tripp and Grueber 2011). The methodology of the study considered six major economic sectors including (i) bioinformatics computer programming services, (ii) testing medical and diagnostic labs, (iii) biologics and diagnostic substances, instruments and equipment, (iv) analytical laboratory instrument, (v) genomics biotech and scientific R&D services, and (vi) devices, drugs, and pharmaceuticals. Based on this information, the study report concluded that the full impact is just beginning to unfold. It is already clear that approximately the $3.8 billion investment into the human genome research project generated about $796 billion in economic impact and created over 300 000 jobs. Such national economic impact represents more than 200-fold return on investment, furthermore launching the genomic revolution medicine with much further impact on new treatments and better chances of improving health.

University Graduate Entrepreneurship. Several major studies highlighted the entrepreneurial impact of leading research universities. The studies primarily focused on the role of alumni in launching new companies. The data are quite compelling that many alumni not only create tremendous wealth but also the rate of launching new businesses is accelerating. Apparently, the economic impact of alumni could not be achieved without the entrepreneurial culture, educational impact, and research innovation competence of university faculty.

As an example, two comprehensive studies focused on the Massachusetts Institute of Technology (MIT) and the creativity of its alumni (Roberts and Eesley 2009; Roberts et al. 2015). It is estimated that MIT alumni launched more than 30 000 currently active companies that employ about 4.6 million people and produce nearly $2 trillion of revenue. Between the 1960s and 2014, the entrepreneurial growth has been particularly strong in the areas of health, medicine, pharmaceutics, biotech, and medical devices.

The comparable numbers from Stanford University in California are equally impressive (Eesley and Miller 2012). The companies launched by alumni include Agilent Technologies, Jawbone, Intuitive Surgical, Google, Facebook, and thousands of others. In the aggregate, the output of companies launched by Stanford alumni would be the equivalent of the world's 10th largest independent nation economy.

Diverse Messages About Economic Impact at Different Levels

Undoubtedly, the economic impact is the important third dimension of most prominent scientific discoveries. It is worth noting that federal research agencies present a much more nuanced and economy-oriented picture of the societal impact of science then realized by the researchers whom they support (UMR 2013).

Speaking about the economic impact of scientific research is necessary for congressional discussions about budget allocation. The economic impact is also acceptable for discussion in big picture presentations of institutional leaders. Considering the treacherous road to societal impact, it is not surprising to hear at different levels mixed messages regarding what is truly desirable in the outcomes of the research enterprise.

In 2013, presidents and chancellors of 165 universities representing all 50 states urged President Obama and members of the 113th Congress to close the innovation gap and increase investments in research and education: "Ignoring the innovation deficit will have serious consequences: a less prepared, less highly skilled U.S. workforce, fewer U.S.-based scientific and technological breakthroughs, fewer U.S.-based patents, and fewer U.S. start-ups, products, and jobs" (Association of American Universities 2018).

On the other hand, these leaders returned to universities where promotion and tenure systems largely ignore breakthroughs, patents, startups, products, and jobs created. Instead, most academic reward systems emphasize traditional indicators of research productivity like the number of peer-reviewed articles, citation numbers, and competitive research grants received. The promise is the economic return on research investment, but the university and academic community rewards are for something else.

Often, speaking about business implications and economic impact is viewed as distraction in scientific discussions. At scientific conferences and in research review panel discussions, mentioning economic impact can be hazardous and counterproductive.

Many scientists view such discussions as an indication of hidden business motivations and powerful source of bias in research projects. Due to these risks, researchers should exercise extreme caution in making references to the economic impact of their research in project proposals and scientific publications.

In reality, learning about the basics of business and, in certain situations, discussing the ultimate mission of science including three-dimensional research success should help researchers in positioning their project plans and projecting the triple value of anticipated results to peer reviewers.

The economic potential of life sciences research should be used as wisely as possible. There are many successful examples of developing a culture of innovation, translational research, technology transfer competence, and risk-taking in science. The opportunities for economic impact are challenging but also major reasons why we should be excited about the potential of science in improving lives.

References

Association of American Universities (2014). Close the innovation deficit. https://www.aau.edu/key-issues/close-innovation-deficit-infographic-sheet (accessed 30 June 2018).

Barboza, D. (1998). Loving a stock, not wisely but too well: the price of obsession with a promising start-up. *The New York Times* (20 September).

Berkrot, B. (2011). Gilead could have had Pharmasset cheap: founder. *Reuters* (22 November).

Bernanke, B.S. (2015). How the fed saved the economy. *The Wall Street Journal* (4 October).

Bornmann, L. (2013). What is societal impact of research and how can it be assessed? A literature survey. *Journal of the American Society for Information Science and Technology* 64 (2): 217–233.

Carreyrou, J. (2015). Hot startup Theranos has struggled with its blood-test technology. *Wall Street Journal*.

Christensen, C.M. (1997). *The Innovator's Dilemma: When New Technologies Cause Great Firms to Fail*. Boston, MA: Harvard Business School Press.

Cohen, S.N. (2013). DNA cloning: a personal view after 40 years. *Proceedings of the National Academy of Sciences* 110 (39): 15521–15529.

Collins, F.S. (2015). Exceptional opportunities in medical science: a view from the National Institutes of Health. *JAMA: the Journal of the American Medical Association* 313 (2): 131–132.

Cookson, C. (2014). Raymond Schinazi fled Nasser's Egypt to become pioneer in antivirals. *Financial Times* (27 July).

Diabetes Control and Complications Trial Research Group (1993). The effect of intensive treatment of diabetes on the development and progression of long-term complications in insulin-dependent diabetes mellitus. *The New England Journal of Medicine* 1993 (329): 977–986.

Eesley, C.E. and Miller, W.F. (2018). Impact: Stanford University's economic impact via innovation and entrepreneurship. *Foundations and Trends® in Entrepreneurship* 14 (2): 130–278.

Feldman, M.P., Colaianni, A., and Liu, C.K. (2007). Lessons from the commercialization of the Cohen-Boyer patents: the Stanford University Licensing Program. In: *Intellectual Property Management in Health and Agricultural Innovation: A Handbook of Best Practices 1797, 1806* (ed. A. Krattiger and R.T. Mahoney). MIHR, PIPRA, Oswaldo Cruz Foundation, and *bio*Developments-International Institute.

Forlenza, G.P., Buckingham, B., and Maahs, D.M. (2016). Progress in diabetes technology: developments in insulin pumps, continuous glucose monitors, and progress towards the artificial pancreas. *Journal of Pediatrics* 169: 13–20.

FTC (2016). FTC approves final order prohibiting "Ultimeyes" manufacturer from making deceptive claims that the app can improve users' vision (23 February). https://www.ftc.gov/news-events/press-releases/2016/02/ftc-approves-final-order-prohibiting-ultimeyes-manufacturer (accessed 28 December 2017).

Herper, M. (2010). The World's most expensive drugs. *Forbes* (22 February).

Kanakis, S.J., Watts, C., and Leichter, S.B. (2002). The business of insulin pumps in diabetes care: clinical and economic considerations. *Clinical Diabetes* 20 (4): 214–216.

Khan, R. (2017). Theranos' $9 billion evaporated: Stanford expert whose questions ignited the Unicorn's trouble. *Forbes* (17 February).

Kilpatrick, K.E. and Brownson, C.A. (2008). *Building the Business Case for Diabetes Self-Management: A Handbook for Program Managers*. St. Louis: Diabetes Initiative, National Program Office at Washington University School of Medicine.

Ligon, B.L. (2002). Biography: Louis Pasteur: a controversial figure in a debate on scientific ethics. In *Seminars in Pediatric Infectious Diseases* (vol. 13, no. 2, pp. 134–141). Philadelphia: WB Saunders.

Moses, H., Matheson, D.H., Cairns-Smith, S. et al. (2015). The anatomy of medical research: US and international comparisons. *JAMA: the Journal of the American Medical Association* 313 (2): 174–189.

News in Brief (2011). End of the Lipitor era. *Nature Reviews Drug Discovery* 10: 889.

NIH (2016). Impact of NIH research. https://www.nih.gov/about-nih/what-we-do/impact-nih-research/our-society (accessed 26 May 2016).

Ohio Research (2012). Kopchick receives Ohio patent impact award. *Research Communications* (3 April). https://www.ohio.edu/research/communications/patentimpact.cfm (accessed 17 September 2016).

Paradis NA. (2016). The rise and fall of Theranos. *The Conversation* (22 April).

Quelch, J.A. (2015). *Consumers, Corporations, and Public Health: A Case-based Approach to Sustainable Business*. New York: Oxford University Press.

Roberts, E.B. and Eesley, C. (2009). *Entrepreneurial Impact: The Role of MIT*. Boston, MA: MIT.

Roberts, E.B., Murray, F., and Kim, J.D. (2015). *Entrepreneurship and Innovation at MIT Continuing Global Growth and Impact*. Boston, MA: MIT.

Scandinavian Simvastatin Survival Study Group (1994). Randomised trial of cholesterol lowering in 4444 patients with coronary heart disease: the Scandinavian Simvastatin Survival Study (4S). *The Lancet* 344 (8934): 1383–1389.

Smith, K.A. (2012). Louis Pasteur, the father of immunology? *Frontiers in Immunology* 3: 68.

Stokes, D.E. (1997). *Pasteur's Quadrant: Basic Science and Technological Innovation*. Washington, DC: Brookings Institution Press.

Topol, E.J. (2004). Failing the public health—rofecoxib, Merck, and the FDA. *New England Journal of Medicine* 351 (17): 1707–1709.

Tripp, S. and Grueber, M. (2011). *Economic Impact of the Human Genome Project*. Battelle Memorial Institute. http://www.battelle.org/docs/default-document-library/economic_impact_of_the_human_genome_project.pdf?sfvrsn=2 (accessed 21 June 2018).

Turner, M. (2016). Drug discovery: a life of tumult and triumph. *Nature* 530 (7589): 157–158.

UMR (2013). *Profiles of Prosperity: How NIH-Supported Research Is Fueling Private Sector Growth and Innovation*. United for Medical Research. http://www.unitedformedicalresearch.com/advocacy_reports/profiles-of-prosperity-how-nih-supported-research-is-fueling-private-sector-growth-and-innovation (accessed 21 June 2018).

UPI (1984). Stanford gets genetic engineering patent. *UPI Archives* (28 August).

Wetterstrand, K.A. (2016). *DNA Sequencing Costs: Data from the NHGRI Genome Sequencing Program (GSP)*. http://www.genome.gov/sequencingcostsdata (accessed 3 July 2016).

Yamaner, M. (2016). Federal funding for research increases by 6% in FY 2014; total federal R&D up 4%. *NSF Info Brief* (April).

Zerhouni, E. (2008). *Testimony on the Fiscal Year 2009 Budget Request Before the House Committee*. NIH Director. https://www.nih.gov/about-nih/who-we-are/nih-director/fy-2009-directors-budget-request-statement (accessed 5 March 2008).

Part Two

Headwinds of Research Innovation

5

Slowdown and Erosion

Recession is when your neighbor loses their grant.
Depression is when you lose your grant.

Andrew Fire (2007)[1]

In his influential report on science strategies, Vannevar Bush highlighted his first and most important principle: stability of funding for research over long periods of time (Bush 1945). Ever since, many award-winning scientists commented on the deficiencies of stability, unexpected obstacles, and concerning trends of research funding.

Regrettably, the research funding system is under stress. Several Nobel Prize-winning scientists also have been exposed to controversies surrounding research funding. Robert Edwards, the discoverer of *in vitro* fertilization (IVF), had to deal with immense ethical concerns that caused great difficulties in conducting and funding his studies. Columbia University professor and Nobel laureate Eric Kandel (2014) pointed out that the changing funding environment is becoming enormously inhibitory.

With the growing recognition of prevalent non-repeatable research results, concerns regarding the effectiveness of the research enterprise are further magnified. Several authors highlighted the need to increase value and reduce waste in biomedical research (Ioannidis 2005; Chalmers et al. 2014).

The funding challenges coincide with the rapid expansion of the scientific workforce, professionals engaged in the creation and production of new knowledge, products, processes, methods, or systems and the management of the projects (including postgraduate PhD students in R&D). Based on US Census Bureau (2017) and UNESCO (United Nations educational, scientific, and cultural organization) Institute for Statistics (2017), Figure 5.1 shows the growth

---

1 Fire, A. (2007). Nobel winner Andrew Fire on ethics, politics, and science. https://www.youtube.com/watch?v=xtfYjL3F-MU (accessed 12 July 2017).

Innovative Research in Life Sciences: Pathways to Scientific Impact, Public Health Improvement, and Economic Progress, First Edition. E. Andrew Balas.
© 2019 John Wiley & Sons, Inc. Published 2019 by John Wiley & Sons, Inc.

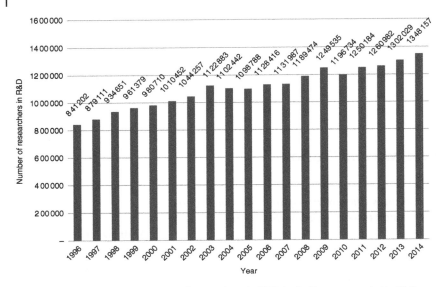

Figure 5.1 Growth of the number of researchers in R&D, including postgraduate PhD students, in the United States. *Source:* Data from US Census Bureau (2017) and UNESCO (United Nations Educational, Scientific, and Cultural Organization) Institute for Statistics (2017).

of the number of professionals in R&D; nearly 3% average annual growth or more than 500 000 researchers added to the workforce over 18 years.

The funding controversies of natural sciences spill over to perceptions of arts as well. Many academics share the passion for Indiana Jones as their favorite professor in fiction and movies. His dedication to finding the truth, open sharing of knowledge, and personal bravery in field trips are truly admirable. What baffles most admirers is his ability to fund high-priced expeditions without ever apparently spending time on writing grant applications and dealing with the overwhelming paperwork of project management. Just two tickets to the airship Zeppelin would have cost more than $15 000 in today's money. It is reasonable to assume that his exciting scientific endeavors were funded through Hollywood connections, not available to anyone else in academia.

Signs of Slowdown in Research Innovation

There appears to be a growing gap between research spending and biomedical innovation that is expected by society. Historical comparisons show that 16 new antibiotics were approved by the FDA between 1983 and 1987, but only

5 new antibiotics were approved between 2003 and 2007 (Boucher et al. 2009). Similarly, 12 classes of new antibiotics were discovered between 1940 and 1962, but only 2 new classes of antibiotics were developed in the following 50 years (Coates et al. 2011).

Many more reports highlighted significant *innovation slowdown* in biomedical research innovation. In 2012, a White House report underscored that R&D productivity is declining as the new drug output is nearly flat. The President's Council of Advisors on Science and Technology prepared this analysis with the aim of propelling innovation in drug discovery, development, and evaluation (PCAST 2012). While the report recognized the tremendous progress in developing new treatments, it also highlighted symptoms that R&D productivity is falling behind expectations. Many companies were exiting the anti-infective market. The challenges were magnified by the "patent cliff": a significant number of drugs with annual sales in the billion-dollar range were coming off patents, resulting in a massive loss in sales to generic substitutions that no longer support research and development.

Along these lines, the annual new molecular entity and new biologic entity approvals reflect a worrisome trend. The number of Food and Drug Administration approved new molecular and biologic entities peaked in the mid-1990s and had been declining until 2005 with some increase afterward (PhRMA Research 2016) (Figure 5.2). Essentially, there was a valley of approvals in 2005. Now there is some acceleration but still below the level of the mid-1990s. This level of new *molecular and biologic entity approval*s is in contrast with the growing spending on research and development by the pharmaceutical industry as reported by the Pharmaceutical Research and Manufacturers of America (PhRMA) member companies.

There appears to be a knowledge gap between basic research and commercial projects. Notably, the following two challenges need to be recognized: (i) predicting the efficacy and toxicity of drug candidates in bench research and (ii) increasing the number of proteins that become druggable targets as opposed to the currently limited selections.

Due to the significant ratio of non-reproducible research results, there may also be a need to improve the rate of trustworthy and consequential research in basic sciences. Another major issue is the inefficiency of clinical trials as they tend to be enormously complex, resource intensive, and heavily regulated.

The general conclusion is that the process of translation should be accelerated. The nearly 20 000 or so FDA-approved pharmaceutical products interact with only about 2% of human proteins (Stockwell 2011).

In developing wide-ranging recommendations, the White House report highlighted the goal of doubling the output of innovative new medicines to

(a)

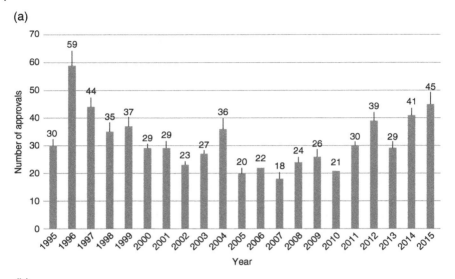

(b)

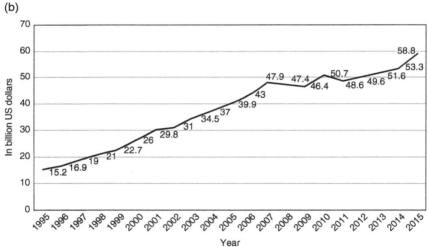

Figure 5.2 (a) Annual new molecular entity and new biologic entity approvals by FDA. (b) Research and development expenditure of total US pharmaceutical industry (PhRMA member companies) *Source:* Data from PhRMA Research (2016).

address major unmet medical needs (PCAST 2012). To meet this goal, there is a need to improve the drug discovery and development process including the drug evaluation process, monitoring and communication about benefits and risks, FDA management and approval process, and economic incentives that could increase innovation and the chances of success.

John Sulston was a British biologist and former director of the Sanger Centre during the most dynamic years of the Human Genome Project. He was the corecipient of the Nobel Prize in Physiology or Medicine 2002 for discoveries concerning the genetic basis of organ development and programmed cell death.

Sulston's research was focused on the cell lineage and genome of the nematode *Caenorhabditis elegans* as a model organism. He mapped the development of nervous system cell lineage and demonstrated its invariability, i.e. it always followed the same sequence of cell division and differentiation. Some cells appear to be programmed to die as the fertilized egg develops into an adult organism. The next breakthrough was the sequencing of the *C. elegans* genome and identification of the genes responsible for the development and programmed death.

In response to the question about sources of funding, Sulston emphatically highlighted, "Much Nobel work would not have got past peer review – that is the essence of Nobel research! Extra funding would need to be rather nonspecific. This is in line with the comment that protected time as a result of getting a basic salary, with less grant hassle, is very important" (Tatsioni et al. 2010).

Later in his career, Sulston's attention turned to the broader social issues of science. Currently, he serves as chair of the Institute for Science, Ethics, and Innovation at the University of Manchester. His scholarly endeavors are based on the recognition that science starts with the overpowering will to understand. His book titled *The Common Thread: A Story of Science, Ethics and the Human Genome* discusses the many controversies of public–private collaborations and accentuates his belief in the free exchange of scientific information (Sulston and Ferry 2002).

Obstacles to Research Innovation

The discrepancy between innovation expectations and limited funding was also highlighted by the Institute of Medicine report titled Neuroscience: Biomarkers and Biosignatures (Altevogt et al. 2008). "The number of innovative medical therapies that have reached the market has been disappointing, given the escalating research investment." One potential reason for the slowdown was the shortage of suitable biomarkers that could make the clinical testing of treatment candidates better focused.

In 2017, Tom Insel, former head of the NIH National Institute of Mental Health, made the following revealing comments about why he moved from government to the private sector: "I had left academia to go to the government because I thought I could have a greater impact on public health measures – morbidity and mortality. I spent $20 billion of taxpayer money over 13 years, and I hadn't budged the suicide numbers, and I hadn't budged any numbers on morbidity... It still felt sometimes like I had gone to war with the wrong army... What I was doing was providing lots of additional funding to people whose major goal was to get a paper in Nature and get tenure" (Piller 2017).

To assess the impact of funding policies on the production of meaningful outcomes, one may look at selected indicators of the novelty of scientific results produced. A New Molecular Entity (NME) is a drug that is without precedent among regulated and FDA-approved drug products. In other words, the new drug is not a version or derivative of a previously investigated and approved substance.

A study evaluated 1453 NMEs based on the original publication and the subsequent FDA approval (Patridge et al. 2015). The ratio of academic institutions as sources of publications about NME discovery was dominant before 1960 but fell below 50% afterward. By publishing 10 or more NMEs, the most prolific academic institutions were the University of Michigan, National Institutes of Health, University of Wisconsin, Harvard University, Columbia University, Johns Hopkins University, Yale University, and the University of Berlin (accounting for all author contributions to the first publications).

Science lagging behind societal needs is illustrated in other examples as well. The United States is experiencing an unprecedented opiate crisis. It is estimated that 1000 emergency department visits are triggered by opiate use every day in the United States. In response to the crisis, Elsevier (2017) developed a graphic summary of research referencing opiates. As measured by the number of publications, research referencing opioids increased 54 times between 1950 and 2015.

In spite of this massive increase of research, the *opioid crisis* has never been greater than now. About 20–30% of patients get an opioid prescription for chronic pain, and about 10–12% develop an opioid use disorder. Nearly half of them transition to heroin use. Prescription opioid misuse is the primary entry point to heroin addiction.

Especially at the time of limited resources, less than meaningful outcomes of research projects can be used to question the societal usefulness of funding them. Federal and state budget discussions are enormously competitive. There is a constant search for lines to be cut or eliminated and free up resources for other, more politically pressing purposes. Such unwanted attention can also be attracted by scientists who publicly invite funding for their research based on unsubstantiated future promises.

One of the most common but often meaningless conclusions of research articles is that "our study documents the need for more research in this area." Regardless of results or area of expertise, most research studies can easily justify such self-serving but otherwise vague supposition. Similarly, when federal grants are used to training and development of research skills, it is questionable when the promised ultimate return on investment is getting more research grants as opposed to accomplishments of scientific innovation or societal benefits.

There are huge variations in the use of research resources as well. *Mapping the human genome* became a race of time and money in the 1990s. There were two competitors at the forefront: the government-funded Human Genome Project, which completed its task in 15 years with more than $3 billion in taxpayer money, and a private company, Celera Genomics, which was financed with $100 million and took less than a decade. Obviously, the two projects benefited each other's progress as well.

In the race for resources, scientists also have to compete with impressive sounding but often ineffective technology promises. New and complex technology cannot and should never be equated with beneficial innovation. In an article for the *New York Times*, Ezekiel Emanuel, professor of medical ethics and health policy at the University of Pennsylvania, highlighted the problem of fake and ineffective innovations in health care (Emanuel 2012). One cited example was a surgical robot once viewed as a breakthrough technology but, in fact, lacking credible evidence of effectiveness and improvements in outcomes. Complicated, expensive, but ineffective "pseudo-innovations" drive up the cost of health care without adding value to it.

A particular challenge is the unpredictability of scientific discoveries. By paraphrasing John Wanamaker, entrepreneur and former postmaster general, one could say: half the money we spend on research is wasted; the trouble is we do not know which half. Charles Kelman, inventor of phacoemulsification, could not find the desired technique in his lengthy and NIH grant funded studies. However, he discovered the crucial technology innovation necessary for lens replacement after learning about the use of his dentist's ultrasonic probe.

One of the increasingly highlighted potential obstacles is the concentration of funding. Analysis of the relationship between the common productivity measures and funding levels for researchers receiving NIH support indicated a gradual concentration of funding between 1985 and 2015 (Katz and Matter 2017). The metrics included bibliometric indices, like publication and citation numbers, and patent counts. These measures are often considered simple proxies in the assessment of *scientific productivity* and *societal impact*. The analyses suggested that a small segment of investigators and research institutes get an ever-increasing proportion of the funds. This funding inequality has been rapidly rising.

The possibility of decreasing marginal *returns from research funding* gets heightened attention at a time of funding shortage for biomedical research (Lauer et al. 2017). In this case, there have been questions regarding the relationship between the amount of NIH funding and research productivity as measured by the Relative Citation Ratio of resulting publications (i.e. RCR compares an article's citations with other papers in the same field). Data on grants held by 71 500 individual investigators over nearly 20 years suggested that productivity goes up to three R01 grants received by an investigator, but afterward more funding appears to be associated with a flattening of the productivity curve. The recommendation was made that capping grant funding of 6% of lead investigators would free up resources for funding 1600 more researchers (Kaiser 2017a). As a follow-up to the studies, NIH decided to introduce a cap on grant funding, but the adverse reaction led to retraction shortly after the announcement (Kaiser 2017b).

Disregarding scientific evidence in the policymaking process not only represents a societal loss in benefiting from what we already know but also threatens public appreciation and support for scientific research. More specifically, when government policies run contrary to scientific evidence, frustration with outcomes and significant financial losses can be expected. Correspondingly, there is also a great need to strengthen the role of scientific evidence in the policymaking process.

Between 1998 and 2004, the US government spent nearly one billion dollars on the National Youth Anti-Drug Media Campaign. The stated goals of the program were educating and enabling young people to reject illegal drugs, preventing people from initiating the use of drugs, and convincing occasional drug users to stop (Hornik et al. 2008b). In spite of the vast amount of resources spent, it is unlikely that the campaign had any favorable effects (Hornik et al. 2008a).

Increasing Costs of Regulatory Compliance

Erosion of funding happens not only by a decrease in the purchasing value of research support but also by the increase of *regulatory and compliance burden* on the research enterprise. As described in the chapter on red tape, the bureaucratic burden on research has been steadily increasing over the past decades. This burden is reflected not only by the escalating number of rules, required approvals, audits, and forms to be filled out but also by the substantial financial costs associated with the increased paperwork.

The Faculty Standing Committee of the Federal Demonstration Partnership conducted a web-based survey to estimate the administrative burdens placed on faculty. Responses from more than 6000 faculty members suggested that 42% of the time spent on federally funded research was on administrative tasks related to the project (Rockwell 2009). The most time-consuming administrative tasks included grant progress report submission, project budget management, equipment and supply purchases, personnel hiring, IRB protocol approvals, training personnel, and students.

In a report to the National Research Council Committee on Research Universities (NRC 2011), one public university highlighted that monitoring certification requires one-quarter of the time of nine separate full-time employees at an estimated cost of $117 000 per year. Another public university estimated that its cost of monitoring certifications for the effort reporting system exceeded $560 000. Another university estimated that more than 18 000 effort reports were processed and 60–90 minutes was spent on each effort report. This investment in time included issuing instructions, completion by faculty and staff, administrative review, tracking, and storing.

In 2012, the National Science Board released a report on the trends and challenges at public research universities. It highlighted diminishing funding and rising expectations. In 2008–2009, public research institution spending on research and development was about 16% of their budget. While the percentage budget share of university research remained in the same range, the trend also reflects erosion due to inflation of costs. Increased research expenditures for many institutions were also driven in large extent by compliance costs. Correspondingly, the National Research Council urged that federal and state policymakers should eliminate regulations that are redundant, ineffective, or inappropriately applied to higher education.

The process for acquiring and using federal research funds is increasingly bureaucratic. In studies of *administrative burden*, the term regulation includes not only the laws but also the general and permanent rules published in the Federal Register by the federal government. In 2015, Vanderbilt University coordinated a study that reviewed the federal regulatory compliance expense based on data from 13 universities. The calculation included all administrative resources needed to acquire and manage research projects, including the full range of associated personal expenses, equipment, office space, and so forth.

The study estimated that the regulatory compliance burden on universities ranges between 7 and 11% of the budget, the vast majority being for research-related compliance. In the totality of research expenditures, compliance was found to be in the range of 11–25%. For Vanderbilt University, the estimate of total compliance expenses was about $146 million annually with research compliance being far the largest component. From a budget of nearly $500 million in federally supported research, $117 million was spent on complying with research regulations.

There have been controversies surrounding the NIH reimbursement for facilities and administrative (F&A) expenses that are often referred to as indirect costs. Such support is supposed to pay for physical infrastructure and other general services in support of research projects. According to a Government Accountability Office (GAO 2013) report, indirect cost reimbursements from NIH to universities varied between 40 and 70% of the direct research expenses (i.e. the percentage of the direct cost added to the award amount). The GAO report also found that the indirect cost increase was faster than the direct costs between 2002 and 2012.

2017 Nobel laureate **Jeffrey Hall** was born in Brooklyn, New York, and raised in the Washington, DC, area. He studied at Amherst College in Massachusetts, the University of Washington in Seattle, and, as a postdoctoral researcher, the California Institute of Technology. Subsequently, he became a faculty member at Brandeis University (Waltham, Massachusetts), where he worked from 1974 until retirement.

Conducting joint research with Michael Rosbash of Brandeis University, they identified the genetic basis of the circadian rhythm, the day–night cycle of sleep and wakefulness. Their breakthrough research involved bioassays of behaviors and molecular genetic studies in Drosophila. Understanding the circadian rhythm is essential not only for the study of most living organisms but also for the interpretation of the wide-ranging impact on human diseases.

Later in his career, Jeffrey Hall became an outspoken critic of the research funding system. In *Current Biology* (2008), he provided an explanation for retiring from research: "I admit that I resent running out of research money... This means, for instance, that recent applications from our lab have had their lungs ripped out, often accompanied by sneering, personal denunciations." "The [typical faculty] job starts with some 'set-up' money for equipping the lab, but next to no means are provided to initiate that 'research program' and to sustain it during the years to come."

After a family trip to the Gettysburg battlefield in 1983, Jeff Hall became not only an avid learner of the nineteenth-century history but also an increasingly appreciated expert about this turning point of the civil war. At Brandeis University, he taught a course on the subject in the history department and later wrote a textbook about the Battle of Gettysburg (Hall 2009). His polymath talents in multiple disciplines remind us to the long-lost universality of science, where success starts with curiosity and resolve.

From a taxpayer's perspective, spending on administration and paperwork is not particularly appealing. In the federal budget development, the desire to save money for other priorities and cutting unattractive administrative expenses are primary motivations of a budget cut. Not surprisingly, the F&A cost-cutting proposal triggered vigorous opposition by research universities and a wide range of professional associations, especially in the light of unchanged regulatory requirements. As a result, this budget cut has been tabled currently.

The Gamble of Science and "Short-termism"

Congressional debates about the funding for science struggle with the major budgetary expenses, while the benefits appear to be less predictable. When President Truman announced the first military use of nuclear power, his words were more than revealing: "We have spent 2 billion dollars on the greatest scientific gamble in history - and won" (Truman 1945).

To people not familiar with the intricacies of scientific research, assessing the future value and guesstimating reasonable returns can be daunting challenges. Scientists need to recognize that the tax-paying public is often uninformed, skeptical, and disoriented when it comes to research. Today, scientists asking for money have to convince a lay audience that their research is the right kind of "gamble" that will become reasonable investment and produce sound outcomes to benefit society.

According to an analysis of the funding and demographic shifts of research supported by the NIH National Heart, Lung, and Blood Institute, the trend line strikingly changed in the beginning of the twenty-first century (Charette et al. 2016). Historically, there was a steady growth in the budget of this institute and correspondingly in its support for research projects. The ratio of applications successfully receiving grant funding was nearly 30% (so-called pay line).

In 2006, the *percentage pay line* crash occurred, and the success rate of grant applications dropped to about 15%. Afterward, there has been a steady and stagnant budget period with continued hyper-competition for grants. Simultaneously, another demographic trend emerged with more established investigators increasing their access to funding, while mid-career and junior investigators lost some grounds. In an attempt to halt these trends, NIH implemented a more protected, new investigator and early-stage investigator policy.

With the multiple funding challenges, the speed of science attracts increased scrutiny. At a time of limited resources, concerns are growing that vaccine development is not keeping up with pandemics. To meet the need and accelerate vaccine development, the Coalition for Epidemic Preparedness Innovations (CEPI) was launched with financial support from the Bill and Melinda Gates Foundation, Britain's Wellcome Trust, and the governments of Germany, Japan, and Norway.

When attempting to enter real world practice, the published or disclosed results of academic laboratories enter into a "shark tank" of skeptics, investigations, and rivalries (Figure 5.3):

- *Replication* by the original researchers and subsequently by competing research laboratories and clinical sites. If others cannot replicate the major conclusions of a particular research study at some points, the promising sounding research results are essentially worthless and lost.
- *Prototype development* to provide proof of concept and exemplify how the results of scientific research can be turned into a practical solution or technology. The prototype may go through several iterations of refinement and improvement, often requiring a dedicated, large, or start-up company resources.

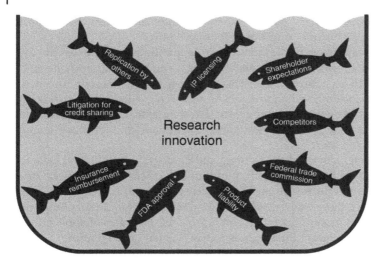

Figure 5.3 Shark tank of exposures in innovation commercialization.

- Compelling *business plan*, successful venture capital, or large corporation financing are needed to support the development and validation of the new technology based on the results of academic research.
- *Validation of the prototype* through demonstration of successful functioning, safety, and efficacy. Typically, such validation involves multiple institutions, individuals, and laboratories.
- *FDA approval* of a particular technology for commercialization and use in patient care and by people in general. Most new technologies need some FDA safety and efficacy clearance before routinely used in the practice of patient care.
- *Insurance reimbursement* is essential for financially sustainable health-care services. In health care, power is not about the Bill of Rights, but about the right to bill. Medicare, Medicaid, and private insurance reimbursements are usually dependent on successful technology assessment and FDA approval.
- In recent year, the Federal Trade Commission has been playing an increasingly vigorous role in the fight against unsubstantiated claims of health and healing. The FTC enforces *truth-in-advertising* laws by looking especially closely at advertising claims that can affect consumers' health or their pocketbooks.
- When success comes, so are the debates and litigation over credit for the discovery and *intellectual property rights.* Various competitors are eager to explore every possible opportunity to take credit for the achievement and gain a substantial share of the revenues.

- *Product liability* is the last shark that can attack the successfully transferred research discoveries after introduction into widespread practice. Any unfortunate, rare, or long latency adverse effect can make customers unhappy triggering lawsuits and huge compensation awards.

To manage uncertainties and control expenses, many funders of life sciences research stay away from long-term commitments and try to support short-term projects with more predictable outcomes. While the approach looks reasonable from a sponsor perspective, the revolving door of short-term projects requires that researchers have to constantly apply for funding and long-term research goals may not be very helpful in the process. Many accomplished researchers need decades of projects before the most significant discoveries. Their frustration with funding was summed up by Adam Smith, Editorial Director of Nobel Media, when he talked about short-termism.

Sputnik Moment: The Need to Accelerate Research

According to a study by Research!America (2016), investment in medical and health research showed growth between 2013 and 2015. The majority of research funding came from industry (64.7%), and the federal government accounted for about 22.6% in 2015. However, recent years have shown an erosion of public support for life sciences research.

Between 1994 and 2004, the total US funding for biomedical research increased 6% per year. Afterward, the rate of growth significantly declined to 0.8% per year (Moses et al. 2015). Such minimal growth of research budgets could not compensate for price increases and inflation. The NIH budget lost almost 25% of its purchasing power over the last decade (Collins 2015). In the past, NIH funded one out of three research proposals, and now the ratio is closer to one out of six.

Meanwhile, international comparisons also showed a relative decline in US government research funding of the global total, from 57% in 2004 to 50% in 2012. Between 1999 and 2009, Asia's share in global R&D expenditures increased from 24 to 32%. During the same period, the US share of life science patents has remained large but slightly declined from 57% in 1981 to 51% in 2011.

A diminished rate of growth in biomedical research funding can be observed from many public and private sponsors in the United States. Notably, several pharmaceutical companies shifted funding priorities to late-phase clinical trials and away from target discovery and validation. Most likely, the complexity, length of time, and cost make late-stage clinical trials a higher budgetary priority.

Such funding challenges are in sharp contrast with the facts that, for example, NIH-funded research has been mostly behind the declining death rates from cardiovascular disease by more than 70% since 1963. The $3.8 billion initial federal investment in the Human Genome Project resulted in nearly $1 trillion in economic activities (Collins 2015).

Some encouraging signs are also emerging. There is growing recognition that investment in biomedical research has made wonders. In recent years, there have been several highly publicized attempts to cut the NIH budget. However, every time Congress came back and actually added money and increased the budget of the NIH.

Sputnik moment is a reference to the launch of the first Earth-orbiting artificial satellite by the Soviet Union that also skyrocketed American public support for scientific research. The plans of chief rocket scientist Sergei Korolev started with larger and more ambitious scientific payloads, but eventually Sputnik 1 was a small polished metal sphere filled with nitrogen, with a radio transmitter, power supply, fan, and antennae. Its beep could be detected by radio receivers, including ham radio operators, all over the world. Sputnik's launch triggered concerns of the US public and Congress. At the same time, the Sputnik moment of 4 October 1957 also generated unprecedented public support for science and jump-started basic and applied research in new areas. When enthusiasm for science is faltering, many scientists talk about the need for a new Sputnik moment that could galvanize society today.

Funding has been viewed as essential for success in research, particularly in natural sciences. A study by Tatsioni et al. (2010) analyzed the funding sources of foremost scientific papers of Nobel Prize winners between 2000 and 2008. In areas of medicine, physics, and chemistry, among the 93 selected papers of 70 Nobel Laureates, 70% reported some funding source, but a substantial portion of the award-winning work appeared to be unfunded. In the follow-up survey, 13 Nobel Laureates responded and offered their views on funding. Apparently, Nobel Prize-winning results can also come from *unfunded research*, especially when institutions offered salary support and protected environment for scientists.

The feedback from Barry Marshall was particularly informative: "No funding. I had no grant experience, and this was a new area of work, so no-one else had it in their research domain." With no support for *Helicobacter pylori* studies, Marshall decided to test the theory on himself and ingested a cultured sample of *H. pylori* after performing baseline assessments. Subsequent assessments proved the growth of the bacteria and tissue damage. "The extreme skepticism of my colleagues led me to believe that I might never be funded to perform the crucial trial of antibiotics" (Marshall 2005). He said later in his Nobel Lecture that he did not expect to become as ill as he had. However, he finally had proof from his own cultures and treatment, and he was able to get funding to research alternative treatments to gastric ulcers and, ultimately, decrease the risk of stomach cancer caused by gastritis.

Institutional or university budget is often unmentioned but, in fact, plays a noteworthy role in supporting research in life sciences. The institutional support typically manifests in allocated free research time, funding for conference travel, and purchase of supplies and equipment from the university budget. There are also significant and valuable opportunities to work on research with PhD students, residents, graduate students, fellows, undergraduate students, and others.

The precious institutional opportunities can supplement initial monetary supports called start-up packages. Considering the benefits and also the magnitude of institutional research funding, today's biomedical researchers have to be skilled in using these resources to their maximum extent while also creating valuable educational experiences for people they mentor. Simultaneously, sponsors of research should continue exploring innovative ways of supporting projects and getting inventions and discoveries through the process more rapidly.

Scientific and technological breakthroughs improve public health and strengthen economic growth. With increased and also better structured budget support, research will undoubtedly continue learning more and more in an accelerating rate. By bridging the gap between the lab and daily life, life sciences research is producing tremendous payoff in terms of human health and also in terms of economics. We are at a remarkable moment of scientific opportunities.

References

Altevogt, B., Hanson, S., and Davis, M. (eds.) (2008). *Neuroscience Biomarkers and Biosignatures: Converging Technologies, Emerging Partnerships: Workshop Summary*. Washington, DC: National Academies Press.

Boucher, H.W., Talbot, G.H., Bradley, J.S. et al. (2009). Bad bugs, no drugs: no ESKAPE! An update from the Infectious Diseases Society of America. *Clinical Infectious Diseases* 48 (1): 1–12.

Bush, V. (1945). *Science, the Endless Frontier: A Report to the President*. Washington, DC: United States Government Printing Office.

Chalmers, I., Bracken, M.B., Djulbegovic, B. et al. (2014). How to increase value and reduce waste when research priorities are set. *The Lancet* 383 (9912): 156–165.

Charette, M.F., Oh, Y.S., Maric-Bilkan, C. et al. (2016). Shifting demographics among research project grant awardees at the National Heart, Lung, and Blood Institute (NHLBI). *PloS One* 11 (12): e0168511.

Coates, A.R., Halls, G., and Hu, Y. (2011). Novel classes of antibiotics or more of the same? *British Journal of Pharmacology* 163 (1): 184–194.

Collins, F.S. (2015). Exceptional opportunities in medical science: a view from the National Institutes of Health. *JAMA: the Journal of the American Medical Association* 313 (2): 131–132.

Elsevier (2017). *The Opioid Epidemic in America*. https://www.slideshare.net/ ElsevierConnect/infographic-the-opioid-epidemic-in-america (accessed 22 November 2017).

Emanuel, E.J. (2012). In medicine, falling for fake innovation. *New York Times* A15.

Government Accountability Office (GAO) (2013). *Biomedical Research: NIH Should Assess the Impact of Growth in Indirect Costs on Its Mission*. GAO-13-760. http://www.gao.gov/products/GAO-13-760. Published 24 September 2013. Released 31 October 2013.

Hall, J.C. (2008). Jeffrey C. Hall. *Current Biology* 18 (3): R101–R102.

Hall, J.C. (2009). *The Stand of the US Army at Gettysburg*. Bloomington: Indiana University Press.

Hornik, R., Jacobsohn, L., Orwin, R. et al. (2008a). Effects of the National Youth Anti-Drug Media Campaign on youths. *American Journal of Public Health* 98 (12): 2229–2236.

Hornik, R., Jacobsohn, L., Orwin, R. et al. (2008b). Effects of the national youth anti-drug media campaign on youths. *American Journal of Public Health* 98 (12): 2229–2236.

Ioannidis, J.P. (2005). Why most published research findings are false. *PLoS Medicine* 2 (8): e124.

Kaiser, J. (2017a). Critics challenge NIH finding that bigger labs aren't necessarily better. *Science* (7 June). doi: 10.1126/science.aan6941.

Kaiser, J. (2017b). Updated: NIH abandons controversial plan to cap grants to big labs, creates new funds for younger scientists. *Science* (8 June).

Kandel, E.R. (2014). Transcript from an interview with Eric R. Kandel. Nobelprize.org. Nobel Media AB 2014. http://www.nobelprize.org/nobel_prizes/medicine/laureates/2000/kandel-interview-transript.html (accessed 23 November 2017).

Katz, Y. and Matter, U. (2017). *On the Biomedical Elite: Inequality and Stasis in Scientific Knowledge Production*. https://dash.harvard.edu/bitstream/ handle/1/33373356/BKC_Report_KatzMatter2017.pdf (accessed 8 November 2017).

Lauer, M.S., Roychowdhury, D., Patel, K. et al. (2017). Marginal returns and levels of research grant support among scientists supported by the National Institutes of Health. *bioRxiv* 142554. https://doi.org/10.1101/142554.

Marshall, B.J. (2005). Helicobacter connections. In: *Les Prix Nobel. The Nobel Prizes 2005* (ed. K. Grandin), 2006. Stockholm: Nobel Foundation.

Moses, H., Matheson, D.H., Cairns-Smith, S. et al. (2015). The anatomy of medical research: US and international comparisons. *JAMA: The Journal of the American Medical Association* 313 (2): 174–189.

National Science Board (NSB) (2012). *Diminishing Funding and Rising Expectations: Trends and Challenges for Public Research Universities*. Arlington, VA: National Science Foundation.

NRC (2011). *Regulatory and Financial Reform of Federal Research Policy Recommendations to the NRC Committee on Research Universities* (21 January). https://energy.gov/sites/prod/files/gcprod/documents/RFIRegReview_Council GovtRelationsAppendix_03212011.pdf (accessed 17 November 2017).

Patridge, E.V., Gareiss, P.C., Kinch, M.S., and Hoyer, D.W. (2015). An analysis of original research contributions toward FDA-approved drugs. *Drug Discovery Today* 20 (10): 1182–1187.

PCAST (President's Council of Advisors on Science and Technology) (2012). *Report to the President on Propelling Innovation in Drug Discovery, Development, and Evaluation*. Washington, DC: PCAST.

PhRMA Research (2016). *Research and Development Expenditure of Total U.S. Pharmaceutical Industry from 1995 to 2015 (in Billion U.S. Dollars)* (May). PhRMA Biopharmaceutical Research Industry Profile.

Piller, C. (2017). They had power and status on the East Coast. So why did these health care gurus decamp for California? To change the world. *STAT* (6 November). https://www.statnews.com (accessed 7 November 2017).

Research!America (2016). *U.S. Investments in Medical and Health Research and Development 2013–2015*. https://www.researchamerica.org/sites/default/files/2016US_Invest_R&D_report.pdf (accessed 8 November 2017).

Rockwell, S. (2009). The FDP faculty burden survey. *Research Management Review* 16 (2): 29.

Stockwell, B.R. (2011). The quest for the cure: The science and stories behind the next generation of medicines. Columbia University Press.

Sulston, J. and Ferry, G. (2002). *The Common Thread: A Story of Science, Ethics and the Human Genome*. London: Bantam.

Tatsioni, A., Vavva, E., and Ioannidis, J.P. (2010). Sources of funding for Nobel Prize-winning work: public or private? *The FASEB Journal* 24 (5): 1335–1339.

Truman Announces Hiroshima Bombing 220643-10 (1945). https://www.youtube.com/watch?v=sDfuau2SfVE (accessed 13 July 2017).

U.S. Census Bureau (2017). *As Reported in the DataBank World Development Indicators*. Washington, DC: The World Bank Group.

UNESCO (United Nations Educational, Scientific, and Cultural Organization) Institute for Statistics (2017). *As Reported in the DataBank World Development Indicators*. Washington, DC: The World Bank Group.

Vanderbilt University (2015). *The Cost of Federal Regulatory Compliance in Higher Education: A Multi-Institutional Study*. https://news.vanderbilt.edu/files/Cost-of-Federal-Regulatory-Compliance-2015.pdf (accessed 13 November 2017).

6

Non-reproducible Research

The seeker of the truth…to make himself an enemy of all that he reads, and, applying his mind to the core and margins of its content, attack it from every side. He should also suspect himself as he performs his critical observation, so that he may avoid falling into either prejudice or leniency.

Ibn al-Haytham (1025)[1]

Reproducibility is regarded as the very essence of good science in addition to generalizability. No one would want to take medication, sit on an airplane, or turn on the light if any of these actions would generate non-reproducible, unpredictable outcomes.

Scientific studies are inherently variable and innovative. Many research methodologies are unique, esoteric, or unprecedented, and the results are frequently unpredictable. Correspondingly, there is an extraordinary need for rigor and perfectionism in attempts to achieve replicable research results.

In recent years, the enormous boulder of *non-reproducible research* presented itself on the way of scientific progress. It appears to be an almost insurmountable obstacle of science, accurate assessment of research performance, biotechnology industry development, and, most importantly, public trust in research results. The problem is not entirely new, but the size and magnitude of this obstacle became much more apparent in recent times.

Several pharmaceutical companies reported stunning inability to reproduce research results that appeared in high-impact, prestigious, and peer-reviewed scientific journals. Bayer scientists were able to reproduce only 21% of 67 target validation projects published in oncology, women's health, and cardiovascular medicine (Prinz et al. 2011). Reproducibility did not show improved correlation with journal impact factor, the number of publications, or the

---

1 Ibn al-Haytham (1025). *Doubts Concerning Ptolemy* (cited by Abdelhamid I. Sabra (2003). Ibn al-Haytham − Brief life of an Arab mathematician: died circa 1040. *Harvard Magazine* (September–October).

Innovative Research in Life Sciences: Pathways to Scientific Impact, Public Health Improvement, and Economic Progress, First Edition. E. Andrew Balas.

number of independent groups among the author affiliations. In a study by Amgen, only six (11%) verified studies were found reproducible in a series of attempts to replicate 53 landmark studies in oncology and hematology (Begley and Ellis 2012).

Quality failures happen in all kinds of production activities, not just in research. In the WWI battle of the Somme, 30% of the 1.5 million British shells fired were duds, and much of the rest had limited or worthless effect due to dysfunctional design. Today, defective manufacturing products, health-care errors, and high academic dropout rates are well-known quality indicators of performance that drive corrective actions.

With the recognition of prevalent non-reproducible research results, uncertainties regarding the effectiveness of the research enterprise are escalating. Obviously, the mainstream of non-reproducible research comes from various biases and not from irregular or criminal behavior.

Not surprisingly, shortcomings of reproducibility and reliability of research are attracting funding agency and scientific scrutiny (Collins and Tabak 2014). With the 50%, 70%, or in some cases 80% estimated non-reproducible research result rates, the concepts of continuous quality improvement need to be fully embraced by the research enterprise.

The Boulder of Non-reproducible Research

Numerous analyses regarding the reproducibility of research have yielded disappointing results. In target-based drug discovery programs, it has long been suspected that potential efficacy is often assessed by the very same researchers who are advocating for the advancement of their compound/target into clinical testing. Obviously, there is a danger of selectively used and overemphasized information supporting the advancement of their research products (MacCoun 1998).

Studies showed that *lack of efficacy* had been the single biggest reason why pharmaceutical compounds fail in clinical trials (43% of all failures). Correspondingly, accurate preclinical assessment of efficacy is vitally important for successful treatment development (Schuster et al. 2005). It is possible that the shift to target-based discovery in recent years may have further increased the effects of bias and contributed to the decline in clinical success rates (Lindner 2007).

A self-report study at MD Anderson found that 50% of researchers were unable to reproduce published research at least once and were often met with resistance when exploring the cause with the original author (Mobley et al. 2013). In another study, 40–89% of interventions were found to be non-replicable due to poor descriptions of methodologies, omissions or changes to primary outcomes, and poor research design (Glasziou et al. 2014). According to another reanalysis, five of seven largest molecular epidemiology cancer studies did not classify patients better than chance (Wacholder et al. 2004).

According to a comparative analysis of HIV behavioral surveillance survey (BSS), only 45% were classified as reproducible based on reported methodology and population-based sampling. Surveys undertaken by NGOs were significantly less likely to be reproducible, and only a few surveys used internationally accepted HIV indicators by gender or age further limiting generalizability (Spiegel and Le 2006).

According to some estimates, the United States spends $28 billion per year on non-reproducible preclinical research, assuming 50% ratio of such research; the global spending is probably twice as much per year (Freedman et al. 2015). Obviously, a large variety of federal agencies, private foundations, and commercial companies sponsor research in life sciences. The cited number considers all funding sources combined. Several authors highlighted the need to increase value and reduce waste in biomedical research (Ioannidis 2005; Chalmers et al. 2014).

Different results and different conclusions need to be distinguished. Reproducible research is expected to lead essentially to the same results every time the experiment is repeated. On the other hand, different researchers coming up with different conclusions are not a new phenomenon in science.

Sometimes, new studies and more accurate measurements can redefine previous understandings, clarify ranges where previous assumptions are still valid but identify new relationships outside those ranges. The evolution of science also changes conclusions not infrequently. In the light of new scientific information, previous assumptions may need to be revised or entirely discarded.

The reproducibility of factual results and simultaneous variations in interpretations should only strengthen our trust in science. For example, the restrictive salt consumption guidelines of the US government have been questioned by a large-scale international study that suggested a closer relationship between salt consumption and elevated blood pressure only in the range of higher salt intake (Mente et al. 2014). Such differences should be interpreted as signs of progress and not irreproducible results in science.

The unacceptably high rate of non-reproducible research shortcoming affects the lives of patients, the practice of health care, the effectiveness of biomedical research, and our ability to generate societal support for research. There is a pressing need to meet the challenges at universities and in research laboratories.

Implications of Non-reproducible Results

When essential conclusions of a research project cannot be replicated by subsequent studies applying the same methodology, one of the two fundamental requirements, reproducibility and generalizability, for scientific validity is violated (Figure 6.1). The negative consequences of non-reproducible research are multiple, wide ranging, and potentially very damaging.

Publication of irreproducible research results causes pollution of information and can cause further *downstream mix-up* in various follow-up research studies. Irreproducible but published research results can mislead systematic reviews, meta-analyses, and clinical practice guidelines. They can generate distraction in scientific exploration and trigger further futile studies and thereby produce more wasted research.

Generating non-reproducible results represents *misuse of human subject's* participation and creates exposure to potential harm. Promising but, in fact, non-reproducible preclinical research results can trigger clinical follow-up studies, including randomized controlled trials with human subject involvement. Such trials not only represent a further waste of resources but also hinder human subject protection and ethical safeguards of research.

Publication of irreproducible research can stuff up academic resumes and lead to false recognition and undeserved academic promotions. In fact, the old

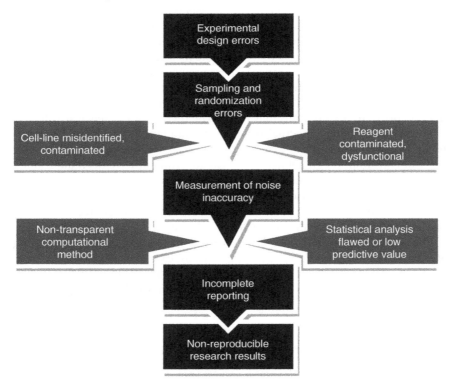

Figure 6.1 Flowchart of non-reproducible results in preclinical research.

"publish or perish" principle is blamed in large part for the rush to generate many publications even at the expense of quality and rigor. Most university promotion and tenure policy manuals set thresholds of scientific productivity partly based on the number of publications.

Production of non-reproducible research is a *waste of research funding* on ultimately useless experiments. The worldwide expenditures on biomedical research are nearly US$100 billion every year. It is estimated that up to 85% can be wasted, primarily due to quality deficiencies in research design and methods, adequacy of publication practices, and reports of research (Chalmers and Glasziou 2009). At the national level, the losses are at the multibillion-dollar level. Furthermore, such research often wastes animal resources in flawed experiments and research studies.

Highly promising but, in fact, untrue discoveries can mislead health-care professionals and the general public. Non-reproducible research is a source of false hopes and major disappointments to patients with various illnesses and their families. When the research is non-reproducible, there is no chance for effective use of the research results to improve individual and community health. Decades ago, James McConnell experimented with conditioning planarians and concluded that the conditioned memory was stored in a molecule that could be transferred by ingestion (Weigmann 2005). Today, it is obvious that memory is not transferable in this way, but funny speculations started about the merit of "eat your professor" and the police state brainwashing people through tap water.

Non-reproducible research can trigger *ineffectual spending* on patenting and commercialization attempts of nonfunctioning compounds or technologies. When research institutions are persuaded to commercialize non-reproducible research, the limited technology transfer resources are misused, and ultimately prestige is damaged. Patenting and subsequent commercialization will not be able to sell nonfunctioning compounds and technologies resulting from non-reproducible research. Trustworthiness of research impacts not only the general credibility of science but also economic development.

The current high rate of non-reproducible research creates *impediment of biotechnology innovation* and successful technology transfer. Significant research funds and industry investments relying on published scientific discoveries turn out to be unsuccessful. The reputation of academic institutions can be damaged as they turn out to be the sources of useless and misleading results. The problem of waste in buying science is not recognized easily as many companies do not want to upset the academic investigators.

Non-reproducible results are *weakening public trust* in the validity of science. In fact, non-reproducible research is not only unscientific but also useless for any purpose. At the time of limited research funding, there is nothing more dangerously appealing to policymakers and state legislatures than limiting or cutting funding for research that produces mostly non-reproducible results. Dramatizing implications of inconsequential, non-reproducible research has explosive political potential.

The lack of repeatability can have harmful effects on forensic applications and lead to *miscarriage of justice*. In 1986, Bennie Starks was convicted on two counts of attempted aggravated sexual assault and aggravated battery. Based on forensic semen analyses and bite marks, he was sentenced to 60 years in prison. Twenty years later, DNA reanalysis of the evidence showed that Bennie Starks could not be the person who committed the rape and the forensic analysis of the original trial was based on non-reproducible flawed results (Gabel 2014).

In some cases, non-reproducible results can lead to potential investigation and article retraction. As illustrated by several high-profile cases of questionable research, funding agencies and journal editors may revisit already published papers at a later point. In cases of significant departure from research practices or misrepresentation of data on which the conclusions were based, the editors may unilaterally retract the paper even in the absence of approval by the authors. In extreme cases, non-reproducibility can cause major *damage to research careers* and also to the institutional reputation.

Overall, it needs to be fully recognized that the overwhelming majority of non-reproducible research results come from well-intended studies of hard-working researchers. Science is an inherently complex endeavor, and unforeseen mistakes can happen all the time. Attempts to improve quality need to work with researchers collaboratively and not against them in achieving change. The chapter on quality improvement in science provides further information about the many opportunities to progress.

Alexis Carrel was an accomplished surgeon who is credited for many pioneering innovations that effectively laid the foundation of modern vascular surgery, including vascular suturing techniques, arteriovenous anastomoses, and the Carrel–Dakin method of wound antisepsis. In 1912, he was awarded the Nobel Prize "in recognition of his work on vascular suture and the transplantation of blood vessels and organs." Later with the aviator Charles Lindbergh, he invented the first perfusion pump to keep tissue alive *in vitro* for several weeks (Dutkowski et al. 2008).

The first problems came with his studies on cellular senescence. According to his statements, tissue from an embryonic chicken heart was maintained in a living culture for over 20 years with regular supplies of nutrient, far beyond the normal life expectancy of chicken. In spite of the great media and scientific attention, no one has been able to replicate the study and later the results appeared to be inconsistent with well-established limits of cell divisions.

The next chapter of his scientific adventures became even more problematic. In his book titled *Man, The Unknown* (Carrel and Rodker 1935), he started advocating eugenics, including endorsement of gassing the mentally ill people. In many ways, he became one of the early promoters of the Nazi eugenics. Obviously, he did not have any experience in genetics and had little regard for principles of humanism and ethics. His far-reaching and appalling pseudoscientific statements represented a gross misuse of his scientific credentials. His active support for the Vichy regime drew criticism, and his death in 1944 prevented putting him on trial.

Sources of Non-repeatable Research Results

The first and most important task is understanding that in the vast majority of cases, the driving factor of non-reproducible research is not intentional fraud or fabrication. The estimated relative contributions of various errors to preclinical research irreproducibility are the following: biological reagents and reference materials (36.10%), study design (27.60%), laboratory protocols (10.80%), and data analysis and reporting (25.50%) (Freedman et al. 2015). The very rare and unusual cases of misconduct have little if anything in common with the much larger, broader, and more pervasive quality issue of non-reproducible research.

Flawed Design and Statistical Analysis. In his much-cited landmark study, Stanford Professor John Ioannidis (2005) showed that most published research results are false largely due to faulty statistical analysis and low positive predictive value of studies. As the Nobel Prize-winning economist Ronald Coase also pointed out, "if you torture the data long enough, it will confess to anything" (cited by Varian 2010).

The causes of flawed design and inadequate sampling can be highly variable. The classic sources of failure like insufficient power, missing power calculation, and non-representative sample remain major challenges. There are too many publications reporting results based on small sample preclinical research (e.g. the prevalent five mouse study). The well-recognized methods of blinding and randomization are much less accepted practices in preclinical research. Such omissions can easily become a major source of bias and non-replicability.

In biomedical research, there has been increasing concern about the failure of preclinical (animal) data to translate to human clinical trials. This often relates to inadequate attention to experimental design and a failure to consider interspecies variations.

As it was highlighted with the headline, every cell has sex (Pardue and Wizemann 2001). The same is true for animal models. Gender differences can bias the experimental results when the sample is not representative of the target population. The choice of species, strains, ages, and sex profoundly influences the likelihood of replicability in research studies. Currently, a high number of preclinical studies still rely on rodent-based animal models with a predominant male emphasis.

Well-designed research also needs defined endpoints before starting the experimentation. Without such plans, experiments can stop at conveniently selected spectacular but in fact random results. Similarly, prospective plans are needed to deal with missing data or outliers.

For example, several studies claimed growing trends in narcissism and significant increase of self-enhancement among students. However, reanalysis of data showed no evidence that level of self-enhancement, as defined by the discrepancy between perceived intelligence and academic achievements, increased among high school students from 1976 to 2006 (Trzesniewski et al. 2008).

In a typical scientific publication, researchers provide access to transformed information, results of statistical analyses. Meanwhile, the source data remain undisclosed and often discarded afterward. It has also been a long-standing complaint of meta-analysis researchers that reports are inconsistent and incomplete in disclosing details essential for further analysis and the original data remain unavailable.

A study showed that sampling errors undermined the validity of genetically modified organism (GMO) analyses (Esbensen et al. 2013). In a group of studies on the standard murine model of amyotrophic lateral sclerosis (ALS), most measurements appeared to record only noise instead of effect (Scott et al. 2008). Reanalysis of recovered data from Minnesota Coronary Experiment showed that incomplete publication at the time of the study led to a subsequent overestimation of the benefits of diet-heart hypothesis (Ramsden et al. 2016).

Ignorance of modeling and unscrupulous use of statistical methods can degrade any experimental study into a kind of cherry dance. First, measuring a lot of things and subsequently testing countless statistical hypotheses without adjusting thresholds of significance. As a result of multiple statistical tests, there will be several seemingly significant results, the cherries. Based on these cherries, speculation starts with explanations to make the random results seemingly logical and useful.

Of course, the fallacy of cherry picking is that the model is created after observing the false positives from a single experiment and the results can be just noise from random variation. Chances of being challenged by someone else measuring the same thing and finding something different are often not sufficiently high. Historically, 0.05 became the threshold of significance, but most such results are not significant, neither statistically nor regarding scientific or patient outcomes.

Assuming that 2.7 million scientific articles are published in one year and half of them included one single statistical tests at the level 0.05, there would be

67 500 false-positive results presented (Fricker 2016). Obviously, the number of false positives is much greater as most experimental research studies present many more statistical tests.

The workshop and report of the Committee on Applied and Theoretical Statistics of the Board on Mathematical Sciences and Their Applications of the National Academies of Sciences, Engineering, and Medicine provides an excellent overview of the statistical challenges of reproducibility (Schwalbe 2016).

Contaminated Cell Lines. Preclinical research often moves from isolated cells or tissue samples to animal research. Obviously, such research is highly vulnerable to problems with cell line identification. Drugs tested on the wrong cell line can give very misleading results and confuse the development of new therapies.

Since the late 1960s, more and more questions have been raised regarding the experimental cell lines of cancer research. Many widely used cell lines turned out to be entirely different and contaminated lines questioning the results produced with their use. In a study of cell lines from diverse head and neck tumor sites, 43% turned out to be misidentified or cross-contaminated (Zhao et al. 2011).

Data from the United States, Europe, and Asia suggests that about 15% of cell lines are misidentified or contaminated (Lorsch et al. 2014). When a research associate brings the great news that free cell lines can be received from a neighboring laboratory, the leader of the research laboratory should be very cautious. Major problems can emerge when the cells are misidentified as it has happened many times.

High frequency of mistaken identities questions the very foundation of cell line-based cancer research. A study of thyroid cancer projects found that 17 out of 40 cell lines have been cross-contaminated (Schweppe et al. 2008). Other studies suggested 36% contamination rate. Another study found significant misidentification, duplication, and loss of integrity of endometrial and ovarian cancer cell lines (Korch et al. 2012). In some cases, the cells are infected with viruses, bacteria, or fungi. A study suggested that 5–10% of cell culture studies used cells contaminated with mycoplasma (Armstrong et al. 2010).

Without external contamination, cells can undergo chromosomal rearrangements and mutations significantly degrading their value for cancer research. Furthermore, even when the cells were correctly identified and had been protected from infection, passing cells from one generation to the other over a prolonged period can change their characteristics and lose representation of the original cell line. Such risks highlight the importance of protecting cell line integrity.

Prudent use of cell lines makes sure that the correct identification is provided based on genetic markers, PCR test, or another objective method. There is a need for validation and accurate reporting of cell line identity (Lorsch et al.

2014). One method is using short tandem repeat (STR) analysis to identify DNA sequences and compare to available online STR fingerprints. The cell line needs to be protected from infection and checked for sterility. New cell lines need to be regularly purchased to avoid deterioration over multiple generations.

Originally, the cell line microarray drug sensitivity signatures were introduced to predict patient response (Bonnefoi et al. 2007). This technology was ranked as one of the top 100 breakthroughs in 2006 (Coombes et al. 2007). However, the results could not be reproduced in large clinical trials later (Baggerly and Coombes 2009). In a published assessment of 18 published microarray studies, only two were reproducible (Ioannidis 2011).

Dysfunctional Reagents. In animal models, genetic variations, differences in species, or inbred versus outbred strains can cause inconsistencies and non-repeatability. Sometimes the laboratory model animal has some characteristics that prevent replicability and relevance to human conditions. For example, Amgen researchers reported that due to a genetic flaw in a commercially available mouse model, their team and an academic lab erroneously reported that Gpr21 influences body weight and glucose metabolism (Wang et al. 2016).

Another study found HIV contamination of commercial PCR enzymes raising concerns in in-house genotypic HIV drug resistance tests (Monleau et al. 2010). A large percentage of murine leukemia virus contamination was found in commercial RT-PCR reagents and mouse DNA contamination in commercially available human DNAs (Zheng et al. 2011).

A study tested the functioning of over 200 antibodies raised against 57 different histone modifications, in *Caenorhabditis elegans, Drosophila melanogaster*, and human cells (Egelhofer et al. 2011). By dot blot or Western blot, over 25% of antibodies failed specificity tests. Among specific antibodies, over 20% failed in subsequent chromatin immunoprecipitation experiments. Apparently, rigorous testing of histone-modification antibodies is necessary before use of such antibodies.

The Undisclosed Becomes Non-repeatable

Lack of Computing Transparency. The research and development progress of computing applications and artificial intelligence represent a particular challenge to reproducibility. Such decision support and artificial intelligence systems are increasingly found in clinical trials and other research applications as well (e.g. computerized clinical decision support for warfarin management (Woller et al. 2015), continuous glucose monitors for insulin dosing (Shapiro 2017).

Often, the complex algorithms are proprietary or automated to the extent that they are not transparent to peer reviewers or other users. Supposedly, machines and computer programs are auditable: the code can be made open and available to readers. In reality nobody likes to read someone else's program, and replicating the functions is often nearly impossible or impractical task.

When a computer system does not have a transparent explanation for its decisions, the opportunity for replication by independent investigators is nonexistent. Beyond research, this is becoming a general societal concern as people are facing mistakes of approvals and rejections generated by enigmatic algorithms. In recent years, frustration has been growing with the unexplained decisions of artificial intelligence systems (e.g. turned down credit applications, being labeled as potential criminal, or being wrongly evaluated in job performance).

Not surprisingly, there have been several court cases challenging the nontransparent computer-generated decisions. The particular problem for life sciences research is that peer-reviewed articles often describe the structure and practical use of complex software systems but do not provide sufficient information to replicate the results. Consequently, an essential requirement of trustworthy science cannot be met.

With the increased use of artificial intelligence and sophisticated computer systems, the push for reproducibility and transparency will likely increase. Several countries are developing legal requirements for transparency and accountability of artificial intelligence systems (e.g. United Kingdom, Germany). The US Food and Drug Administration (FDA) is developing guidelines for appropriate premarket assessment of such experimental computer systems and requirements for introduction into routine patient care.

António Moniz was a Portuguese neurologist, professor, and dean of the Medical School at the University of Lisbon. Besides medicine his passion was politics, and he was elected to the parliament, during the First World War, served as ambassador to Spain and afterward as Minister for External Affairs of Portugal. After retiring from politics, he returned to medicine and medical research (Tan and Yip 2014).

His most important scientific breakthrough was the X-ray contrast visualization of blood vessels in the brain. After several failed experiments, Moniz was the first to develop a successful cerebral angiogram. This pioneering diagnostic technique made possible the diagnosis of internal carotid occlusion, among others, and subsequently made him world renowned. In recognition, he was twice nominated for the Nobel Prize (Jansson 2014).

The next major development by Moniz was the leucotomy surgical procedure that is today more widely known as lobotomy. In an attempt to treat schizophrenia, the procedure was cutting nerve fibers connecting the frontal lobe with other parts of the brain. After the first set of surgeries, Moniz reported seven cures, seven improvements, and six unchanged cases. However, he also noted that patients who had deteriorated from the mental illness did not gain much.

This procedure was spreading for a few years with tens of thousands of surgeries performed in various countries. Moreover, Moniz got the Nobel Prize for his development of prefrontal leukotomy in 1949. Soon afterward, there were growing concerns with the questionable outcomes, and the discovery of drug treatment for schizophrenia made the procedure largely obsolete.

The controversies of prefrontal leukotomy should warn everyone that, prospective, well-designed, long-term studies of therapeutic outcomes are essential to assess the benefits of new treatments before widespread practical implementation (Jansson 2014).

Research Reporting Deficiencies. Compounding the problem of non-reproducible results, various reporting defects can add further confusion. Unreported results, lack of data transparency, and selective reporting are among the frequently mentioned shortcomings.

Unreported results by definition are not replicable. Unfortunately, several studies have documented the very low reporting rates from well-funded clinical trials. Among 4347 interventional clinical trials, the ratio of clinical trials with results presented within 24 months of study completion ranged from 16.2 to 55.3% across 51 academic medical centers (Chen et al. 2016). Within the same timeframe, the ratio of result reporting on ClinicalTrials.gov was less, between 1.6 and 40.7%. Certainly, non-publication of results defeats the effort of trial participants, funding from the sponsors including taxpayers, and the scientific opportunity recognized at the time of designing the trial.

Selective reporting of significant results has been a long-recognized danger in research. However, pictures can be unintentionally manipulated as well. The typical research publication displays the best or most successful picture based on the subjective judgment of the researchers. Often, the picture is selected to "best illustrate" the point the author wants to make. In the age of printed scientific journals, limitation to the one best picture was perhaps justified. With the gradually emerging electronic and multimedia technologies, complete disclosure is technically feasible and also desirable. Greater transparency and more objective reporting could be achieved by disclosing all pictures from reported experiments.

Data transparency also needs improvement. In many studies, the presented average can come from clustering of the majority of data points that can also be pulled by one or two major outliers in a different direction. Disclosure of data would answer questions and also make reanalysis possible. One may also argue for more opportunities to reuse experimental data by other researchers. Along these lines, there are calls that the FDA should make all data available from submitted clinical trials. Of course, sharing data also requires full disclosure of the methodology that generated them to interpret meaningfully.

A particular challenge of open sharing is that disclosure of research data can go against informed consent even in cases when anonymity is protected. When a patient signs up to participate in an experiment, he or she does not sign up for future unspecified and unforeseen research on the data or specimen. Again, the concept of informed consent is in crisis as it is becoming almost unethical to limit use for a single published study. Perhaps in cases when the risk of re-identification remains below a certain level, such broader use could be reasonably assumed. If open data sharing becomes routine, the entire process of human subject approval and informed consent needs to be updated.

Rare but Troubling: Pathological Science and Fraudulent Research

At the far end of the spectrum of non-reproducible research, there are two distinct entities, *pathological science* and *fraudulent research*. They sound related occurrences, and their outcome is similarly useless non-reproducible result, but, in reality, their roots are quite different.

Pathological science is a product of extreme wishful thinking, illusions generated by marginal differences and subjective biases. Not infrequently, pathological science is generated by highly respected, indeed accomplished scientists. There are many examples of pathological life science. In 1988, an immunologist published a paper in the prestigious journal *Nature* and stated that the transmission of immune system information could be related to the molecular organization of water. In spite of great media attention, the findings have never been corroborated by anyone (Davenas et al. 1988).

In 1989, two accomplished and respected scientists reported energy-producing nuclear fusion at room temperature, so-called cold fusion (Fleischmann et al. 1989). However, the findings could never be corroborated, and apparently some non-reproducible observations were grossly overinterpreted. Apparently, the interpretation was more influenced by hopeful thinking than reliance on actual reproducible measurements.

Uncommon but grave turn of research is intentional scientific misconduct. It can range from negligent research to purposefully fake data that undermines trust in the results of science. For example, a psychologist researcher manipulated or faked data in more than 50 studies of social psychology in Europe. Some of his studies on fabricated data were published in the high-impact journal *Nature* (Callaway 2011).

Purposefully fraudulent research is extremely rare but highly corrosive problem. In many ways, it is just another version of criminal behavior well known in many other areas of life. For such people, the prestige of being recognized as scientist and the opportunity to get millions of dollars in grant funding become too attractive. The typical fraud is built on the fabrication of data, falsified experiments, and doctored images.

Another university researcher in America became famous in the area of HIV vaccine development. He faked his research by spiking rabbit blood samples with human HIV antibodies to make the vaccine effect more impressive. His research generated millions of dollars in NIH grant support. Eventually, the fraud became recognized, and the researcher was not only barred from receiving further funding but also sentenced to 57 months of prison time after a felony conviction (Reardon 2015).

In 2007, it was discovered that a psychology professor, long considered a leading expert in animal and human cognition, engaged in misconduct. He published about one peer-reviewed paper per month for several years (Ledford 2010). One of his most famous experiments led to the report that one of the smallest monkeys, the cotton-top tamarin, can recognize themselves in a mirror. However, the experimental results could not be replicated by others, and later the author could not replicate them either. A broader investigation by the NIH Office of Research Integrity (ORI) found that he falsified data and made false statements about research methodologies in six federally funded studies. Ultimately, the psychology professor resigned from his faculty position and left academia.

In some extreme cases of fraudulent research, emotional machinations, political connections, and celebrities are involved to multiply support for nonsense theories and meaningless results. In Asia, a veterinarian researcher specialized in stem cell studies and published sensational articles on human embryonic stem cell cloning in some of the most prestigious journals in the world. At some point, he was even named supreme scientist, an honor accompanied by extremely generous funding support in his country. Later, it turned out that his stem cell research was faked, and he embezzled funds in the range of millions of dollars (Gottweis and Triendl 2006).

One of the most infamous and politically prominent mega-frauds of history was Trofim Lysenko's theory of environmentally acquired inheritance and his denial of the Mendelian inheritance in Stalin's Soviet Union. The pseudoscientific and oppressive *Lysenkoism* not only caused major damage to the progress of biology and genetics in his country but also led to the defamation, firing, arrest, and death of many dissenting scientists.

The Way Out: Modeling, Transparency, and Continuous Quality Improvement

Knowing the dedication, tireless effort, and good intentions of the vast majority of researchers, one may wonder about the system-wide problems of research that need fixing. Corrective design actions should bring the reliability of the scientific enterprise up to standards and public expectations.

The words of Ignaz Semmelweis (1861), the pioneer of antiseptic surgery, should encourage our action for the prevention of errors in science: "No matter how painful and oppressive such recognition may be, the remedy does not lie in suppression. If the misfortune is not to persist forever, then this truth must be made known to everyone concerned."

Science is about the search for the truth. Various errors and incorrect assumptions are everyday experiences and part of the usual process in research projects. When a life sciences study is completed, careful cross-checking and *scientific rigor* are needed to verify the accuracy of facts and data published by the research laboratory. Balas and Ellis (2017) recommended a three-point plan for reproducibility that includes transparency, replication, and *triangulation* before publication (see the quality improvement chapter).

Contrary to popular assumptions, *self-replication* does not require repeating the entire project as most research is characterized by a large number of failed experiments, and only the successful ones deserve replication at a fraction of the total cost of the project. Being thoroughly familiar with the process, having all the equipment and resources in place, the self-replication should be an affordable and straightforward task.

In the process of cross-checking, the unexpected, non-reproducible, or unexplained outliers can also become major turning points on the road to discovery. As illustrated by John Gurdon's frog gene, early recognition of non-reproducibility and careful follow-up investigation can yield exceptionally valuable information, in his case Nobel Prize-winning discoveries (see more details in Chapter 16).

To move forward, constructive and practical messages are needed to benefit research producers as well as users of biomedical research discoveries. Suggestions to improve research quality include broader data-sharing requirements and greater monitoring from funding agencies and journal publishers (Chalmers et al. 2014; Glasziou et al. 2014).

NIH is implementing a policy for data sharing of research projects above certain thresholds. Among other benefits, sharing data opens scientific inquiry, promotes new research, helps the testing of new or alternative hypotheses, facilitates the education of new researchers, and expedites the translation of research results into knowledge, products, and procedures.

According to Peng et al. (2006), availability of the following information is needed for full replicability of published research results generated with the use of complex software systems: analytical dataset, computer code underlying figures and tables, software environment to execute the code, adequate documentation of the data and software environment, and effective methods of distribution.

Particularly, the pressure is increasing on academic institutions to improve practices and change the disorienting incentives that appear to encourage production of untrustworthy results. Some analysts talk about institutional expectation bias that is provoking non-reproducible results. Several factors are suspected being institutional incentives of non-reproducible research results:

the focus on isolated original research studies, the constant counting of publications particularly in high ranked journals, and the use of extramural funding as a measure of scientific productivity in evaluation, promotion, and tenure deliberations. There is a great need for accountability for broader societal outcomes and impact of research institutions.

Reliance on the underfunded editorial offices of scientific journals and overloaded peer reviewers is probably not a very effective prevention strategy. With an automotive analogy, the best place to improve new product manufacturing quality is the place of production, not in the dealer's showroom.

Quality improvement experts highlight that every defect is, in fact, an opportunity for improvement and the discussions should focus on positive strategies. It should also be recognized that criminalizing unintentional biases is likely to alienate many researchers who actually should be our best allies in quality improvement efforts. Naming and shaming generate only enemies and committed defenders of status quo. Progress needs allies to effect change.

Universities and research institutions increasingly recognize the relationship between reliability of research and innovation success benefiting society. The chapter on quality improvement provides more information about opportunities and processes to achieve improvement. One of the most promising opportunities of reducing non-repeatable research is taking institutional responsibility.

Well-thought-out actions are more durable than the products of haste. Every defect should be viewed as an opportunity to do better than others. Scientists striving to conduct quality research can get ahead by better understanding the root causes of non-reproducible research and by applying the best practices of preventing such distractions.

Ultimately, scientific rigor and sound quality management are laying the foundation for a myriad of novel initiatives and significant discoveries. With continuous improvement, remarkable progress will be achieved in advancing the science and treatment of major diseases. This is the most opportune time for anybody who is interested in the mysteries of life and prevention of illness.

References

Armstrong, S.E., Mariano, J.A., and Lundin, D.J. (2010). The scope of mycoplasma contamination within the biopharmaceutical industry. *Biologicals* 38 (2): 211–213.

Baggerly, K.A. and Coombes, K.R. (2009). Deriving chemosensitivity from cell lines: forensic bioinformatics and reproducible research in high-throughput biology. *The Annals of Applied Statistics* 3 (4): 1309–1334.

Balas, E.A. and Ellis, L.M. (2017). Preclinical data: three-point plan for reproducibility. *Nature* 543 (7643): 40–40.

Begley, C.G. and Ellis, L.M. (2012). Drug development: raise standards for preclinical cancer research. *Nature* 483 (7391): 531–533.

Bonnefoi, H., Potti, A., Delorenzi, M. et al. (2007). RETRACTED: validation of gene signatures that predict the response of breast cancer to neoadjuvant chemotherapy: a substudy of the EORTC 10994/BIG 00-01 clinical trial. *The Lancet Oncology* 8 (12): 1071–1078.

Callaway, E. (2011). Report finds massive fraud at Dutch universities. *Nature* 479: 15.

Carrel, A. and Rodker, J. (1935). *Man, the Unknown*, 207. New York: Harper.

Chalmers, I. and Glasziou, P. (2009). Avoidable waste in the production and reporting of research evidence. *Obstetrics & Gynecology* 114 (6): 1341–1345.

Chalmers, I., Bracken, M.B., Djulbegovic, B. et al. (2014). How to increase value and reduce waste when research priorities are set. *The Lancet* 383 (9912): 156–165.

Chen, R., Desai, N.R., Ross, J.S. et al. (2016). Publication and reporting of clinical trial results: cross sectional analysis across academic medical centers. *British Medical Journal* 352: i637.

Collins, F.S. and Tabak, L.A. (2014). NIH plans to enhance reproducibility. *Nature* 505 (7485): 612.

Coombes, K.R., Wang, J., and Baggerly, K.A. (2007). Microarrays: retracing steps. *Nature Medicine* 13 (11): 1276–1277.

Davenas, E., Beauvais, F., Amara, J. et al. (1988). Human basophil degranulation triggered by very dilute antiserum against IgE. *Nature* 333 (6176): 816–818.

Dutkowski, P., De Rougemont, O., and Clavien, P.A. (2008). Alexis carrel: genius, innovator and ideologist. *American Journal of Transplantation* 8 (10): 1998–2003.

Egelhofer, T.A., Minoda, A., Klugman, S. et al. (2011). An assessment of histone-modification antibody quality. *Nature Structural & Molecular Biology* 18 (1): 91–93. http://doi.org/10.1038/nsmb.1972.

Esbensen, K.H., Pitard, F., and Paoletti, C. (2013). Sampling errors undermine valid genetically modified organism (GMO) analysis. *TOS Forum* 1 (1): 25–26.

Fleischmann, M., Pons, S., and Hoffman, R. (1989). Measurement of γ-rays from cold fusion. *Nature* 339: 29.

Freedman, L.P., Cockburn, I.M., and Simcoe, T.S. (2015). The economics of reproducibility in preclinical research. *PLoS Biology* 13 (6): e1002625. https://doi.org/10.1371/journal.pbio.1002165.

Fricker, R.D. (2016). False positives are statistically inevitable. *Science* 351 (6273): 569–570.

Gabel, J.D. (2014). Realizing reliability in forensic science from the ground up. *Journal of Criminal Law and Criminology* 104 (2): 283.

Glasziou, P., Altman, D.G., Bossuyt, P. et al. (2014). Reducing waste from incomplete or unusable reports of biomedical research. *The Lancet* 383 (9913): 267–276.

Gottweis, H. and Triendl, R. (2006). South Korean policy failure and the Hwang debacle. *Nature Biotechnology* 24 (2): 141–143.

Ioannidis, J.P. (2005). Why most published research findings are false. *PLoS Medicine* 2 (8): e124.

Ioannidis, J.P. (2011). Improving validation practices in "Omics" research. *Science* 334: 1230.

Jansson, B. (2014). Controversial psychosurgery resulted in a Nobel Prize. Nobelprize.org. Nobel Media AB 2014. http://www.nobelprize.org/ nobel_prizes/medicine/laureates/1949/moniz-article.html (accessed 24 November 2017).

Korch, C., Spillman, M.A., Jackson, T.A. et al. (2012). DNA profiling analysis of endometrial and ovarian cell lines reveals misidentification, redundancy and contamination. *Gynecologic Oncology* 127 (1): 241–248.

Ledford, H. (2010). Harvard probe kept under wraps. *Nature* 466: 908–909.

Lindner, M.D. (2007). Clinical attrition due to biased preclinical assessments of potential efficacy. *Pharmacology & Therapeutics* 115 (1): 148–175.

Lorsch, J.R., Collins, F.S., and Lippincott-Schwartz, J. (2014). Fixing problems with cell lines. *Science* 346 (6216): 1452–1453.

MacCoun, R.J. (1998). Biases in the interpretation and use of research results. *Annual Review of Psychology* 49: 259–287.

Mente, A., O'Donnell, M.J., and Yusuf, S. (2014). The population risks of dietary salt excess are exaggerated. *Canadian Journal of Cardiology* 30 (5): 507–512.

Mobley, A., Linder, S.K., Braeuer, R. et al. (2013). A survey on data reproducibility in cancer research provides insights into our limited ability to translate findings from the laboratory to the clinic. *PLoS One* 8 (5): e63221.

Monleau, M., Plantier, J.C., and Peeters, M. (2010). Short communication HIV contamination of commercial PCR enzymes raises the importance of quality control of low-cost in-house genotypic HIV drug resistance tests. *Antiviral Therapy* 15: 121–126.

Pardue, M.L. and Wizemann, T.M. (eds.) (2001). *Exploring the Biological Contributions to Human Health: Does Sex Matter?* Washington, DC: National Academies Press.

Peng, R.D., Dominici, F., and Zeger, S.L. (2006). Reproducible epidemiology research. *American Journal of Epidemiology* 163 (9): 783–789.

Prinz, F., Schlange, T., and Asadullah, K. (2011). Believe it or not: how much can we rely on published data on potential drug targets? *Nature Reviews Drug Discovery* 10 (9): 712–712.

Ramsden, C.E., Zamora, D., Majchrzak-Hong, S. et al. (2016). Re-evaluation of the traditional diet-heart hypothesis: analysis of recovered data from Minnesota coronary experiment (1968–73). *British Medical Journal* 353: i1246.

Reardon, S. (2015). US vaccine researcher sentenced to prison for fraud. *Nature* 523 (7559): 138–139.

Schuster, D., Laggner, C., and Langer, T. (2005). Why drugs fail – a study on side effects in new chemical entities. *Current Pharmaceutical Design* 11 (27): 3545–3559.

Schwalbe, M. (ed.) (2016). *Statistical Challenges in Assessing and Fostering the Reproducibility of Scientific Results: Summary of a Workshop*. Washington, DC: National Academies Press.

Schweppe, R.E., Klopper, J.P., Korch, C. et al. (2008). Deoxyribonucleic acid profiling analysis of 40 human thyroid cancer cell lines reveals cross-contamination resulting in cell line redundancy and misidentification. *The Journal of Clinical Endocrinology & Metabolism* 93 (11): 4331–4341.

Scott, S., Kranz, J.E., Cole, J. et al. (2008). Design, power, and interpretation of studies in the standard murine model of ALS. *Amyotrophic Lateral Sclerosis* 9 (1): 4–15.

Semmelweis, I.F. (1861). *The Etiology, Concept, and Prophylaxis of Childbed Fever* (No. 2). Cited by University of Wisconsin Press, 1983.

Shapiro, A.R. (2017). FDA approval of nonadjunctive use of continuous glucose monitors for insulin dosing: a potentially risky decision. *JAMA: The Journal of the American Medical Association* 318 (16): 1541–1542.

Spiegel, P.B. and Le, P.V. (2006). HIV behavioral surveillance surveys in conflict and post-conflict situations: a call for improvement. *Global Public Health* 1 (2): 147–156.

Tan, S.Y. and Yip, A. (2014). António Egas Moniz (1874–1955): lobotomy pioneer and Nobel laureate. *Singapore Medical Journal* 55 (4): 175–176. doi: http://doi.org/10.11622/smedj.2014048.

Trzesniewski, K.H., Donnellan, M.B., and Robins, R.W. (2008). Do today's young people really think they are so extraordinary? An examination of secular trends in narcissism and self-enhancement. *Psychological Science* 19 (2): 181–188.

Varian, H.R. (2010). Computer mediated transactions. *The American Economic Review* 100 (2): 1–10.

Wacholder, S., Chanock, S., Garcia-Closas, M. et al. (2004). Assessing the probability that a positive report is false: an approach for molecular epidemiology studies. *Journal of the National Cancer Institute* 96 (6): 434–442.

Wang, J., Pan, Z., Baribault, H. et al. (2016). GPR21 KO mice demonstrate no resistance to high fat diet induced obesity or improved glucose tolerance. *F1000Research* 5: 136.

Weigmann, K. (2005). The consequence of errors. *EMBO Reports* 6 (4): 306–309.

Woller, S.C., Stevens, S.M., Towner, S. et al. (2015). Computerized clinical decision support improves warfarin management and decreases recurrent venous thromboembolism. *Clinical and Applied Thrombosis/Hemostasis* 21 (3): 197–203.

Zhao, M., Sano, D., Pickering, C.R. et al. (2011). Assembly and initial characterization of a panel of 85 genomically validated cell lines from diverse head and neck tumor sites. *Clinical Cancer Research* 17 (23): 7248–7264.

Zheng, H., Jia, H., Shankar, A. et al. (2011). Detection of murine leukemia virus or mouse DNA in commercial RT-PCR reagents and human DNAs. *PLoS One* 6 (12): e29050.

7

Red Tape and Litigation

Creativity in science, as in the arts, cannot be organized. It arises spontaneously from individual talent. Well-run laboratories can foster it, but hierarchical organizations, inflexible, bureaucratic rules, and mountains of futile paperwork can kill it.

Max Perutz (2002)[1]

The scientific enterprise has never been so large and never involved so many people as it is today. Research and development of new technologies receives unprecedented public support as people all over the world enjoy the many benefits of scientific achievements. Much longer life expectancy, elimination of many diseases, unprecedented speed and comfort of transportation, revolutionized communication and entertainment, and new energy sources are just some of the examples that benefit people almost universally.

In this process of professionalization, the huge number of researchers, the skyrocketing research literature, and the vast amounts of money spent on research invite proliferation of rules, administrative structures, managers, compliance officers, performance systems, and paperwork. Today, the resulting bureaucracies help to manage the research enterprise, protect intellectual property, and eliminate many sources of waste and abuse. No one wants to bring back the times when researchers did not have to seek human subject review before starting their project or when people could get appointed to leading academic positions without scientific credentials.

Bureaucracies solve some problems, but they also create many others. Scientists and creative thinkers increasingly complain about the bureaucratic red tape, especially in life sciences. The best scientific ideas come from the wondering mind that is unobstructed by bureaucratic rules. The paperwork

---

1 Perutz, M.F. (2002). *I Wish I'd Made You Angry Earlier: Essays on Science, Scientists, and Humanity.* Oxford University Press, New York.

Innovative Research in Life Sciences: Pathways to Scientific Impact, Public Health Improvement, and Economic Progress, First Edition. E. Andrew Balas.

and regulatory pressure on researchers appear in various shapes and forms but ultimately distracting attention, wasting precious time, and costing a lot of money.

As the societal value of research increases, scientific and financial interests collide periodically. With the escalating number of rules and regulations, interpretation can become contentious. Some of these conflicts play out in the court of law. Particularly, there are two hot-button issues, the right to intellectual property and acceptable research conduct. This chapter provides an overview of the escalating red tape and court cases of research innovation.

Bureaucratization of Research

The term red tape comes from the time when red tapes were used to tie official documents centuries ago. Today, it is the phrase for unnecessary formality and paperwork that delays results. Red tape can include excessive filing and certification requirements, requiring multiple people or committee approvals, too much reporting, investigations and inspections, extensive enforcement practices, and bureaucratic hair splitting or foot dragging. Oppressive procedures take preference over results and rules over creative initiatives. It is all about living off the public purse without meaningful results.

In connection with an extensive analysis of trends in academic research policies, Bozeman (2015) suggested the following definition of bureaucratization as "the deployment of the institution of bureaucracy and its attendant rules, regulations, and procedures, as the means of social organization of work." Additionally, Bozeman defines red tape as "rules, regulations, and procedures that require compliance but do not meet the organization's functional objective." The top-down approach of the government and closed-mouth bureaucracy and hoarding of information can indeed slow down research.

Harvard economist Andrei Shleifer conducted a revealing study about the regulation of entry of startup companies in 85 countries (Djankov et al. 2002). The lowest number of procedures was required in Canada and Australia, only two. In lots of countries, the required number of procedures was between 10 and 21. As another reflection of the bureaucratic burden, the time to obtain legal status to operate a startup company varied between 2 and 152 business days. Similar variations were observed for the total cost of getting legal status for the new startup.

Based on the huge variations and disproportionately large regulatory requirements in many countries, some general conclusions can be reached. More regulatory requirements do not appear to serve the public interest. Democratic countries have fewer regulatory requirements, and more regulatory requirements appear to serve only the needs of bureaucrats and politicians.

Red tape is known to stifle competition and increase costs. Effectively, it becomes a disincentive of innovation, and researchers can drown in the paperwork. The other danger is when the number of rules is growing, and complexity increases, many institutions and researchers find out how the system can be gamed, and impression of high productivity can be made up without real value to society. Certainly, production of scores of inconsequential and non-reproducible research publications is partly blamed on the bureaucratization and bean counting that likes to measure productivity by the number of publications, sometimes in high-impact journals.

PI Administrative Burden. There are many symptoms of increasing paperwork. Planning research projects and applying for funding involves a considerable amount of paperwork. The NIH estimates that it takes about 22 hours of paperwork to complete a grant application, not including the research plan that is the essential core of the entire application (NIH 2016).

Over the years, there has been a tremendous increase in the administrative requirements of research, reducing the time devoted to science and increasing the amount of time demanding administration and compliance. A $500 000 proposal in the 1980s was about 15 pages long, and today it can easily exceed 100 pages. Federal grant applications are quite often hundred or several hundred pages long.

According to a survey of more than 6000 US researchers conducted by Decker et al. (2007), over 42% of time related to federal grants was spent on pre- and post-award administrative activities, not directly relevant to the actual research. There are alarming signs that cumulative research technicalities create a significant distraction from creativity and humanistic mission by overwhelming the time and attention of biomedical researchers.

According to the 2012 FDP Faculty Workload Survey, an average of 42% of the principal investigator research time assigned to federally funded projects is spent on meeting requirements rather than conducting active research (i.e. proposal preparation, pre-award administration, post-award administration, report preparation) (Rockwell et al. 2012). The results are very similar to those found in the 2005 FDP survey, suggesting little change over time.

Further increasing the burden of writing-intensive applications, there has been a gradual decrease in NIH pay lines, the percentage of grant applications that get funding. In other words, the average investigator needs to write many more grant applications to have the same chance of winning support as just a few decades ago. National research leadership started talking about hypercompetition as a result of decreasing pay lines. In a hypercompetitive culture, learning grant writing skills appear to be more important than actually conducting research that responds to scientific opportunities and societal needs (Alberts et al. 2014).

Bureaucratization is becoming emblematic of scientific conference hallway discussions about papers published and grant applications being developed. Many scientists complain that the first question that they hear is about grant funding generated instead of the substance of their research. Many researchers receive email advertising about grant writing course offerings on a weekly basis. Internationally, research training is often simplified to scientific publication writing skills. For some people, research starts with paper and ends with paper.

Obviously, there is an urgent need to refocus the discussions on good science: identifying important questions about nature and society, building models, emphasizing basic mechanisms, and carefully looking at the evidence. Most scientific papers are not read by a general audience, only by the specialists of the same or comparable professions. If scientists want to break out of the bureaucracy of research, they need to look at the actual intellectual and practical impact of their research and make sure that interactions with other scientists focus on the substance of unanswered questions and not on the paperwork surrounding research.

The Massachusetts Institute of Technology sent a survey to 921 active principal investigators (PIs) in 2015. According to the PI administrative burden survey, the most frustrating aspects of research are managing grant/contract expenditures, proposal preparation, effort reporting, personnel administration, and contract negotiations (MIT Office of Sponsored Programs 2015). Recommendations of surveyed PIs pointed toward more and better software tools in addressing the crisis of administrative burden.

Product Approval. In the health biotechnology sector, the *regulatory burden* is particularly significant as the risk of harm is disproportionately larger. Commercialization of intellectual property is particularly challenged by requirements of regulatory approval and health-care reimbursement issues. As a result, there is a burden of delays and inconsistency in efforts to convert genomics and other life sciences research into social and economic benefits.

When people talk about bureaucratic approvals in health care, the first example that is often mentioned is the approval by the Food and Drug Administration (FDA). Indeed, the requirements are very high, the process is very lengthy, and the costs are enormous. At every stage of new drug development, the testing for efficacy and safety is expensive, the risk of failure is high, and the approval process is burdensome.

On the other hand, new drug approval is the critical point where the rubber hits the road, and the evidence of benefits is checked in the public interest. There have been many cases when the public was or should have been protected by the scrutiny of approval. One of the most famous ones was thalidomide, the sleeping pill that was introduced in Europe in the 1950s. It turned out to be responsible for causing major malformations of limbs in newborns when taken by pregnant mothers. The mortality rate of these children was also excessively high, about 40–50%.

In the United States, FDA reviewer Frances Kelsey resisted pressure from the manufacturer and also the fact that 20 countries already approved the drug and insisted on the need for complete testing before approval (Bren 2001). The extra time and testing revealed the devastating side effects. Undoubtedly, her action saved thousands of children from death and malformations. In recognition of her work and accomplishments, Kelsey received the President's Award for Distinguished Federal Civilian Service from John F. Kennedy.

Quite appropriately, the FDA approval is layered, and different levels of scrutiny apply to products that have different risk profiles. Pharmaceutical treatments are among the most regulated in the world. Apparently, a white pill that is entirely judged based on what is written on the box can be highly beneficial and toxic, sometimes simultaneously. Customers have no way to judge the claims until it is too late, and consequently there is total dependence on prior testing and regulatory approval.

The regulatory burden is much less or nearly negligible when the diagnostic or therapeutic intervention is more transparent and can be better judged by the users. For example, the regulatory requirements for prosthetic limbs, first aid kits, and operating tables are less stringent. On the other hand, implantable devices like pacemakers and closed-loop insulin pumps are again more rigorously regulated. Regardless of the level of regulatory burden, the approval process must be factored in when new product development plans are designed.

To some extent, the increased regulatory and administrative burden is blamed for the escalating expenses of new drug development. Notably, the cost of drug development rose by a factor of 2 roughly every nine years apparently following the so-called Eroom's law as opposed to Moore's law of the cost of semiconductor development that projects decrease by a factor of 2 every two years (Scannell et al. 2012).

Max Perutz was an Austrian chemist who graduated from Vienna University and subsequently worked at the Cavendish Laboratory of Cambridge University in most of his life. He arrived in Cambridge at the time when his parent's business was seized by the Nazis.

Perutz was a pioneer of molecular and structural biology. Among his particular accomplishments is the use of X-ray crystallography for the analysis of larger molecules. He had a special interest in the structure and function of hemoglobin. He developed several methodologies and studied the three-dimensional structures of macromolecules. For his studies of the structures of globular proteins, he got the Nobel Prize in Chemistry jointly with John Kendrew in 1962.

His book titled *I Wish I'd Made You Angry Earlier: Essays on Science, Scientists, and Humanity* is not just a great memoir of his life and accomplishments but also a unique window into the world of many renowned contemporary scholars (Perutz 2002). His book is an excellent source of insightful comments about the role of creativity and ethical considerations in scientific discoveries. Particularly, the chapter on Fritz Haber, the chemist who made nitrogen fertilizers a reality for most of the food production all over the world, is particularly insightful. In the book, Perutz proved that he was not only a brilliant scientist and astute observer but also an elegant and skilled writer.

Informed Consent. South African surgeon and head of the Department of Experimental Surgery Christiaan Barnard performed the first human-to-human heart transplant operations at Groote Schuur Hospital. Between 1967 and 1983, more than 50 heart transplant operations were carried out by his team.

At that time, the Consent of Operation form signed by the patient was less than one page: "I the undersigned consent to the performance of an operation on myself and the administration of an anesthetic for the purpose. I understand that the operation will be a heart transplant, but I agree to the surgeon extending the scope thereof or to his carrying out additional or alternative measures if during the operation he considers such procedure necessary. I fully appreciate the nature, scope, risks, and probable consequences of the operation which have been explained to me by Professor Barnard and I accept any risks associated with and consequences arising out of such operation."

Today, informed consent forms are much longer and more detailed, quite appropriately. However, some of the excesses are startling, like the 29-page consent form in a recent cancer clinical trial. Providing written explanation is well justified, but exceeding human attention span and burdening patients with unnecessary details do not seem to add value to the process of informed consent.

Not surprisingly, studies show evidence of significant misconceptions of participants and providers suggesting the low quality of the informed consent process (Joffe et al. 2001). Fortunately, effectiveness can be significantly improved by the use of multimedia, enhanced form, and extended discussion (Nishimura et al. 2013). One of the major side benefits of early heart transplant operations was the emerging definition of brain stem death, a valuable and enduring concept of patient care (Hoffenberg 2001).

Regulatory Burden on Creativity

Institutional review boards have the meritorious mission of protecting subjects of research and consequently the integrity of research as well. Following recommendations of the Helsinki Declaration, the Common Rule governs Institutional review boards and research involving human subjects since 1981. The Common Rule sets requirements for compliance by research institutions, for obtaining and documenting informed consent and institutional review board (IRB) structure and function.

While no one argues with the need and critical mission of *human subject protection* by IRBs, concerns are growing regarding the expanding bureaucracy of the process. Some argue that IRBs tend to substitute bureaucratic ethics of documentation for professional ethics (Heimer and Petty 2010). University of Michigan Professor Carl Schneider did not mince words in describing the signs of misregulation in the IRB system, the problem of trying to do more good than harm and the often arbitrary, capricious decisions (Schneider 2015).

In 2016, NIH released a new policy on the use of a single institutional review board for multisite research (Hodge and Gostin 2017). The much-anticipated elimination of repetitive IRB reviews was an important step forward in reducing the unnecessary administrative burden and inefficiencies without actually diminishing human subject protections. In other words, duplicative and triplicate reviews were required for more than 30 years, a practice that is discontinued by the new policy.

Testing the safety and efficacy of new treatments requires order, precision, hard data, and orderly procedures. These rules have been designed to protect participants of experiments and also the general public from any potential harm. However, these rules can put patients into rather impossible situations, and therefore flexibility and willingness to adapt are recommended (Fauci 2016).

Anthony Fauci, director of the National Institute of Allergy and Infectious Diseases, recalled a particularly memorable situation at the beginning of the AIDS epidemic in 1984. There were very few drugs available to prolong patient survival, measured only in months at that time. They just started experimenting with ganciclovir to treat the cytomegalovirus infection causing blindness among AIDS patients. Clinical trial rules required that patients could use only one drug during testing. In reality, these patients also needed to take another medication to prolong their lives, particularly AZT. By revisiting the rules, the trial investigators could manage an experiment that also accepted patients with multiple drugs. Ultimately the more flexible approach greatly accelerated the approval of an important new treatment for a major complication of AIDS.

Being responsive to opportunities for innovation is becoming a particular challenge for large organizations with significant research and development arms. In many organizations, looking for disruptive thinkers runs against rule-based thinking and committed defenders of status quo. There is a sense of urgency to rebalance bureaucratization and innovation, i.e. the creation and enforcement of bureaucratic rules versus behaviors that break rules, and promote innovation and entrepreneurship. Organizations that fail to promote innovation put themselves into a competitive risk and, in some cases, to the risk of extinction.

To address the crisis of bureaucratization, the National Academies of Sciences produced a 260-page report urging a new regulatory framework for the twenty-first century (NAS 2016). The report recognizes that effective regulation is essential to the success of the research enterprise (e.g. performance accountability, scientific integrity, stewardship of federal funds, and well-being of the people and animals involved in research). On the other hand, continued expansion and ever-growing requirements are starting to diminish effectiveness.

When regulations are inconsistent, duplicative or unclear universities and researchers are burdened further. The report suggests harmonization of human subject protections, conflict of interest statements, single IRB for multisite studies, simplified processing of minimal risk research, and revised informed consent form for biospecimen collection. The report recommended the creation of a new public–private forum and a designated government official to promote the adequate conception, development, and harmonization of research policies.

To estimate the bureaucratic burden, Vanderbilt University conducted a study of spending in 13 institutions in FY2014 (Vanderbilt University 2015). As mentioned before, the research-related compliance costs range from 11 to 25% of research expenditures. Obviously, institutions with larger research portfolio have proportionally higher compliance costs. The primary drivers of administrative research costs were grants and contracts, human subjects, environmental health and safety (research), animal research compliance, research misconduct, conflict of interest, export compliance, and technology transfer. Extrapolating findings to all institutions of higher education in the United States, the estimated total cost of federal compliance is about $27 billion.

Georgios Papanicolaou was a Greek-American scientist who was born on the island of Euboea, Greece. He completed medical school studies at the University of Athens and got his PhD in zoology at the University of Munich in Germany. After graduation, he married Andromache Mavroyeni, and before the First World War, the couple immigrated to the United States. This move brought hardship and many difficulties to the couple initially.

Papanicolaou's real passion was research, and eventually, he got a job at Cornell University Medical College. Initially, he was interested in studying sex chromosomes and reproductive cycles. Soon he learned to distinguish normal cells from cancerous cells in vaginal smears (Tan and Tatsumura 2015). In 1943, he published, jointly with Dr. Trout, the landmark book titled *Diagnosis of Uterine Cancer by the Vaginal Smear*. It showed that the simple procedure of examining smears taken from the vagina and cervix under the microscope could be used for the correct classification of normal and abnormal cells.

Georgios Papanicolaou was a hardworking, curious, but modest scientist, fully immersed in his research. His wife was an unparalleled supporter to the unusual extent that she was his experimental subject, sometimes for daily collection of vaginal smears. Certainly, Papanicolaou revolutionized the early detection of cervical cancer. His name was too complicated for the otherwise grateful next generations. This is the reason why we know the test with the simple abbreviation, Pap smear.

For his discoveries, Georgios Papanicolaou received several honorary doctorates from various universities internationally: the Albert Lasker Award for Clinical Medical Research from the American Public Health Association and the Medal of Honor from the American Cancer Society.

Protection of Intellectual Property

Protection of intellectual property is an essential step toward commercialization of discoveries. Ability to defend invention as intellectual property is a fundamental societal function, and the system of IP protection serves a significant meritorious mission (Table 7.1).

On the other hand, the system of intellectual property protection also generates considerable administrative burden. In patenting life sciences discoveries, frequently the first challenge is to identify what is naturally occurring and what is not, only the latter being patentable. According to Title 35 of the United States Code, laws of nature or physical phenomena are not patentable.

Understandably, the human genome project generated an entirely new set of complex legal, ethical, and business interest issues regarding what is patentable.

Table 7.1 Side-by-side comparison of ownership and recognition levels associated with various IP protections.

	Peer-reviewed publication	Patent	Research copyright	Trade secret
What is protected?	Description of the study and its results	Products, services	Protects the way facts, ideas, systems, or methods are expressed	Know-how that is not made publicly known
What is the creator of the idea called?	Author	Assignee/inventor	Author	Typically, unknown
Is there a name recognition of the individual creator/s?	Yes	Yes	Yes or no	No
How long does the name recognition exist?	Forever	Utility and plant patents – 20 yr, design patents – 14 yr	Life of the author plus 70 years	None
Revenues – does the author receive residuals from his creation?	No, usually the publisher	Usually assigned to the institution, % decided based on policies	Decided on a case by case basis	Typically, held by the corporation
Right to use and further develop idea	Complete control by author	Ownership resides with the institution	Ownership resides with the institution	Ownership resides with the institution but typically unknown

The book by Robert Cook-Deegan (1994) provides an outstanding early overview of the evolution of conflicts and concepts.

The Association for Molecular Pathology v. Myriad Genetics was a court case challenging the validity of gene patents covering isolated DNA sequences. The patents owned or controlled by Myriad Genetics were intended to be used to diagnose cancer risks by looking for mutated DNA sequences and to identify drugs using isolated DNA sequences. In addition to isolated DNA sequences, the case also included synthetically created exons-only strands of nucleotides otherwise called as complementary DNA (cDNA).

The case went through a very lengthy legal process with appeals and eventually ended up in the Supreme Court. The decision of the Supreme Court was based on the majority opinion that "a naturally occurring DNA segment is a product of nature and not patent eligible merely because it has been isolated, but cDNA is patent eligible because it is not naturally occurring" (Supreme Court of the United States 2012).

The patenting system has been extensively studied regarding administrative burden. It can take several years of work, considerable paperwork, and about \$10–\$30 000 expense to prosecute a patent in the United States. Subsequently, maintenance fees are due 3½, 7½, and 11½ years after granting the utility patent to secure a continuation of the protection (\$1600, \$3600, and \$7400, respectively).

There are various ways to measure the value of innovation in comparison to the expenses and administrative burden of protection:

- Presumably, the simplest measure is counting the number of patents filed or owned by an institution. Unfortunately, such numbers are not particularly useful as patents vary greatly in private and social value. Many patents are criticized for absurdity (e.g. bird diaper, a device for simulating a "high five"). Furthermore, most patents are never licensed, and many of them are not maintained over the years. Perhaps most importantly, a patent is just one way to protect intellectual property, while many creative projects use different channels of protection (e.g. copyright, trade secret).
- Licensing revenues represent one valuation approach. Many deals are not disclosed, and the revenues may be subject to a variety of business considerations, unrelated to the intrinsic value of innovative results. The particular difficulty of this measurement is that you also need to deal with estimates of future projected revenues, and such numbers are not reliable.
- Litigation is a sign that several parties compete for the ownership of the patent. Obviously, litigation is also an expression of perceived value as it should greatly exceed the legal costs involved for each party involved. However, litigation affects only a small number of patents, and many well-functioning licensing revenue-generating patents are never litigated.
- Failure to pay maintenance fees is an assessment by the owner that the relatively small cost of continuing the patent protection is more than the actual

value of the patent. Obviously, licensed or litigated patents are unlikely to fall into this category. For these reasons, the ratio of non-renewed patents is considered as one of the most informative indicators of useless protection and waste in the system.

The US Patent and Trademark Office receives nearly 600 000 utility patent applications annually. The number of applications tripled between 1995 and 2015. About half of the applications get a patent granted. Meanwhile, there are 6000–8000 patent disputes/litigation annually. In other words, the vast majority of patents are never litigated. This low ratio may be consistent with the fact that many patents are defensive measures to serve as deterrents blocking others, to create a patent thicket as a defense against competitors, or to be used solely for signaling purposes.

A major empirical study by Moore found that 53.71% of patentees allow their patents to expire by not paying the maintenance fees (Moore 2005). It appears that renewal rate data are better predictors of patent value because they capture many forms of private value (i.e. defensive, deterrent, signaling, or licensing revenue generation). Indeed, the majority of patents become non-renewed and therefore can be considered worthless. These numbers also reflect the administrative burden that does not generate value.

Conflicts over Credit and Case Law of Inventor Recognition

Intellectual property and research innovation have great emotional, societal, and also financial value. Several high-profile conflicts provide additional insight and have effectively generated case law defining boundaries of inventor recognition.

The innovation-related court cases demonstrate how miscalculations by inventors and institutions can be expensive and counterproductive to the innovative discovery process. The collected 20 cases of university v. inventor and inventor v. university litigation illustrate the challenges to all parties involved (Table 7.2).

In an academic research environment, contentious and at times lengthy litigation can arise when revenue sharing or lack of sharing is viewed unfairly by one of the parties. In some cases, the frustrated author/inventor takes matters into his or her own hands and protests the institution's IP policies by filing for patent/copyright. It is important for researchers to understand and agree with the policies of their institutions before embarking on research that may result in disclosable IP.

A classic *credit denial dispute* was a university v. research assistant inventor case in 1989. A student research assistant inventor developed breakthrough

Table 7.2 Inventors and universities in the court.

University v. inventor	Inventor v. university
While being employed at a university, filing a patent in the inventor's name, against institutional policies (Fenn v. Yale University 2005)	Changing revenue sharing IP policies retroactively (Singer v. Regents of the University of California 1997; University of Pittsburgh v. David Townsend; Ronald Nutt; CTI Molecular Imaging, Inc. and CTI PET Systems Inc., No. 3:04-cv-291 2007)
Removing IP-related materials against institutional policies (lab notes, results, samples, etc.) (Taborsky 1995)	Failure to enforce faculty fiduciary responsibility for their assistants and students, faculty have a responsibility to mentor and nurture students and to instruct them on how to protect their IP (Grimshaw 2001)
Publicly disclosing/reporting technology that is commercialized as a trade secret by the institution (Mayo Clinic v. Peter Elkin 2011)	Excluding substantial co-inventor researchers from successful IP protection (Grimshaw 2001; Stern v. Trustees of Columbia University 2006)
Using patented, copyrighted technology without compensating the patent, copyright owner (Cook Biotech Inc. and Purdue Research Foundation v. ACell, Inc., Stephen F. Badylak, and Alan R. Sprievack 2005; University of Pittsburgh v. David Townsend; Ronald Nutt; CTI Molecular Imaging, Inc. and CTI PET Systems Inc. 2007)	Failure to follow institution's IP policies resulting in failed production and lawsuits (Shearer 2010; Stanford 2010)
Divergent interpretations of university IP policies (Shaw v. Regents of University of California 1997; University of West Virginia 2003)	Variable or unsupportive institutional IP policies (Shaw v. Regents of University of California 1997)
Suing innovator researchers at an excessive legal and public relations expense (Duke University v. Madey 2003; Shearer 2010)	Excessive and exasperating tax/profit taking institutional policies regarding licensing revenues (Platzer 1992)

research on clinoptilolite and the methodology to enhance its ability to absorb ammonium, which is useful in wastewater treatment. He had conducted the research on his own time and with permission from the dean of research. However, the university determined that he was not entitled to any compensation for the discoveries. Subsequently, the research assistant inventor refused to turn over laboratory notebooks for what the university claimed was proprietary and confidential. He also filed for and received three patents for his work, which he refused to reassign to the university. As a result, he was arrested for theft and after serving jail time returned to overseas where he earned his PhD and continued his research (Taborsky Peter v. State of Florida 1995).

Patenting under your name while being employed at a university can also be illustrated by the case of a researcher who invented the electrospray mass spectrometry technique, which later earned him the Nobel Prize (Fenn v. Yale University 2005). The university stated that the researcher deliberately downplayed the importance of his invention and stole the patent from the university by filing it for himself even though the university had indicated they desired to maintain ownership of the patent. The researcher claimed that he completed work on the discovery after his forced retirement from the university at the age of 70. He was also unhappy with the university's royalty divisions: the first $100 000 of royalties is divided 50–50 between the inventor and university, with 40% of the second $100 000 royalty and 30% of anything above $200 000 going to the inventor. The researcher's patent has already earned more than $5 million in royalties as of 2005. The court ultimately sided with the university. The researcher continued to fight the ruling in subsequent cases until his death in 2010 (Childs 2011).

Sudden nonconsensual declaration of a trade secret by an employer can severely limit academic freedom. In a major academic health center, a medical researcher began working on neurolinguistic programming (NLP) to develop electronic medical records management for the clinic system. The researcher claimed ownership of the software because he began working on it before coming to this academic health center. The academic health center disagreed and claimed ownership through the institution's IP policies.

Later, the academic health center licensed the software as a trade secret to a health information system vendor. Upon leaving the institution, the medical researcher gave presentations on his software at a professional conference, which the health information system vendor and major academic health center contended in court were proprietary and decreased the value of the program. The medical researcher was sued by the institution, and the court ordered him to pay $1.9 million in lawyer's fees. Upon appeal, the judgment against him was reversed, and detailed accounting was requested from the institutions instead (Mayo Clinic v. Peter Elkin 2011).

Bending licensing compensation can also lead to prolonged litigation. In a case, university researcher and colleagues invented new magnetic resonance imaging (MRI) techniques while working at the university. As a condition of their employment, the faculty had signed agreements assigning the patents on all discoveries to the university, in return for a 50% share of the net royalties and licensing fees. The university negotiated a minimal royalty rate with the purchasing company for the license, while a research funding agreement was also negotiated by a different office in which the company provided over $20 million in research funding. The faculty inventors claimed that this money should be considered licensing fees as well, rather than merely research funding as the university claimed (Schwartz 2012). The inventors sued, and the court agreed that the inventors were due an increased share, but only awarded them $4 million, less than the amount they had sought (Schwartz 2012).

Another case illustrates the hazards of bending licensing compensation and changing institutional policies unilaterally. A faculty member was hired by the university to conduct research and teach. At the time of his hiring, the university policy included assignment of research patents and inventions to the university, with a 50% share in net royalties and fees from licensing (Shaw v. Regents of University of California 1997). The university later reduced this flat percentage to a sliding scale. The researcher with two other university faculty members disclosed an invention of six new strawberry plants. The university informed the researcher that the inventions would be handled under the current patent policy, not the one he originally signed. The researcher refused and later sued and won, with the courts agreeing that he should be compensated according to the policy in place at the time of his appointment (Shaw v. Regents of University of California 1997; Singer v. Regents of the University of California 1997).

Unsurprisingly, biomedical research in life science is heavily regulated to protect the public from credible sounding but unsubstantiated and misleading claims. The Federal Trade Commission (FTC) works in the prevention of fraudulent, deceptive, or unfair business practices and provides information to recognize and avoid them. The FTC is also a member of the National Prevention Council that coordinates prevention, wellness, and health promotion at the federal level. In advancing its mission, the FTC is making sure that advertising of health claims is substantiated by scientific evidence.

Aftermath of Research IP Disputes. The long-term impact of protracted legal battles on institutional reputation, research productivity, and innovation environment can be devastating for both the institution and the innovative researcher.

For many scientists, the product of their lifelong research creates a serious, long-term psychological attachment. They often fight for years and sometimes decades for the intellectual rights to their inventions, even unto their deaths. Tenacious defense of research ideas and property is a hallmark of many IP disputes.

The legal expense generated by litigation can be immense, and often, the payoff is not as great as the amount of resources dedicated to the legal process. One faculty member involved in an IP dispute with his former employer was drawn into a lengthy legal process. In the end, the university was forced to pay $2.5 million instead of the $300 000 the professor initially expected.

Negative publicity lives long in the digital age, as simple searches can easily dredge up unfortunate incidents for future researchers and students to find (Chou 2001). Related implications involve reduced donations from the very same faculty and alumni involved in research. One former associate pointed out in discussing a case: the university had lost out on millions of dollars from alumni donations because of the way the former employees had been treated (Kesan and Shah 2004).

Perhaps the most troubling effects are on the faculty and innovators themselves. Being placed into a metaphoric corner can impinge on clear thought and decision making and leave the professor pursuing justice on his or her own. One researcher spent six years in jail rather than turn over his patents to the university where he was formerly employed. When offered a pardon, he refused, because it came with the condition that he turns over his patents (Taborsky Peter v. State of Florida 1995).

Several cases mention that years of productivity and advancement of research were delayed by the litigation, putting discoveries seemingly at the low end of the priority scale (Curators of the University of Missouri v. Suppes et al. 2009; Shearer 2010). Unfortunately, all the time, emotional and physical efforts, as

---

John Fenn was an American chemist who developed the soft desorption ionization methods for mass spectrometric analyses. Born and raised on a farm in Delaware, he later recalled his grandmother's noteworthy observation: "A single chestnut tree in the front yard, so large and prolific that the nuts from that one tree paid the taxes on the farm every year."

Analysis by mass spectrometry revolutionized chemistry, but its use for large and complex molecules was limited by fragmentation. Dr. Fenn's breakthrough was to use the strong electric field to disperse charged droplets that subsequently evaporate into smaller droplets until they contain a singly charged protein. When the proteins are separated, scientists can employ the usual mass spectrometry to accelerate the molecules and measure their mass (Fenn et al. 1989). In his Nobel Prize lecture, Fenn proudly declared that his method gave "electrospray wings to molecular elephants" (Fenn 2002).

Fenn's plan has wide-ranging applications with extraordinary precision, particularly in the analysis of large and complex molecules of biological interest (e.g. proteins, polyethylene glycol oligomers). He successfully used the method to study the antibiotic gramicidin S and the immunosuppressant cyclosporine A. Today, the applications are wide ranging, including proteomics, drug discovery efforts, cancer diagnosis, doping analysis in sports, analyzing environmental pollution, and many others.

Dr. Fenn was a longtime professor at Yale University, and he developed the new technology after retirement as emeritus professor. He patented the electrospray ionization under his name, instead of assigning it to Yale. The long and drawn-out legal battle over the patent rights ended with the ruling of a federal judge ordering Dr. Fenn to pay Yale more than $1 million. This order was based on the finding that he "misrepresented the importance and commercial viability of the invention" (Childs 2011).

For his work, he received the Award for Advancements in Chemical Instrumentation from the American Chemical Society. In addition to receiving the Nobel Prize in chemistry, Dr. Fenn was also an elected member of the National Academy of Sciences.

well as financial resources spent on the defense and claims to intellectual property, limit the amount of creativity and effort put into continued research. In these cases, professors are also forced to become legal experts, rather than continuing pursuit of groundbreaking research.

Getting much-deserved credit for accomplishments is a deeply personal and intensely emotional matter for most people. The mother's pride in her children is legendary. Often, similar sentiments are expressed by people with significant achievements. Not infrequently, researchers and inventors can feel the same way about their scientific discovery or revolutionary new technology. The feeling of missed credit or demand for the much-deserved but not received credit can last a lifetime. Giving credit and denying credit are serious matters, and all parties involved should understand the lasting consequences.

Not just among scientists but also in other professions, there are many examples of credit denial and subsequent prolonged search for recognition in a large variety of professions. A particularly illuminating case was the WWI occupation of Fort de Douaumont, the largest defensive fort protecting Verdun, France. It fell when a group from the German Brandenburg Regiment, led by Lieutenant Radtke, penetrated and captured the crew. However, a latecomer, von Brandis, met the telephone group and reported the Fort was "firmly in our hands," which was forwarded to the high command as the fort was "firmly in the hands of Lieutenant von Brandis." Von Brandis became the hero of Douaumont and received the highest German order of merit. No less than 50 years later, Radtke was still searching for the missed recognition and went to Verdun commemorations in a wheelchair to protest, but officials prevented his anti-Brandis declaration. Similarly, there have been several cases of scientists demanding credit for discoveries over several decades.

Accountability for Reproducibility

In today's research institutions, the pressure to publish and obtain grant funding is tremendous. If you do not get your grant, you may lose your job that pays your mortgage, the schooling for your children, and the family vacation. You may be forced to look for and move to another job. Too prevalent opinion is that to get the next grant, you have to publish fast and many papers.

In the scientific community, the usual practice is *measuring productivity* based on the number of published peer-reviewed articles and the impact factor of the journals. Such measurements of productivity serve as the principal guide in grant review and continuation grant awards. Most researchers try to meet these requirements with hard work and dedication.

The pressure to publish perhaps can stimulate research productivity but can also become the source of superficial research. Sometimes, well-intended and reasonably good researchers inadvertently slip into trouble. Often, the publishing race can lead to cutting corners, picking the best results and pushing out too many articles. Some outside observers may notice an unexplained spike in the number of publications by a researcher, right before the major grant resubmission.

One of the most exciting opportunities of new information and communication technologies is the growing number of respectable electronic journals like *PLOS* and many others. These are typically open-access journals that charge a fee for publication. They are useful additions to the traditional section of peer-reviewed journals.

The excellent technical opportunities are abused by some *predatory publishers*. These are essentially protruding operations without meaningful peer review for scientific standards. In 2013, Bohannon conducted an informal study about the acceptance of a bogus research paper. After submitting 304 versions of a report on presumed wonder drug, more than half of journals accepted it for publication in spite of the many fatal flaws in the manuscript. As always, researchers are strongly advised to check journal quality in reputable sources (e.g. the list of journals indexed by the National Library of Medicine).

With the rapidly growing attention to non-repeatable research, accusations of irreproducible research are becoming corrosive on someone's integrity and reputation. The mere statement of inability to replicate published results can generate concern of researchers. The authors of such publications are forced into defending the value of their research. Honest errors have to be distinguished from irregular research.

Publishing results without successful replication by the originating or another laboratory is becoming an invitation for trouble. By publishing non-repeatable research results, the skeptics, adversaries, and competitors get a gold mine for questioning the validity of results at best or integrity as a scholar at worst. Even the highest-ranking academic leaders are not immune from such scrutiny.

There are many telltale signs of likely non-reproducibility. The results look too good to be true. The findings give perfect scientific evidence to one side in a highly contentious political debate. Curiously, the number of published manuscripts skyrockets just before the renewal of a major research grant. Several pictures look the same or slightly altered in the publication.

At these points, the journal editor may decide that retraction is the only logical next step. Subsequently, institutional investigations start and committees jump into action. The researcher and the results become focal points of investigations. Accuracy and completeness of experimental records are examined. No deficiency of methodology or unsubstantiated conclusion will escape attention. When the investigating committees come to unfavorable conclusions, it is hard to prevent or limit the damage. Not infrequently, retraction of articles is initiated, and, in extreme cases, termination of employment is recommended. The professional career of an aspiring researcher can come to an unflattering end.

In the case of proven *scientific misconduct* on NIH-funded projects, the federal government is involved. According to 42 CFR Part 93 guidance, "research misconduct is defined as fabrication, falsification, or plagiarism in proposing, performing, or reviewing research, or in reporting research results." At this phase, more and more legal professionals step into the frontline of the process on both sides. Often, a settlement is reached, and the case with retraction of publications, a censure of conduct, and prohibition of federal funding for a certain period are implemented.

When the pride of the researcher is wounded and livelihood is threatened, the case can quickly progress to the court system. The likelihood of becoming a court case is greatly heightened by the fact that scientific conduct review committees have no appeals process, and their judgment quickly becomes the incontestable reference point for all further institutional actions. The court of law becomes the only suitable point of recourse.

All these troubles can be completely prevented by research rigor and careful documentation of methods and results. Transparency of reporting and also open access to research data can further reassure all readers and users of the results. Carefully navigating the paperwork requirements of research and eliminating such requirements whenever possible should bear fruits.

In summary, flying into the headwinds of research is becoming daily reality of modern research (i.e. non-reproducibility, bureaucratization, and erosion of support). On the other hand, skills of maneuvering such headwinds also represent a major opportunity to get ahead of the competition. There are many challenges to face, but they are all addressable in life sciences.

At major research universities and institutions, innumerable laboratories demonstrate how talented scientists are finding new and better ways of doing things. As a result, researchers are getting inventions and discoveries through the complex process more rapidly. Ultimately, the reproducible scientific and technological breakthroughs generated by research should have a tremendous, much-appreciated effect on our lives in the years to come.

References

Alberts, B., Kirschner, M.W., Tilghman, S., and Varmus, H. (2014). Rescuing US biomedical research from its systemic flaws. *Proceedings of the National Academy of Sciences* 111 (16): 5773–5777.

Bohannon, J. (2013). Who's afraid of peer review. *Science* 342 (6154): 60–65.

Bozeman, B. (2015). Bureaucratization in academic research policy: perspectives from red tape theory. *Paper prepared in connection with keynote address for: 20th International Conference on Science and Technology Indicators Università della Svizzera Italiana in Lugano, Switzerland* (2–4 September 2015).

Bren, L. (2001). Frances Oldham Kelsey: FDA medical reviewer leaves her mark on history. *FDA Consumer* 35 (2): 24–29.

Childs, M. (2011). John Fenn: Nobel Prize-winner whose work speeded up research into new drugs. *The Independent* (28 January).

Cook Biotech Inc. and Purdue Research Foundation v. ACell, Inc., Stephen F. Badylak, and Alan R. Sprievack, No. 4:03-CV-0046 (United States District Court for the Northern District of Indiana, LaFayette Division 2005).

Cook-Deegan, R.M. (1994). *The Gene Wars: Science, Politics, and the Human Genome.* New York: WW Norton & Company.

Curators of the University of Missouri v. Suppes et al. (Missouri Western District Court 2009).

Decker, R.S., Wimsatt, L., Trice, A.G., and Konstan, J.A. (2007). *A Profile of Federal-Grant Administrative Burden Among Federal Demonstration Partnership Faculty. A Report of the Faculty Standing Committee of the Federal Demonstration Partnership.* http://www.iscintelligence.com/archivos_subidos/usfacultyburden_5.pdf (accessed 12 June 2018).

Djankov, S., La Porta, R., Lopez-de-Silanes, F., and Shleifer, A. (2002). The regulation of entry. *The Quarterly Journal of Economics* 117 (1): 1–37.

Duke University v. Madey, 539 U.S. 958; 123 S. Ct. 2639; 156 L. Ed. 2d 656; 2003 U.S. LEXIS 5045; 71 U.S.L.W. 3799 (Supreme Court of the United States 2003).

Fauci, A. (2016). Embracing flexibility as a matter of scientific principle. https://www.youtube.com/watch?v=KHkE7kvczq4 (accessed 7 December 2017).

Fenn, J.B. (2002). Electrospray wings for molecular elephants. *Nobel Lecture* (8 December).

Fenn, J.B., Mann, M., Meng, C.K. et al. (1989). Electrospray ionization for mass spectrometry of large biomolecules. *Science* 246 (4926): 64–71.

Fenn v. Yale University (United States District Court for the District of Connecticut 2005).

Grimshaw, K. (2001). A victory for the student researcher: Chou v. University of Chicago. *Duke Law & Technology Review* 1 (1): 1–7.

Heimer, C.A. and Petty, J. (2010). Bureaucratic ethics: IRBs and the legal regulation of human subjects research. *Annual Review of Law and Social Science* 6: 601–626.

Hodge, J.G. and Gostin, L.O. (2017). Revamping the US Federal Common Rule: modernizing human participant research regulations. *JAMA: The Journal of the American Medical Association* 317 (15): 1521–1522.

Hoffenberg, R. (2001). Christiaan Barnard: his first transplants and their impact on concepts of death. *BMJ: British Medical Journal* 323 (7327): 1478.

Joany Chou v. The University of Chicago and Arch Development Corporation, and Bernard Roizman, and Aviron Company, 254 F. 3rd 1347;20021 US App. Lexis 15028; 59 USPQ2D (BNA) 1257 (United States Court of Appeals for the Federal Court 2001).

Joffe, S., Cook, E.F., Cleary, P.D. et al. (2001). The quality of informed consent in cancer clinical trials: a cross-sectional survey. *The Lancet* 358 (9295): 1772–1777.

Kesan, J.P. and Shah, R.C. (2004). *Illinois Public Law and Legal Theory Research Papers Series*. Research Paper No. 04-07 (18 March).

Mayo Clinic v. Peter Elkin, Civil No. 09-322 (DSD/JJK). 2011 U.S. Dist. LEXIS 95049 (United States District Court for the District of Minnesota 2011).

MIT Office of Sponsored Programs (2015). *PI Administrative Burden Survey-Results*. Boston, MA: MIT.

Moore, K.A. (2005). Worthless patents. *Berkeley Technology Law Journal* 20 (4): 1527–1552. Article 3.

National Academies of Sciences, Engineering, and Medicine (2016). *Optimizing the Nation's Investment in Academic Research: A New Regulatory Framework for the 21st Century*. Washington, DC: National Academies Press.

NIH (2016). Paperwork burden. https://grants.nih.gov/grants/paperwork-burden. htm (accessed 10 July 2017).

Nishimura, A., Carey, J., Erwin, P.J. et al. (2013). Improving understanding of the research informed consent process: a systematic review of 54 interventions tested in randomized control trials. *BMC Medical Ethics* 14 (1): 28.

Perutz, M.F. (2002). *I Wish I'd Made You Angry Earlier: Essays on Science, Scientists, and Humanity*. New York: Oxford University Press.

Platzer, E. (1992). Dr. Karl Welte and Dr. Roland Mertelsmann, Plaintiffs-Appellants, v. Sloan-Kettering Institute for Cancer Research, Defendant-Appellee. No. 92-1280. United States Court of Appeals, Federal Circuit (10 November).

Rockwell, S., Shaver, K., and Brutkiewicz, R. (2012). Faculty Workload Survey. http://sites.nationalacademies.org/cs/groups/pgasite/documents/webpage/pga_087667.pdf (accessed 30 June 2018).

Scannell, J.W., Blanckley, A., Boldon, H., and Warrington, B. (2012). Diagnosing the decline in pharmaceutical R&D efficiency. *Nature Reviews Drug Discovery* 11 (3): 191.

Schneider, C. (2015). *The Censor's Hand: The Misregulation of Human-Subject Research*. Boston, MA: MIT Press.

Schwartz, D. (2012). Draw lines carefully when licensing technology to sponsored research partners. *Tech Transfer eNews Blog* Vol. 2014. http://techtransfercentral.com/2012/08/29/draw-lines-carefully-when-licensing-technology-to-sponsored-research-partners (accessed 12 June 2018).

Shaw v. Regents of University of California, 58 Cal. App. 4th 44 [67 Cal. Rptr. 2d 850] (Superior Court of Yolo County 1997).

Shearer, L. (2010). Drug's inventor, UGA Foundation settle. *Athens Banner-Herald* (3 April).

Singer v. Regents of the University of California, 997 Cal. App. Unpub. LEXIS 3, 26 November 1997 (San Francisco Superior Court no. 950381 1997).

Stanford (2010). Stanford-Roche patent fight draws U.S. Supreme Court review (1 November).

Stern v. Trustees of Columbia University, No. 05-1291 (Fed. Cir., 17 January 2006)

Supreme Court of the United States. Association for Molecular Pathology et al. v. Myriad Genetics, et al. 569 U.S. (2012).

Taborsky, Peter v. State of Florida, 659 So. 2nd 112, 7 July 1995 (Court of Appeal of Florida, Second District 1995).

Tan, S.Y. and Tatsumura, Y. (2015). George Papanicolaou (1883–1962): discoverer of the Pap smear. *Singapore Medical Journal* 56 (10): 586.

University of Pittsburgh v. David Townsend; Ronald Nutt; CTI Molecular Imaging, Inc. and CTI PET Systems Inc., No. 3:04-cv-291. 2007 U.S. Dist. LEXIS 56860, 30 March 2007 (United States District Court for the Eastern District of Tennessee 2007).

University of West Virginia, Board of Trustees, Plaintiff-Appellee, and West Virginia University Research Corporation, Third Party Defendant-Appellee, and James Earl Smith and Integral Concepts, Inc., Third Party Defendants, v. Kurt L. Vanvoorhies, Defendant/Third Party Plaintiff-Appellant, 342 F.3D 1290 (United States Court of Appeals for the Federal Circuit 2003).

Vanderbilt University (2015). The Cost of Federal Regulatory Compliance in Higher Education: A Multi-Institutional Study. An Assessment of Federal Regulatory Compliance Costs at 13 Institutions in FY 2013-2014. https://news.vanderbilt.edu/files/Cost-of-Federal-Regulatory-Compliance-2015.pdf (accessed 12 June 2018).

Part Three

Boosters of Research Productivity

8

Humanism for Innovation

Science can improve lives in ways that are elegant in design and moving in practice.

Harold Varmus (1996)[1]

Impressive stories of healing have always been prominent in recognizing the impact of biomedical research over the years. Celebrations of great medical discoveries highlight millions whose lives have been saved and improved. In recognizing that there are high stakes for both risk and benefit to the human condition when undertaking research, it is impossible not to see humanism as the core mission of scientific investigation.

Life sciences should serve the needs of people and humanity in general. Humanism is often defined as respectful and compassionate service of human beings to one another that is sensitive to the values and cultural backgrounds of others. In scientific research, this means creation and discovery of new knowledge that will universally benefit both people and society.

In the twenty-first century, biomedical research is facing unprecedented challenges. The frequency of safety and quality failures as well as non-reproducible research is unacceptably high (Begley and Ellis 2012; Prinz et al. 2011). The innovation productivity of biomedical research is behind expectations. There is also confusion in society as to what intended or desirable outcomes are being sought in the life sciences. When error occurs repeatedly many systems and organizations will pause to examine their process, reconsider their goals, and refocus on their mission. Perhaps it is time for the research community to do the same.

In defining humanism, the Gold Foundation highlights integrity, compassion, excellence, altruism, respect, empathy, and service as critically important values in the growth and development of clinicians. The Gold Humanism

---

1 Varmus, H. (1996). Harvard University commencement address (6 June).

Innovative Research in Life Sciences: Pathways to Scientific Impact, Public Health Improvement, and Economic Progress, First Edition. E. Andrew Balas.

Honor Society recognizes students, residents, and faculty who are exemplars of compassionate patient care and who serve as role models, mentors, and leaders in health care. The history and tradition of science and the impact resulting from so many great scientific discoveries illustrate the importance of similar ethical grounding in scientific discovery for the good of humanity. Perhaps it is time for the research community to more openly embrace the values of humanism in similar fashion.

Talking about the humanistic *mission of life* science is a necessary step on the path to defining important corollaries like social responsibility in supporting the progress of science, improvements in public health, and economic development. Humanism has also been the traditional driving force of many significant research discoveries. Without including humanistic values, it would be difficult, if not impossible, to have a meaningful conversation about the mission of science or the measurement of scientific performance.

In his 1940 speech dedicating the National Institutes of Health (NIH) in Bethesda, President Franklin Roosevelt stated vigorously that "...the National Institutes of Health speak the universal language of humanitarianism. It has been devoted throughout its long and distinguished history to furthering the health of all mankind, in which service it has recognized no limitations imposed by international boundaries; and has recognized no distinctions of race, creed or color."

This chapter will explore the significance of humanism in biomedical research and propose innovative strategies to reduce waste and improve outcomes in support of the greater mission of improving human lives and welfare. Proposed are six compelling arguments that humanism should take a more central role in biomedical research (Figure 8.1).

Research Targeting by Values of Humanism

By many estimates unexplored domains of nature vastly outnumber what has already been explored. Physicists point out that 96% of the universe is mostly or entirely unknown (Panek 2011). Life sciences researchers argue that 86% of existing species on Earth and 91% of species in the ocean still await description (Mora et al. 2011). Scientific research has fared no better in seeking to grasp the vast universe of knowledge that still awaits attempting to know the human condition.

In choosing research interests and planning projects, researchers need to prioritize domains of unknown. One logical starting point is to develop a clear understanding of social burden and impact on human life. In many cases, a sense of compassion and empathy can lead the researcher to recognize previously unrecognized problems and effective solutions.

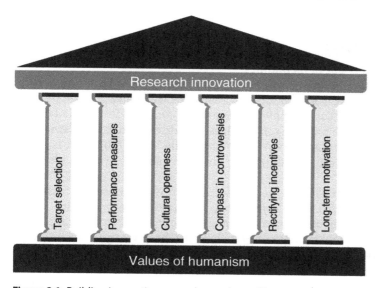

Figure 8.1 Building innovative research on values of humanism.

Empathy has great relevance not only to patient care but also to biomedical research. It is a cognitive attribute that predominantly involves understanding another person's concerns and seeing them from their perspective (Basch 1983). Empathy can also be defined as an understanding of the inner experiences and perspectives of another person, combined with an ability to communicate it to others and positively act on the shared knowledge (Mercer and Reynolds 2002). Empathy can inform as to what may be important research questions and areas worthwhile to investigate.

With humanism as the starting point, specific aims of research can be more clearly stated. Metaphorically, you cannot be a sharpshooter if you do not see the target. There will always be diverse ideas and perceptions of desirable research outcomes in the research community. The goal of relieving societal burden and limiting the impact of diseases on human lives can enrich the discussion and serve to provide unifying guidance in life sciences research.

Understandably, long-range views of science can also inspire basic science research, not just applied studies. For example, basic research on vesicular transport in the cells or gene repair mechanisms has the obvious promises of broad societal value down the road.

In the light of humanism, not all priorities are of equal value, however. Arguably, the ideal research targets relieve major disease burden or develop an understanding of fundamental physiologic processes to address human needs.

Attempts to understand functions of the human body, genetic and environmental impact, and evolution of disease face challenges in attempting to prioritize domains of the unknown. Value can be measured in terms of outcomes achieved per resources expended (Kaplan and Porter 2011). A fitting goal of biomedical research is to improve the value being delivered to people and society as a result of its findings.

An important clarification of what serves people well is the World Health Organization (WHO) Model List of Essential Medicines (World Health Organization 2016). Essential medicines are recommended to be available at all times. These medicines are selected based on disease prevalence, evidence on efficacy and safety, and cost-effectiveness. The WHO model list is effectively used by numerous countries in the development of their essential medicines lists.

In the light of humanism and the above-described WHO concepts of essential treatments, we can also better define what practical significance means: effect size, safety, frequency of potential use, and cost-effectiveness. The difference between *practical and statistical significance* has been recognized for a long time. When two drugs reduce elevated blood pressure, the one with larger effect is likely practically more significant. Similar statements could be made about prevalence, safety, and cost-effectiveness of new treatments. It is also widely accepted that effect size is a major but not the only determinant of practical significance.

Understanding of the human experience has not only overarching general implications but also a much more direct influence on research in areas where people and their surroundings interact. Some of the most prominent areas include human performance studies in complex technical environments and syndemics that focus on the interaction of diseases and social/ environmental factors. Other examples include studies of disabilities and prosthetic devices, certain occupations and associated conditions, or transportation safety.

Humanism can also guide choices about experimental designs and subjects. In its report on chimpanzees in biomedical and behavioral research, the Institute of Medicine (IOM) concluded that the number one principle of assessing research should be that the experiment can be justified if the knowledge gained is necessary to advance the public's health (National Research Council 2011). It is logical to apply this same principle broadly to human and laboratory research.

Humanism also informs and refines cultural considerations in research. For example, studies have shown that weekly religious attendance is associated with longer life expectancy, comparable with regular physical activity or statin-type lipid-lowering agents (Hall 2006). According to Bates and Plog (1990), culture is a system of shared beliefs, values, customs, behaviors, and artifacts that the members of society use to cope with their world and with one another and that are transmitted from generation through learning. In short, culture is a shared way of life that evolves and defines itself through the behaviors, values, and common experiences of those living it.

The pioneer of *in vitro* fertilization, Nobel laureate **Robert Edwards,** worked on solving a problem that most people thought was not a problem, infertility. When he started to research, gynecologists did not pay much attention to infertility. Overpopulation and family planning were popular and accepted concerns, but infertility was not. In reality, infertility affects more than 10% of couples.

The infertility research came out of a combination of his belief in social justice and curiosity as a scientist. A period of crisis followed when his major Medical Research Council grant application was rejected because the reviews emphasized risks and again marginalized potential benefits.

His meticulous triangulation confirmed the robustness and generalizability of his preclinical research results and accurately projected success in human trials. In his studies leading to the discovery of *in vitro* fertilization, Robert Edwards studied the successful *in vitro* maturation of eggs in rats, mice, and hamsters (Johnson 2011). His collaboration with Patrick Steptoe, an obstetrician at the Oldham General Hospital, was particularly transformative in reaching the human application.

After the publication of his discoveries, Edwards was accused of being Nazi doctor and Frankenstein. Throughout his research, he worked hard to make sure that his ethical groundwork was correct. In the ensuing media battles as well as in research, he continued working tirelessly for the infertile patients, the no-hoper without a voice.

Consequently, the humanities have been important allies to the quest for humanism in research. Clinicians know how to save lives but musicians know how to express life. Historically, visual arts have been highly effective in illustrating physiology and pathology, portraying disease processes, and educating researchers. Clearly, peer-reviewed research publications do not represent the full depth and breadth of human experience relevant to life sciences research. Newspapers and the writings in nonscientific journals played a key role in the recognition and investigation of Legionnaires' disease.

Humanities and lay initiatives can portray and visualize what others, including many scientists, do not readily recognize. Grassroots initiatives like the Dutch parent group, Downpride, have been effective in explaining the experience of living with Down syndrome and offer a broader understanding of the consequences of prenatal screening (Lindeman 2015).

Identifying Measures for Performance Improvement

Current assessments of research productivity are often driven by introverted technicalities with limited relevance to the ultimate public benefit and mission of science. Traditional measures include the number of peer-reviewed publications, citations of publications as measured by journal impact factor, competitive grant funding received, or various derivative indices (h-index, g-index, or i10-index). However, there is at best unclear relationship between researchers' citations and actual beneficial impact on public health or individual well-being.

Humanism, its influence and impact, cannot be measured by data alone. Any attempt to simplify, mechanize, or trivialize the application of humanism is doomed to fail. Humanism encompasses the whole human experience and not just measurable parts. We need to sharply distinguish the valuable guidance of humanism in developing *performance measures* from the inherently futile attempts of measuring humanism itself.

In the research process, bioethicists and institutional review boards are the guardians of humanistic practices. There is no such thing as ethics by the numbers, and there can be no justification for experimenting on human beings to save the lives of others – even a lot of them. Most importantly, life sciences researchers themselves should be the best followers of humanistic ideals in choosing their research targets and methodologies.

One area where natural scientists and creative artists often appear to be on the same page is the vehement opposition to measuring performance. As it was passionately stated by Leslie Sage, senior editor in *Physical Sciences, Nature*, about the value of impact factor and other measures, "All of these metrics are an attempt to satisfy bean counters. They are inappropriate, they have no place in science. They only exist to make deans happy" (Sage 2011).

Confusion about measures and outcomes is also pointed out by the IOM report on the Clinical Translational Research Center program of NIH. It concluded that "Although it would be ideal to evaluate the CTSA Program's impact on clinical care and public health, currently this is neither feasible nor realistic given the numerous driving forces that shape the research enterprise." The performance of translational research centers is difficult to evaluate given the "lack of high-level common metrics that are barriers to overall program accountability." Clearly defined and measurable goals are needed to provide such programs with a basis for evaluation, reporting, and accountability for the individual CTSAs and the overall program (Leshner et al. 2013).

In 2012, the NIH National Heart, Lung, and Blood Institute started the Centers for Accelerated Innovations (NCAI) to accelerate the *translation of scientific discoveries* into commercial products that improve health for patients. One may reasonably assume that if you can accelerate innovation, then you can measure the speed and demonstrate improvement according to predefined metrics. However, such metrics of innovation have not yet been apparent at this point.

While humanism and value to humanity cannot be measured, they are essential for the development of measures assessing continuous performance improvement. Since 2013, enhanced NIH biosketch standards have placed increased emphasis on contribution to science as defined by influence of the finding(s) on the progress of science or the application of those finding(s) to health or technology, peer-reviewed publications, or other non-publication research products (can include audio or video products, patents, data and research materials, databases, educational aids or curricula, instruments or equipment, models, protocols, and software or netware).

If the humanistic mission were to be further embraced, measurement of outcomes and progress toward them would likely be more efficient. As pointed out earlier, the ultimate outcomes of research include new scientific models, improvement in health and wellness, and economic development. With such a tripartite humanistic mission of research, important aspects of the process become measurable.

A major threshold of research is rigor and trustworthiness. Scientific misconduct ranging from sloppy research to fake data undermines trust in the results of science (Martinson et al. 2005). For example, researcher Diederik Stapel falsified more than 50 studies in social psychology. Some of his studies based on fabricated data were published in the prestigious journal *Nature*. The loss of rigor can range from misconduct to innocently looking mistakes. With the focus on the humanistic mission of science, we can have a standard that surpasses all technically limiting definitions of quality research.

Based on the tripartite humanistic mission, the quality improvement chapter of this book offers several measures of research performance (e.g. the ratio of publications and IP disclosures, per faculty licensing). These few measures are just illustrations of opportunities. We are only at the beginning of a road to quality improvement in biomedical research. Forty years ago, there were only very few outcome-oriented measures of health-care quality. Today hundreds of multiple measurement tools support effective quality improvement in clinics and hospitals across the country.

Creativity Needs Diversity of People, Cultures, and Ideas

The history of science is filled not only with landmark discoveries but also with inspiring chronicles of scientists coming from diverse backgrounds. While exact statistics are hard to come by, even superficial look at major discoveries often finds immigrants, all kinds of cultures, and many miles of travel in the life of innovative scientists.

Science and creativity need the diversity of people, cultures, and ideas. It is often pointed out that research can only be built on *critical thinking* skills: the ability to look at problems from multiple directions and to appreciate many diverse views on the same issue. The critical thinker can understand how others think and synthesize the best points.

Involving intended users and beneficiaries of results in the design, conduct, and publication of research studies makes great sense in living up to the humanistic mission of applied science. Depending on the nature of research, clinicians, faculty administrators, patient advocacy groups, and people living with various conditions are intended to benefit from applied scientific projects. Through their participation, they can add valuable reality check, practical observations, and credibility to the resulting publications.

Diversity is one of the best ways to foster innovation. Conversely, a large number of research failures come from the lack of the diversity of thought. In many circles, diversity is presented only as an issue of fundamental fairness and also a matter of compliance with equal opportunity requirements. These are true and relevant considerations, sufficient justifications on their own. However, biomedical research innovations show something even more important: the power of cultural diversity and the vibrant intellectual environment in producing great new ideas.

Every major step of research needs critical thinking and *cultural diversity*. In conducting literature searches, the primary challenge is finding and using the terms and vocabulary that are associated with the desired articles. In writing scientific articles, the challenge is to express ideas and findings in a way that peer reviewers and readers down the road will understand quickly and easily. Most importantly, the creative scientists of the twenty-first century, sometimes known as "brainiacs," need the unimpeded collision of ideas and cultural diversity.

Many years ago, a health services researcher was invited to visit a health administration department of a major university. The department chair showed him the place and proudly highlighted that every faculty member in the department was an MD. The health services researcher was horrified and asked him: how can you produce anything meaningful in such monotonous environment? The exact response is no longer remembered, but the inquisitive researcher has never been invited back to the department.

One of the most famous research and development projects of all times, the Manhattan project, was triggered by a call from scientists escaping from Europe. In 1938, the immigrant physicist Leo Szilard contacted the White House urging them to look at the potential of nuclear fission. When failed to get attention, he turned to another immigrant, Albert Einstein, and their letter to President Roosevelt led to the creation of the Manhattan project.

The Manhattan project was scientifically greatly accelerated by Enrico Fermi, an Italian immigrant scientist, and at some point, the project also employed several Hungarian-born scientists (John von Neumann, Eugene Wigner, Edward Teller, all graduates of the same Fasor High School in Budapest). Years later, when Fermi asked about extraterrestrial life and Martians, Szilard's response was that "of course, they are already here among us: they just call themselves Hungarians" (Teresi 1993).

Rock Solid Values When the Controversies Mount

Especially life sciences researchers need well-grounded ethical values when controversies start and grow to stay focused and avoid troubles. Emerging new concepts, experiments, or initial results can generate fear, intellectual disagreement or outright hostility in the scientific community or the general public discussion.

Trailblazing science often triggers rejection. In the fog of controversies, the values of humanism and insistence on replicable evidence can guide the researcher to success. The IVF development by Robert Edwards provides one of the most inspiring illustrations of value-driven science. Without it, research appears to be a profession that is often lost in its technicalities.

Curiosity has long been recognized as one of the principal drivers of scientific investigations. It is a major intrinsic motivation to learn that plays an essential role in triggering and pursuing scientific studies. Newer functional magnetic resonance imaging studies could also locate the brain's structures involved (Gruber et al. 2014). Studies revealed that the activity in the midbrain and the nucleus accumbens, the brain's "pleasure center," responsible for reward, motivation, and addiction was enhanced during states of high curiosity.

In recent decades, individual *curiosity-driven science* has been increasingly displaced by a mechanized process that generates and screens vast amounts of correlative data with the hope of getting an answer to life's fundamental questions. High-throughput analysis methodologies, the bean counting of peer-reviewed publications and grant funding in assessing scientific productivity, and a large number of repetitive experiments necessary in many scientific studies add to the pressures of mechanizing research.

No one should expect lasting or impactful results from life sciences research that, though perhaps technically exquisite, fall short in humanistic quality and substance. Barondess (1996) provided a striking historical analysis how one of the most technologically sophisticated and respected medical research enterprises in the world degraded itself by abandoning its humanistic mission in the Nazi era.

In pharmaceutical research, promising compounds are often identified in preclinical research. Another sign of mechanized research is the high-throughput chemical analysis in drug development. The number of compounds needed from promising preclinical discoveries to result in one launched drug is a measure of productivity in drug development. The average ratio of one successful product launch and necessary original compounds varies between 0.5% for Alzheimer's research and 4.6% in MRSA research (Calcoen et al. 2015).

The embellished technical minutia of biomedical research and the exaggerating terminology have generated much confusion about scientific outcomes. Significance in scientific publications often has little relevance to practical application. *Impact factor* does not reflect the practical implications to public health or economic development. Finally, research citation in the form of researchers citing each other has ambiguous usefulness in furthering scientific progress.

Contemporary vaccine research is a good example of the intricate humanistic and societal value judgments in target selection (Madharan et al. 2012). In the past, benefits of vaccination were so obvious that choices regarding development and use of vaccines required only common sense. Given the vast expansion of the vaccine enterprise, the need for prioritization is becoming greater than ever. Today, decisions regarding research and development, implementation, and delivery of vaccines are becoming increasingly sophisticated. Important considerations include health, economic, demographic, scientific and business, programmatic, policy, public concerns, and intangible values.

Currently, most research educational programs overwhelmingly focus on the technology of research. Limited attention is given to the ultimate mission of science, societal goals of research, target selection, stages of technology transfer, and various disclosure options in the process of transfer to maximize beneficial impact.

Being mindful of humanity is equally important in protecting the integrity of researchers and fostering the avoidance of scientific misconduct. The many billions of dollars flowing into research, job security associated with many academic and tenured positions, and a vibrant environment of leading research teams, can create an irresistible attraction to generate desired data in any way possible. A month barely passes without reading about scientific misconduct. Encouraging attitudes and behaviors that are humanistic is one way to ensure responsible research.

Marshall Nirenberg was a biochemist and geneticist working at NIH at the time when the DNA double helix was already discovered, and its role in protein synthesis was also presumed at least in principle. However, what was not known is the codebook of protein synthesis, i.e. the relationship between various DNA base pair combinations and the resulting proteins.

Nirenberg and several other researchers recognized the challenge and concentrated great efforts on decoding the protein synthesis. This recognition effectively launched the coding race. Several ingenious methodologies and testing devices were invented to enable and accelerate the process of decoding. Many NIH scientists put their own research on hold to join the effort and help to earn the first NIH Nobel Prize.

The term scientific discovery tends to include an element of surprise, but sometimes results come from a process similar to industrial production. In such cases, the end result of research comes from tedious, meticulous, and tireless effort.

The concentration of NIH resources and progress was so great that competing teams of researchers quietly left the race and turned to other areas of interest. Ultimately, the Nirenberg effort led to the decoding of DNA transcription and earned the much-anticipated Nobel Prize.

Counteracting Flawed Economic and Social Incentives

The partnership of publicly funded research with private industry has been a major stimulus to diagnostic and therapeutic innovation. The list of successful and innovative products and services is long, benefiting society on a global scale. Life expectancy and quality of life have improved in nearly every modern country of the world.

Meanwhile, points of tension remain increasingly apparent in the world of life sciences revenue generation. The first is a dearth of research on diseases that affect third world countries and those with low income. Developing countries are particularly disadvantaged as a result of market-driven and commodified research that may do more harm than good for their populace.

The huge discrepancy between production costs and market prices of certain medications represents another source of major tension. The profit margin is often portrayed as necessary to fund increasingly expensive product development. However, the margin also generates corporate profit. Regardless of how it is justified, a significant markup in price makes many drugs unavailable to patients in developing countries.

The study of a large number of randomized clinical trials found a weak association between the global *burden of disease* and number of responsive trials (Emdin et al. 2015). Repeatedly, humanistic research finds itself in the position of opposing prevailing financial interests, such as developing cost-effective treatment for rare diseases. It also validates the substantive value of lower expense treatment (Friedberg 2014).

In 1990, the Commission on Health Research for Development estimated that only about 5% of support for health research was applied to the health issues of developing countries, where most of the world's preventable deaths occurred. Adopting the Pareto principle, the Global Forum introduced the term "10/90 gap" to highlight the imbalance between the magnitude of problems of developing countries and resources devoted to addressing them (Horton 2003).

Typically, the strategic plans of universities and major corporations start with a compelling statement of values mission and vision. In the life of an organization, there is no way to predict the myriad of questions and challenges surfacing. However, a clear statement of values can be invaluable in guiding

---

Norbert Hirschhorn, inventor and developer of the oral rehydration therapy, is an Austrian-born American public health physician. In addition to his great scientific work, he is also an award-winning poet.

During his first assignment in the US Public Health Service to Pakistan, he witnessed a devastating cholera outbreak. The IV therapy was expensive and in short supply, triggering a search for an efficient oral solution. When starting the development of oral therapy, Dr. Hirschhorn had to prove that his supervisor did not get the physiology right when he tried a solution but failed and several patients died.

With a better understanding of physiology, Dr. Hirschhorn designed the new oral treatment, drinking water with calibrated amounts of sugar and salt added (Hirschhorn 1982). In the first trial, he slept next to the child patient to ensure safety. It was a complete success – "like seeing Lazarus come back from the dead – a miracle," he said later. Today, the oral rehydration therapy of Dr. Hirschhorn is credited with saving over 50 million lives.

---

decisions and driving future actions. Research is more likely to be successful with firm recognition of humanistic value, defining the mission of particular research and offering a vision of how anticipated results will benefit.

Focus and Motivation for Sacrificial Effort

Research has always been time consuming and effort intensive. Countless hours in the laboratory lead to many failures before success. A large part of research is searching but finding nothing. Nobel laureate Tim Hunt compared research with gardening where most of the time is spent on weeding. "If you don't like the weeding, you don't like gardening" (Hunt 2014).

Aspiring researchers, in order to sustain their momentum, must embrace the realities of low success rates in scientific investigation. Salvarsan, the first modern successful chemotherapeutic agent, was identified as compound 606 because it was the sixth in the sixth group of compounds synthesized for testing. In the 1960s, a scientist searching for treatment of drug-resistant malaria screened an estimated 240 000 compounds without success. Famously, Dolly the sheep was born out of 277 attempts of somatic cell nuclear transfer, and only one survived to adulthood (Campbell et al. 1996).

Research also often involves personal risk taking. Occasionally it is the ultimate test of discovery and acceptance. Nobel laureate Barry Marshall decided to test his gastric ulcer theory on himself and ingested a cultured sample of *H. pylori* to prove the growth of the bacteria and tissue damage in the stomach. After isolating the active anti-malaria compound artemisinin in wormwood,

Youyou Tu tested the medicine on herself to ensure it was safe. In 1976, Peter Piot was part of a team that discovered the Ebola virus and subsequently traveled to Zaire to study and quell the highly dangerous outbreak.

A different kind of risk is taken when scientists go against prevailing views. Before the discovery of the oncogenic role of papillomavirus, it was widely but incorrectly believed that cervical cancer was caused by the herpes simplex virus. When Harald zur Hausen first presented his work, it was greeted with "stony silence" and indifference. One may conclude that indeed landmark discoveries often go against the prevailing opinion.

Arts and life sciences have long inspired each other. Since the work of Andreas Vesalius and Leonardo da Vinci, visual arts have played a quintessential role in illuminating and explaining anatomical structures and their functions. Humanities support whole person understanding, not just a measurable aspect. One may say, scientists know how to save lives but artists know what life is.

Humanities can guide an understanding of general patterns of response to health and illness, offer insight into unique and individual responses, and provide a framework and vocabulary for communications about health and disease (Scott 2000). According to Harold Varmus (2016), "Life is richer if you have a broad exposure to culture... The ideas that come from humanities and the arts play a role in thinking about the meaning of scientific work."

Movies can be uniquely helpful in understanding the psychopathology of mental illness and the interactions of such diseases with daily life, family relations, and society in general (Wedding and Niemiec 2014). Science fiction is not just a source of motifs and narratives in announcements of research discoveries but also serves as a stimulus to the innovative thought process around biomedical research and technology development.

Biomedical research can also learn from many other professions in recognizing the significance of moral values in achieving desirable outcomes. In 2014, the Strategic Studies Institute and US Army War College released a landmark report on the soldier's morality and professional ethic (Snider and Shine 2014). It highlights that the "services can ill afford to lose the irrefutable power of soldiers' personal moralities as they serve in both peace and war, providing an additional motivation and resilience to prevail in the arduous tasks and inevitable recoveries inherent in their sacrificial service."

Thoughtful inquiry also needs to examine the motivation of academic researchers in the arduous tasks of research. Success in science requires major and long-term commitment, as described in the how to succeed guide of NIH (Yewdell 2008). Certainly, humanism and personal moralities can provide the much-needed additional motivation and resilience to prevail in research. Ultimately, it is important to realize that personal moral agency has a potentiating effect on attitudes and behavior in biomedical research.

Numbers Need Stories and Stories Need Numbers

Humanism has tremendous potential in focusing research and encouraging sustained effort toward success. Evidence from diverse sources suggests that compassion, altruism, respect, and empathy are powerful components of good research. The next step is unlocking and strengthening the value of humanism in the development of future researchers (i.e. PhD and postdoctoral programs).

The first opportunity is to *discuss humanism* as a reference point in science and bring this issue up in discussions about research projects and models. Ask, how can this idea benefit patients? What is the number of people affected by the targeted conditions? What is the potential impact of proving that the new model is correct? What are our options to strengthen scientific rigor and make the results repeatable with a high level of confidence?

Historically, the white coat was the laboratory scientists uniform. Physicians followed suit later and for the purpose of gaining trust by being associated with science. The purpose of white coat ceremonies is to mark the transition from student to a health professional and highlight the humanistic mission of patient care. White coat ceremonies are spreading in various health professional education programs. Along these lines, white coat ceremonies of PhD programs and celebration of the humanistic mission of life sciences represent logical extensions of the best traditions.

In positioning for success in research, the profound role of the humanities cannot be overstated. Reading newspaper articles and blogs about the particular disease can give further insight. Appreciation of literature, performing, and visual arts should be part of the scientist's life. Often, artistic expressions of human experiences provide valuable reference points for explaining the practical significance of someone's research. Researchers should have experiences of health needs and exposure to role models of humanism.

The second opportunity is building an *inclusive environment* that is culturally diverse, providing experiences that build compassion and facilite understanding the needs of other people. Inviting people with diverse backgrounds, many different views, and nontraditional ideas can create the environment that fosters innovation. Meeting with patients having the particular condition, traveling to places where the need is greatest, and talking to families and community leaders impacted by the consequences of the disease might be helpful. Listening to others, understanding their concerns, and providing encouragement can build compassion. Standing up for one's own unprecedented views is also an experience that will enrich the personal and professional lives of researchers and the work they do.

In the book of Al Sommer, titled *Ten Lessons in Public Health*, the first chapter advises to travel to places of need and go where the problems are (Sommer 2013). Dr. Sommer is an ophthalmologist and epidemiologist at Johns Hopkins University. His research is most recognized for the discovery that treating severely vitamin A-deficient children with an inexpensive, large dose vitamin

A capsule twice a year reduces mortality by 34% (Sommer et al. 1986). This very first lesson of his book passionately explains the life-changing experience he had in East Pakistan (today Bangladesh).

The third opportunity is to apply, promote, and propagate research performance evaluation based on humanistic ideals. So, what? This was the favorite question of Professor Reed Gardner, the pioneering medical informatician at the University of Utah. In other words, answering the following questions can be helpful: What are your current projects, and what are your goals? Who will benefit from the targeted results? Does the world need this particular research project? What makes us believe that the results are true as opposed to some accidental statistical significance? What would help these patients?

In the hallways of scientific conferences and seminars, asking about the targeted problem should be among the first questions in discussing someone's research. Conversely, asking about someone's peer-reviewed publications and current grant funding should be last. Let first things remain first: discussions should be focused on what the researchers want to understand, and the technicalities of realization will come afterward.

Curiosity, once considered a vice, ultimately became one of the principles and much-appreciated drivers of scientific investigations. Wanting to understand everything became a praiseworthy motivation. The concept of curiosity has evolved in parallel with the progress of empirical research methodologies. Today, researchers can unleash curiosity in many directions. Sometimes curiosity-driven research is in contrast to problem-driven, applied research. Certainly, both drivers have merits, but they are not mutually exclusive. As concerns about effectiveness are growing, scientists need to consider expected outcomes when directing their curiosity.

Humanism as outlined in the six columns of responsible research will help guide the creation of well-rounded and humanistic science for the good of mankind. One can argue that all research takes place in hazy environment that makes the challenges of the work seem more grotesque and larger than they really are. Just as high-functioning organizations are grounded in mission and values, successful researchers will be guided by the ethos of humanism and offer discoveries that are trustworthy and beneficial for many generations to come. Eagerness to serve the broader community, interest in others, and desire to help people are powerful motivators and also useful guides for productive research.

References

Barondess, J.A. (1996). Medicine against society. Lessons from the third Reich. *JAMA: the Journal of the American Medical Association* 276 (20): 1657–1661.

Basch, M.F. (1983). Empathic understanding: a review of the concept and some theoretical considerations. *Journal of the American Psychoanalytic Association* 31: 101–126.

Bates, D.G. and Plog, F. (1990). *Cultural Anthropology*. New York: McGraw-Hill.

Begley, C.G. and Ellis, L.M. (2012). Drug development: Raise standards for preclinical cancer research. *Nature* 483 (7391): 531–533.

Calcoen, D., Elias, L., and Yu, X. (2015). What does it take to produce a breakthrough drug? *Nature Reviews Drug Discovery* 14 (3): 161–162.

Campbell, K.H., McWhir, J., Ritchie, W.A., and Wilmut, I. (1996). Sheep cloned by nuclear transfer from a cultured cell line. *Nature* 380 (6569): 64–66.

Emdin, C.A., Odutayo, A., Hsiao, A.J. et al. (2015). Association between randomised trial evidence and global burden of disease: cross sectional study (Epidemiological Study of Randomized Trials – ESORT). *BMJ* 350: h117.

Friedberg, J.W. (2014). End of rituximab maintenance for low–tumor burden follicular lymphoma. *Journal of Clinical Oncology* 32 (28): 3093–3095.

Gruber, M.J., Gelman, B.D., and Ranganath, C. (2014). States of curiosity modulate hippocampus-dependent learning via the dopaminergic circuit. *Neuron* 84 (2): 486–496.

Hall, D.E. (2006). Religious attendance: more cost-effective than lipitor? *Journal of the American Board of Family Medicine* 19 (2): 103–109.

Hirschhorn, N. (1982). Oral rehydration therapy for diarrhea in children – a basic primer. *Nutrition Reviews* 40 (4): 97–104.

Horton, R. (2003). Medical journals: evidence of bias against the diseases of poverty. *Lancet* 361: 712–713.

Hunt, T. (2014). Interview. Nobelprize.org. Nobel Media AB 2014. http://www.nobelprize.org/nobel_prizes/medicine/laureates/2001/hunt-interview.html (accessed 7 April 2017)

Johnson, M.H. (2011). Robert Edwards: Nobel laureate in physiology or medicine. In: *Les Prix Nobel. The Nobel Prizes 2010* (ed. K. Grandin). Stockholm: Nobel Foundation.

Kaplan, R.S. and Porter, M.E. (2011). How to solve the cost crisis in health care. *Harvard Business Review*, 89(9), 46–52, 54, 56–61.

Leshner, A.I., Terry, S.F., Schultz, A.M., and Liverman, C.T. (eds.) (2013). *The CTSA Program at NIH: Opportunities for Advancing Clinical and Translational Research*. Washington, DC: National Academies Press.

Lindeman R. (2015). Down syndrome screening isn't about public health. It's about eliminating a group of people. *Washington Post* (16 June).

Madharan, G., Sangha, K., Phelps, C. et al. (2012). *Ranking Vaccines: a Prioritization Framework: Phase I: Demonstration of Concept and a Software Blueprint*. Washington, DC: National Academies Press.

Martinson, B.C., Anderson, M.S., and De Vries, R. (2005). Scientists behaving badly. *Nature* 435 (7043): 737–738.

Mercer, S.W. and Reynolds, W.J. (2002). Empathy and quality of care. *British Journal of General Practice* 52 (suppl): S9–S12.

Mora, C., Tittensor, D.P., Adl, S. et al. (2011). How many species are there on earth and in the ocean? *PLoS Biol* 9 (8): e1001127.

National Research Council (2011). *Chimpanzees in Biomedical and Behavioral Research: Assessing the Necessity.* Washington, DC: National Academies Press.

Panek (2011). *Panek's the 4 Percent Universe.* Boston, MA: Houghton Mifflin Harcourt.

Prinz, F., Schlange, T., and Asadullah, K. (2011). Believe it or not: how much can we rely on published data on potential drug targets? *Nature Reviews Drug Discovery* 10 (9): 712–712.

Roosevelt (1940). President Roosevelt's address dedicating the National Institute of Health at Bethesda, MD (31 October).

Sage, L. (2011). How to publish in Nature – Leslie Sage (SETI talks). https://www.youtube.com/watch?v=ys8wHUPd6Vo (accessed 22 June 2018).

Scott, P.A. (2000). The relationship between the arts and medicine. *Medical Humanities* 26 (1): 3–8.

Snider, D.M. and Shine, A.P. (2014). *A Soldier's Morality, Religion, and our Professional Ethic: Does the Army's Culture Facilitate Integration, Character Development, and Trust in the Profession?* DTIC Document. US Army War College Press.

Sommer, A. (2013). *Ten Lessons in Public Health: Inspiration for Tomorrow's Leaders.* Baltimore, MD: JHU Press.

Sommer, A., Tarwotjo, I., Djunaedi, E. et al. (1986). Impact of vitamin a supplementation on childhood mortality. A randomised controlled community trial. *Lancet* 24: 1169–1173.

Teresi, D. (1993). Hungarians think the darndest things. *New York Times* (24 January).

Varmus, H. (2016) Life is richer if you have a broad exposure to culture. Nobel Laureate Harold Varmus. https://www.youtube.com/watch?v=wVKaypBl4aQ (accessed 11 August 2017).

Wedding, D. and Niemiec, R.M. (2014). *Movies and Mental Illness: Using Films to Understand Psychopathology.* Cambridge, MA: Hogrefe Publishing.

World Health Organization (2016). *The Selection and Use of Essential Medicines: Report of the WHO Expert Committee, 2015 (Including the 19th WHO Model List of Essential Medicines and the 5th WHO Model List of Essential Medicines for Children)* (No. 994). Geneva: World Health Organization.

Yewdell, J.W. (2008). How to succeed in science: a concise guide for young biomedical scientists. Part I: taking the plunge. *Nature Reviews Molecular Cell Biology* 9 (5): 413–416.

9

Desire to Understand First

We were both readers. We liked facts. We wanted to find out which facts were correct... We want the answer more then proving that we were right yesterday.

James Watson (1970)[1]

The role of outstanding minds is often recognizable in the history of most important scientific discoveries. In recent years, there have been many systematic studies about very smart people and their role in science. Examining occurrences of surprising originality, geniuses of science either founded new scientific disciplines or revolutionized established disciplines (Simonton 2013). Robert Koch, the founder of modern microbiology, could exemplify the first category, and Charles Darwin can illustrate the revolutionizing impact on a discipline.

Meanwhile, the information about experiences and aspirations of award-winning scientists is also rapidly growing. In partnership with AstraZeneca, the Nobel Prize Inspiration Initiative developed a program focusing on the characteristics of a scientist. Presenting remarks from 16 laurates, the website offers advice about characteristics of a scientist, collaboration, communicating research, getting started, inspiration and aspiration, life after the Nobel Prize, surprises and setbacks, the nature of discovery, and work–life balance (NPII 2017).

Studies suggest that surprising originality produced by a scientific genius appears to be less likely in these days. Due to the rapid growth and increasing complexity of modern science, it may be appropriate to assume that new discoveries are more likely to emerge from large, well-funded collaborative teams of researchers. However, the brilliant mind remains the primary ingredient of success, and one may ask the question: what sort of brain power is needed for the greatest scientific discoveries?

---

1 Watson, J. (1970). Horace Freeland Judson: (RARE) interview with James Watson and Francis Crick [Video file]. https://www.youtube.com/watch?v=NGBDFq5Kaw0 (accessed 11 July 2016).

Innovative Research in Life Sciences: Pathways to Scientific Impact, Public Health Improvement, and Economic Progress, First Edition. E. Andrew Balas.
© 2019 John Wiley & Sons, Inc. Published 2019 by John Wiley & Sons, Inc.

We Will Surpass on the Gray Matter

Exceptional focus on the chosen scientific questions and aggressive pursuit of the answer are certainly hallmarks of successful research. As strategists often note, only door kickers win wars. Such talent includes extraordinary ability to stay away from the ancillary and keep focused on the central questions. Much of research is a kind of detective work. Once David Baltimore said that the virologist is among the luckiest of biologists because (s)he can see into the chosen pet down to the details of its smallest molecules.

There have been many studies about the role of innate talent in producing creative scientific achievements. The long-running Study of Mathematically Precocious Youth (SMPY) has been looking into the progress of gifted children and their ultimate accomplishments. The studies found a correlation between the number of patents and peer-refereed publications produced and their earlier scores on SATs and spatial ability tests (Kell et al. 2013). On the other hand, IQ scores are notoriously inaccurate in predicting success in science and demonstrably fail to recognize several Nobel Prize winners.

Overall the SAT mathematics score at age 13 correlates well with the percentage of students who achieved the following outcomes: any doctorate, STEM publications, STEM doctorates, patents, and income in 95th percentile (Clynes 2016). While the list of these observed accomplishments is important, they provide little insight regarding the genesis of most important scientific discoveries. Studies reveal that people at the very top of the range on the mathematics section of the SAT go on to outperform the others.

Based on a wide-ranging analysis of citations and awards, Albert-Laszlo Barabasi suggested that *scientific creativity* remains virtually unchanged with aging and many great discoveries come from senior scientists (Sinatra et al. 2016). The analysis was based on reconstructing the publication profile of award-winning scientists and the citation of their articles over ten years. The study identified a certain factor Q that appears to play an influential role in defining individual productivity. The factor Q is particular to each scientist and remains largely unchanged during the scientist's career.

The life and work of many successful researchers show examples how they quiet their mind, focus full attention on the scientific problem, and dismiss any distractions that come their way. "We will surpass on the gray matter" are the famous words attributed to Mstislav Keldysh, academician, applied mathematician, and one of the intellectual fathers of the highly successful early space exploration program (Gromov 1999).

Interestingly, the physical activity walking appears to improve creativity, a mutually beneficial interaction. Aristotle founded the Peripatetic (walking) school where the name came from the habit of walking while lecturing. The nineteenth-century German philosopher Friedrich Nietzsche wrote that "Only thoughts which come from walking have any value." (Nietzsche and Large 1998).

A more recent Stanford study on a sample of volunteers concluded that taking a walk increases creativity and walking outside produced the most novel and highest quality analogies (Oppezzo and Schwartz 2014).

In more general terms, becoming an excellent scientist is a journey. We are all discovering what kind of character we possess, what parts of our personality traits we want to develop, and what we want to downsize. Listing aspirations and hallmarks of successful research might assist not only the aspiring researcher in self-assessment but also the research leader in tailoring assistance to junior researchers and institutions in interviewing research faculty candidates (Table 9.1).

Research Talent Needs Scientific Opportunity

Besides intellectual capacities of individuals, the *scientific opportunity* also plays an enormously influential role in achieving significant discoveries. New opportunities in science are stemming from research advances or better research instruments (IOM 1998). The role of opportunities and the timeliness of research are further underscored by the many occurrences of joint or simultaneous discoveries (e.g. the HIV discovery by Robert Gallo and Luc Montagnier independently or CRISPR/Cas9 discovery by several research teams).

In his famous article, Abraham Flexner the founding director of the Institute for Advanced Study in Princeton highlighted the importance of the unobstructed pursuit of useless knowledge (Flexner 1939). As an illustration, he pointed out that Marconi's discovery of the radio and wireless communication was a practically important but scientifically marginal accomplishment in comparison with Maxwell's studies in the field of magnetism and electricity.

Flexner's words capture the essence of creative professions "A poem, a symphony, a painting, a mathematical truth, a new scientific fact, all bear in themselves all the justification that universities, colleges, and institutes of research need or require." In the background of such creativity, there is curiosity, which is probably the outstanding characteristic of the best basic scientists.

On the other hand, glamourizing personalities and believing that bright intellect is everything while scientific opportunity means nothing are assumptions that are invariably doomed to fail. Politicians and journalists often fall into this trap. When science is just not advanced enough, and the necessary knowledge base is not yet available, there is no scientific opportunity even for the most brilliant mind.

The overblown and absurd focus on the individual scientific talent was pervasive in the totalitarian approach to science. Historians note that Hitler was pathologically fearful of cancer, particularly, vocal cord cancer that could threaten his oratory skills central to his power. In reality, vocal cord location is one of the rarest among all cancers. Anyway, he would not have time to develop cancer because he killed not only tenth of millions of people but also himself much sooner.

Table 9.1 Aspirational checklist for successful research in life sciences.

The basics	Yes?
1) Do you have one well-defined research interest and set of questions?	
2) Are you asking new questions and seeking important answers?	
3) Is your research attacking one of the great scientific problems of our times?	

Twelve research skills

4) Can you explain the societal mission and human implications of your research?

5) Do you develop models to understand the natural phenomenon you are studying?

6) Have you learned from population health studies as they relate to your research?

7) Have you ever targeted something that was contrary to assumptions of the majority?

8) Have you learned from nature how it solves the problems you face?

9) Have you studied the success factors of best scientists and your competitors?

10) Do you use latest technologies and clever, creative methodologies in research?

11) Do you conduct engaged research using a broad range of societal and practical sources?

12) Have you been active in multi-institutional and international collaborations?

13) Have you been able to generate significant research funding and grow your team?

14) Can you articulate methods and results of your research to people outside your field?

15) Do you have the curiosity to learn and passion to advance understanding in your field?

Intellectual impact

16) Do you have well-cited publications and/or high-impact journal articles?

17) Do you have several research results that have been replicated by others?

Public health impact

20) Have you ever solved some practical health-care problems using your scientific skills?

21) Are some of your results making a difference in the practice of others?

Economic impact

18) Have you ever worked with industry that was productive and mutually beneficial?

19) Do you have patents that are licensed or commercialized intellectual property?

Nikola Tesla is one of the most celebrated engineers and physicist in the history of science and technology. His unprecedented accomplishments related to the generation and use of alternating current electricity changed our lives profoundly. With its myriad of electronic devices, modern health care is among the most prominent beneficiaries of his inventions.

He was born in a Serbian family and raised in the part of the Austrian Empire that is called Croatia today. After working in several European countries, Tesla came to the United States and for a few years joined Edison's lab. Later, Tesla became affiliated with Westinghouse, and the company purchased several of his patents.

Over the years Tesla designed numerous groundbreaking, highly innovative technologies such as induction electric motor, power generator, rotating magnetic field, Tesla coil, and many others. His pioneering experimentation with X-ray is particularly noteworthy around the same time when Wilhelm Röntgen started.

Stories about scientist sleeping in the laboratory and constantly thinking about new designs got a new meaning with the words of Nikola Tesla: "My method is different. I do not rush into actual work. When I get an idea. I start at once building it up in my imagination. I change the construction, make improvements and operate the device in my mind. It is absolutely immaterial to me whether I run my turbine in thought or test it in my shop. I even note if it is out of balance. There is no difference whatever, the results are the same. In this way, I am able to rapidly develop and perfect a concept without touching anything" (Tesla 1919).

During those years, Nobel laureate Otto Warburg, an expert on metabolism, was among the most highly regarded life scientists in Germany. Leading Nazis of the 1930s believed that Warburg, thanks to his brilliant intellect, can solve the problem of cancer in no time and exempted him from the general persecution of people of Jewish decent. Indeed, Warburg retained the opportunity to conduct research and made some further significant discoveries. While he also did some studies on cancer, he never got even close to understanding the mechanisms or possible treatments. There was simply no scientific opportunity at that time.

Curiosity and Research Agenda

As mentioned, perhaps the most widely recognized hallmark of good basic science is curiosity. It is the drive to study, observe, analyze, develop, and test with the ultimate purpose of full understanding. The dedicated, curious researcher wants to understand how the world works. Brilliance in science can be defined by the speed of producing innovative results.

In the popular perception, the brilliant scientist is preternaturally curious, driven, focused, and creative person. By merging the words brain and maniac, sometimes these people are called brainiacs. Once, Sergei Korolev, the father of practical astronautics, recalled his meeting with the rocket scientist Tsiolkovsky: "All the meaning of my life was the one thing – to get to the stars" (Hartford 1997). Curiosity is based on the ability to concentrate effectively.

In a social context, curiosity or asking questions has an enormously powerful ally. According to a study by Harvard researchers, disclosing information about the self is intrinsically rewarding (Tamir and Mitchell 2012). On average, people devote 30–40% of speech output to informing others of their subjective experiences. Apparently, the reason for such enormous allocation of speech output is the intrinsic neural and cognitive rewards that particularly increased activation in the mesolimbic dopamine system of the brain. In fact, participants in the experiment were willing to forgo money to introspect about the self.

Successful researchers are passionate learners about nature and motivated seekers of solutions. The diverse interactions of engaged research and asking many questions give other people an opportunity for richly rewarded sharing of subjective experiences. If you do not want to know the answer, you will not be a real scientist.

Frederick Sanger, British chemist, was the recipient of two Nobel Prizes in the same field, one in 1958 and one in 1980. He got his first Nobel Prize for sequencing the proteins of the insulin molecule. His second Nobel Prize was for the discovery of a method sequencing the DNA. Sanger later remembered "Rather fundamental questions which were quite mysterious when I started. When I started, there were no sequences known at all either in proteins or in nucleic acids" (Sanger 2014).

The power of focus, creativity, and innovation is among the greatest joys of life and also the most inspiring forces. Famously, when Nobel laureate Warburg received an award from the German government, he sent a letter back that they should mail the award because he has no time to attend award ceremonies. A few years ago, Grigori Perelman, the famous Russian mathematician, turned down the Fields Medal for his breakthrough work on the Poincaré conjecture by saying that the prize was completely irrelevant for him.

Research is not just another line of business. Many young researchers dream about managing very large groups, coordinating the research of others, or becoming directors or deans. Contrary to such beliefs, this will no longer be research but mostly administration. Administrative positions and honorific prizes are not the same as the joy and intellectual value of scientific discoveries.

In the short term, it is possible to advance and get easy recognition by following the right politics in project planning, choosing powerful supervisors, and choosing politically correct publishing strategy. The prevalent tendency is quickly publishing results, often without proper examination, and chasing scientometric numbers.

Especially, young researchers need to understand the benefits of long-term interests and plans. One may say that life can be unfair in the short run but it is fair in the long run. The most successful scientists play the long-term card instead of trying to follow quickly changing political fashions. Only the systematic, determined, and carefully reexamined research will be the one that lasts.

The productive scientist needs an ambitious and coherent *research agenda* to be successful in generating discoveries, getting ahead of competition, and attracting support (Figure 9.1). The foundation of the agenda is a well-informed and focused research interest that drives all subsequent activities. Based on this foundation, extensive data collection, exploration of the scientific literature, and methodology developments can be built. The research also needs to develop its fitting set of networks, including numerous discussions and conference presentations. Development of grant applications can become further opportunities to deepen understanding of the critical issues. Eventually, the multitude of diverse but synergistic activities of the research agenda should lead to grant funding, discoveries, and innovation.

Research agendas can be distinguished according to their focus on (i) a natural or social phenomena by applying a variety of methodologies to gain full

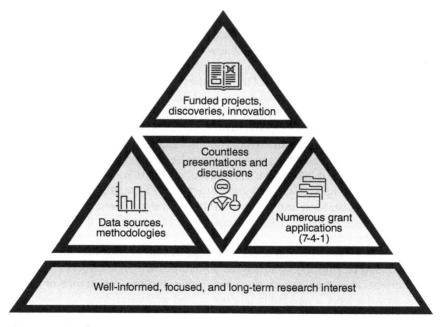

Figure 9.1 Develop your research agenda: pyramid of activities.

understanding on what is happening in nature and society (e.g. transport system of the cell); (ii) a new cutting-edge methodology that will be used to answer unprecedented questions, study a variety of natural phenomena, and effectively generate new knowledge (e.g. genetic sequencing); or (iii) an important practical problem by gathering information and combining technologies from all available sources to meet a major need (e.g. artificial kidney for dialysis).

Research driven by agenda represents an essential quality benchmark. An overarching interest needs to be the multiyear guide of day-to-day activities. With a clear agenda, one research project has a logical relationship with other projects of the same principal investigator or laboratory.

Research institutions and universities have a major interest in pushing people upward on the level of research agenda. Projects without agenda often fall into the category of recording and reporting research by making various measurements and publishing findings with some added speculations about possible reasons. Without long-term agenda, such research is wandering and vulnerable to non-reproducibility.

Michael Brown, professor of the University of Texas Southwestern Medical Center at Dallas, received the Nobel Prize in Physiology or Medicine together with Joseph L. Goldstein "for their discoveries concerning the regulation of cholesterol metabolism." Their wide-ranging studies produced many remarkable results, including identification of the rate-controlling enzyme in cholesterol biosynthesis, the role of low-density lipoproteins in extracting cholesterol from the bloodstream, and the receptor-mediated endocytosis. The direct practical implication of their studies was the development of statin drugs credited for reducing heart disease and stroke, a significant public health improvement.

Many years later, Michael Brown (2014) summarized nine simple steps to get the Nobel Prize. First among them is being curious. Other essential steps include working very hard. The aspiring researcher is to be totally consumed by the task at hand. You have to think about it day and night. It has to be constantly on your mind. You always have to try to figure out what the next step should be. You should always plan for it and think about the next experiment and also think about what went wrong with the last experiment. Do not believe the nay-sayers, if you have the evidence.

Michael Brown's suggested important steps toward getting the Nobel Prize also included picking the right spouse. Sometimes scientific careers end prematurely because their spouse does not understand the importance of their work, why the work is needed in late hours, and why vacation is not taken as planned. Many unexpected questions can come up when research is dictating someone's life.

Being Passionate Without Emotions

Science requires systematic, meticulous work with rigor, attention to detail, and pursuit of accuracy. The ability to do the hard work includes bouncing back from failures that are numerous and frequent in scientific work. There has to be an unwavering dedication to finish long-term projects and to succeed in the long run. "You need to be resilient to unexpected things happening," Nobel laureate Elizabeth Blackburn warned.

Productive research requires focused, tenacious, determined, and hard work. There are many technical and intricate details in scientific work, and projects can take twists and turns. If the researcher is passionate about the subject, hard work comes naturally and without such motivation is impossible to endure.

Passion is the drive, focus, *resilience*, and will to achieve that is needed for success. It is the full mental and emotional commitment to the desired outcome of the scientific endeavor. Research often involves sacrifice with long hours of work and effort dictated by needs of the experiment, not the needs of the experimenter.

Producing an enormous number of failures is usual in the process of discovery and innovation. Occasionally, the resilience of the scientist is called stick-to-itiveness or persistence in research. The life and achievements of many great research programs illustrate that focus, hard work, and patience eventually will pay off with unprecedented discoveries:

- In searching for effective malaria treatment, Nobel laureate Youyou Tu, a researcher at the Academy of Traditional Chinese Medicine in Beijing, collected 2000 candidate recipes and analyzed 380 extracts from 200 herbs before the uniquely effective artemisinin was discovered in 1972. "The work was the top priority so I was certainly willing to sacrifice my personal life," the famously understating scientist later recalled.
- In 1974, the UN-based Special Programme for Research and Training in Tropical Diseases (TDR) was launched (Crump and Omura 2011). Nearly 10 000 compounds, supplied by many pharmaceutical companies and also soil samples, were screened before the highly effective ivermectin was identified.
- The large numbers of failures are also seen in engineering innovation. When Dyson developed the innovative dual cyclone vacuum cleaner, he spent 15 years creating over 5000 versions that failed before the one that worked well (Goodman 2012).

On the other hand, emotions also represent a grave danger to scientific objectivity. Emotional attachment to preconceived notions, favorite methodologies, or expected outcomes can derail the best scientific project. In the research process, such harmful attachments can lead to trivial and non-repeatable research results.

The ups and downs of scientific work can derail when emotions distort the picture or bias interpretation of facts. Therefore, being passionate without emotions is an unsurpassed value. Nobel laureate James Rothman summed it up: "The hardest thing about being a scientist is you have to be prepared to fail most of the time" (Rothman 2013).

The Power of Logic and Abstraction

Rationality or logic is the method of modern science. The evidence-based processes of thinking and reasoning are essential to the scientific method of examining facts and ideas. The starting point is evidence that is obvious to the eye or mind. Subsequently, the sound logic proceeds according to strict principles of validity to make the results replicable and generalizable. Elegant science takes evidence, forms own imaginative interpretation, and develops creative, out-of-the-box solutions that ultimately can be breathtakingly simple (Glynn 2010).

In seeking the truth, scientists need to be open-minded in soliciting advice and honest about the results of understanding. According to Paul Nurse, the ability to think independently, taking evidence, and forming your own opinion are essential skills of good scientists. The best researchers can not only collect and hold specific information but also categorize and generalize the information into abstract knowledge and concepts

Critical thinking is a normative concept that includes skills or abilities of reasoned assessment plus dispositions to value sound reasoning and being guided by such assessments (Siegel 1980). As people often note there are three sides to every story: ours, yours, and the truth. Critical thinking is a concept that involves open-mindedness, fair-mindedness, independent-mindedness, intellectual modesty and humility, inquiring attitude, and respect for others in group deliberation.

Critical thinking involves strong observational skills and ability to locate and use trustworthy information, conduct an objective analysis, and make clear, reasoned judgments while recognizing that very often multiple reasonable views of the same subject coexist. According to Tim Hunt, good scientists have a tremendous respect for the truth. Attention to detail, healthy skepticism, and effort to validate all the main facts from multiple directions are part of reproducible logic.

According to Scheffler (2014), the fundamental trait to be encouraged is reasonableness by liberating the mind from adherence to fashions of thought and dictates of authority. Along these lines, education should encourage asking questions, looking for evidence, seeking and scrutinizing alternatives, and being critical of their own ideas as well as others. Reasonableness can be difficult to achieve, but it is essential in the study of nature and the pursuit of science.

Principles and methods of mathematics play a particularly important role in modern science. As Nikola Tesla summed it up, "There is scarcely a subject that cannot be mathematically treated and the effects calculated or the results determined beforehand from the available theoretical and practical data. The carrying out into practice of a crude idea as is being generally done is, I hold, nothing but a waste of energy, money and time." It is unfortunate tunnel vision of life sciences when the application of mathematical methods is narrowly limited to biostatistics, in fact, only a sub-discipline of mathematics. In the overreliance on biostatistics, we are idealizing averages and disrespecting outliers.

The comprehensive, current 2010 Mathematics Subject Classification in many other sources lists a large number of important mathematical areas with the high potential to benefit life sciences studies. The particularly promising areas include but not limited to combinatorics, linear and multilinear algebra, matrix theory, associative rings and algebras, group theory and generalizations, topological groups, measure and integration, ordinary differential equations, partial differential equations, integral transforms, calculus of variations and optimization, geometry, general topology, probability theory and stochastic processes, numerical analysis, fluid mechanics, and game theory. Interested readers are advised to consider the potential of these and many other areas and consult applied mathematicians interested in life sciences studies.

Born in Minnesota, **Randy Schekman** was a curious child. At some point, he reported his parents to the police for taking his money and not letting him buy a microscope. Fortunately, his parents relented, and he got the microscope to satisfy his interest in microbes. Furthering his skills in biology, he studied at UCLA (BA in Molecular Sciences) and Stanford University with the mentorship of Arthur Kornberg (PhD in DNA replication studies).

After graduation, he took a faculty position at the University of California, Berkeley, where most of his research focused on the machinery regulating vesicle traffic, a major transport system in the cells. The evolution of his models is an educational glimpse into the progress of award-winning research. The biotech industry was quick to recognize the value of this research and yeast culture fermentation to produce commercial quantities of human secreted proteins (e.g. more than one-third of insulin production in the world).

Ultimately, Schekman's discoveries led to many recognitions including the Albert Lasker Award for Basic Medical Research, the Louisa Gross Horwitz Prize from Columbia University, and the Nobel Prize in 2013. Randy Schekman also became respected as the editor of the *Proceedings of the National Academy of Sciences* and a forceful advocate for reforming scientific publication practices. He never patented his discoveries.

The closing words of his Nobel lecture (2014) captured the essence and importance of curiosity-driven research: "I trust the pursuit of basic discovery unconnected to any practical application will continue to motivate young scholars and that the agencies, government and private, that made discovery an adventure for me will continue to do so for as long as we thirst for knowledge of the natural world."

Creativity and Divergent Thinking

The originality of great scientists is their independence and creativity and the power of saying things that have never been said before and designing things that have never been realized before. It is also called *divergent thinking* that builds on the capacity to produce novel ideas and extends into different, unprecedented directions.

According to Wallas (1926), the process of creativity has four stages: preparation, incubation, illumination or insight, and verification. After having struggled with the problem for an extended period of time in the process of incubation, solution often comes suddenly and unexpectedly in the stage of illumination. Interestingly, some studies show that certain autistic traits are associated with high numbers of unusual responses on divergent thinking tasks (Best et al. 2015).

When novel concepts and ideas come up, the reception is not always welcoming. From time to time, researchers encounter resistance and hostility toward their ideas and discoveries. Sometimes, the researcher is ostracized for years before the world accepts what the researcher already knows from evidence.

According to one of the foremost strategists, Donn Starry, "a concept is an idea, thought, a general notion. In its broadest sense, a concept describes what is to be done" (Jensen 2016). In sciences, the operational environment is the set of circumstances surrounding someone's research laboratory, projects, and research goals. It is the aggregate of conditions, circumstances, and influences that affect the use of available resources. According to experts in strategic studies, intelligence is all about understanding the operational environment of the particular task.

Using any and all sources of information to study a natural phenomenon requires engaged research. "An experiment is a question which science poses to

Nature and a measurement is the recording of Nature's answer," according to Max Planck (see autobiography published in 2014). The scientific equivalent of moving from concept to intelligence is discussed in Chapter 12 on engaged research.

Sometimes, divergent thinking becomes a political risk. History shows that many dictators aggressively interfered with science causing great suffering of scientists and to the detriment of progress in their country. In Stalin's Soviet Union, genetics, cybernetics, and sociology were branded as "bourgeois pseudosciences," among others. As a result, many successful intellectual leaders of the highly accomplished Soviet space program of the 1960s emerged from the Stalin era prisons of the Gulag system where they could barely survive.

After World War I, the exaggerated nationalism and fascism created the notion of Aryan physics as opposed to other physics. The murderous hostility toward diverse ideas and groups of people reached unprecedented heights. Ultimately, fascism also had a devastating impact on science and productivity. Before 1933, German Nobel laureates outnumbered US recipients nearly 3 : 1 among all winners. Between 1934 and 2015, this ratio completely reversed: US Nobel laureates outnumbered German recipients by more than 4 : 1 (Kirk 2015).

Winning Science: Who Gets the Award?

When appropriately selected, scientific awards serve a very beneficial purpose not just by recognizing scientists but also by bringing their significant achievements to the attention of the general public. The best and most prestigious scientific awards recognize achievements by developing a consistent, high-quality selection process that identifies internationally renowned prize winners. The well-designed awards are much more than words of recognition, medal, and perhaps prize money. They also provide insight how success can be achieved.

In the field of life sciences, some of the most prestigious *scientific awards* include the Gairdner International Awards (Canada), Grand Medal (France), Shaw Prize (Hong Kong), Dan David Prize (Israel), Wolf Prize (Israel), Balzan Prize (Italy and Switzerland), Kyoto Prize (Japan), Japan Prize (Japan), Nobel Prize (Norway), Kavli Prize (Norway), King Faisal International Prize (Saudi Arabia), Crafoord Prize (Sweden), Queen Elizabeth Medal (UK), Copley Medal (UK), Breakthrough Prizes (US), and Lasker Award (US). Often, memberships in national academies and national honor societies of major professions are also counted as prized scientific recognitions.

Among the proliferating international scientific awards, the Nobel Prize, established by Alfred Noble in 1895, serves as the most ambitious benchmark and has a towering reputation. In his Will, Alfred Nobel directed "prizes to those who, during the preceding year, shall have conferred the greatest benefit to mankind." "It is my express wish that in awarding the prizes no consideration be given to the nationality of the candidates, but that the worthiest shall receive the prize" (Nobel 1895). Far from being perfect, the Nobel Prize remains the most highly regarded award. Many leading research organizations, like NIH, measure their accomplishments by citing their support for Nobel Prize winners.

Typically, scientific awards recognize individuals, no more than three in any given year. This *rule of three* has been working well for over a century but now gradually challenged by the concept of team science. Many great scientific achievements of the recent decades come from the collaboration and competition of a vast number of scientists. The CRISPR/Cas9 gene editing system discovery is one of the most recent and best-known examples. While the very broad footing of major discoveries is increasingly apparent, it does not seem to diminish the role of most influential thought leaders. Individual awards are likely to remain dominant in recognition of scientific excellence.

Production of highly cited articles, landmark discoveries, or a large number of inventions does not guarantee recognition in the form of scientific awards. Most scientific awards represent recognition by peers and occasionally political considerations (e.g. preference for the citizens of a particular country or region). There have been many scholarly attempts to predict recipients of major scientific awards based on measures of research productivity, but none of them have been entirely successful.

Ultimately, understanding nature and helping others are the true worthy goals of scientific research. Many award-winning scientists pointed out that answering important questions is the only meaningful way to success. Award is the recognition but not the purpose of productive research.

Reading Scientific Papers Can Be More Difficult Than Writing Them

As it turns out, not just scientific communications but also scientific awards are vulnerable to the dangers of generating biased or distorted news. The 2016 US election brought the concept of "fake news" to the forefront of public discussions. Recognizing vulnerabilities of scientific awards should not diminish their value but caution against romanticizing either the award or the recipients.

Detection of non-reproducible research results is one of the most fundamental challenges of reading scientific reports and recognizing achievements. Not infrequently, such results originate from exceptionally "productive" researchers regarding their number of publications. Given the high rate of non-reproducible research in the literature, 50–85% according to some estimates, it is not unreasonable to acknowledge that some of it can lead to futile attempts of practical application and also to undeserved recognition.

The most effective researchers recognize 'incredible' numbers long before others do. While numbers tend to look scientific and trustworthy, we are often surrounded by inaccurate and profoundly misleading numbers. Therefore, checking the reproducibility of research results is one of the first and most essential steps of accepting research results and quality award selection.

Political bias represents another hazard of interpreting scientific results. When readers are scratching their heads in trying to find out what the award recipient actually contributed, the credibility of the award is undermined. Obviously, so-called lifetime awards that rely on traditional measures of scientific productivity and popularity among peers can be more vulnerable to rewarding political connections. Membership in national academies and national professional honor societies is certainly in the category of lifetime achievement recognitions.

Careful monitoring and controlling the influence of political bubbles can limit the impact of people rewarding each other for conforming views. Observers highlight the Eurocentric approach of the Nobel Prize selection that appears to disadvantage Asian scientists.

Highly successful but divergent thinkers without political clout may not do well in lifetime recognitions. One of the most famous examples is Nobel laurate and Lasker Award winner Tu Youyou who is not a member of the national academies in her home country. Beyond careful checking of the reproducibility of someone's research results, appropriately weighing the societal benefits realized from someone's work could improve the trustworthiness of lifetime achievement recognitions.

Business relationships tend to attract particularly intense political scrutiny. When Harald zur Hausen received the Nobel Prize, concerns were raised that AstraZeneca, the company with significant interest in the HPV vaccine production, also had business connections to a member of the Nobel committee. Conversely, the 2016 Lasker-DeBakey Clinical Medical Research Award for the Hepatitis C replicon system and drug development did not include Emory researcher Raymond Schinazi, a pivotal contributor to the discovery. There were speculations that the omission happened not for the lack of essential scientific contribution but due to Schinazi's simultaneous business successes in retroviral therapy development.

None of the above concerns are comparable with the ultimate challenge of scientific award selection: finding the greatest and most meritorious discoveries. In the rapidly expanding and increasingly questioned scientific literature, the task can be daunting. No wonder, awards tend to attract each other. For example, the Lasker Award is sometimes described as the antechamber of the Nobel Prize.

The 2003 Nobel Prize in Physiology or Medicine gave an illustration of not giving well-deserved credit for the scientific accomplishment. In that year, Paul Lauterbur and Peter Mansfield got the award "for their discoveries concerning magnetic resonance imaging." However, the earlier and landmark contributions of Raymond Damadian were missed and unrecognized, in spite of the availability of a third slot for sharing the award (Kauffman 2014).

In a world where reading scientific papers can be more difficult than writing those, readers who can quickly absorb large volumes of information and also separate trustworthy, practically useful messages from the non-repeatable and inconsequential publications can enjoy distinct advantages.

Ultimately, the foundation of scientific success is the desire to understand first. Impressive researchers think in unexpected directions, show outstanding creativity, and develop elegant models. The best scientists achieve high-speed execution of ideas that avoids waste, astonishes observers, and produces extraordinary accomplishments.

References

Best, C., Arora, S., Porter, F., and Doherty, M. (2015). The relationship between subthreshold autistic traits, ambiguous figure perception and divergent thinking. *Journal of Autism and Developmental Disorders* 45 (12): 4064–4073.

Brown, M. (2014). How to win a Nobel Prize – 9 simple steps | Dr. Michael Brown | TEDxUTA. https://www.youtube.com/watch?v=MdarocitY6k (accessed 5 March 2017).

Clynes, T. (2016). How to raise a genius. *Nature* 537 (8 September).

Crump, A. and Omura, S. (2011). Ivermectin, "wonder drug" from Japan: the human use perspective. *Proceedings of the Japan Academy, Series B* 87 (2): 13–28.

Flexner, A. (1939). The usefulness of useless knowledge. *Harpers* (179): 544–552.

Glynn, I. (2010). *Elegance in Science: The Beauty of Simplicity*. Oxford: Oxford University Press.

Goodman, N. (2012). James Dyson on using failure to drive success. *Entrepreneur* (5 November).

Gromov, G. (1999). *The History of Computer Science*. http://abcdefgh.livejournal. com/1563118.html (accessed 19 March 2017).

Harford, J. (1997). *Korolev – How One Man Masterminded the Soviet Drive to Beat America to the Moon*. New York: Wiley.

Institute of Medicine (US) Committee on the NIH Research Priority-Setting Process (1998). *Scientific Opportunities and Public Needs: Improving Priority Setting and Public Input at the National Institutes of Health*. Washington, DC: National Academies Press (US).

Jensen, B. (2016). *Forging the Sword: Doctrinal Change in the US Army*. Stanford, CA: Stanford University Press.

Kauffman, G. (2014). Nobel prize for MRI imaging denied to Raymond v. Damadian a decade ago. *Chemical Educator* 19: 73–90.

Kell, H.J., Lubinski, D., Benbow, C.P., and Steiger, J.H. (2013). Creativity and technical innovation spatial ability's unique role. *Psychological Science* 24 (9): 1831–1836.

Kirk, A. (2015). Nobel Prize winners: which country has the most Nobel laureates? *The Telegraph* (12 October).

Nietzsche, F. and Large, D. (1998). *Twilight of the Idols*. Oxford: Oxford University Press (Paperbacks).

Nobel, A. (1895). The will. Nobelprize.org. Nobel Media AB 2014. http://www.nobelprize.org/alfred_nobel/will (accessed 17 Mar 2017).

Nobel Prize Inspiration Initiative (NPII) (2017). http://www.nobelprizeii.org (accessed 11 March 2017).

Oppezzo, M. and Schwartz, D.L. (2014). Give your ideas some legs: the positive effect of walking on creative thinking. *Journal of Experimental Psychology: Learning, Memory, and Cognition* 40 (4): 1142.

Planck, M. (2014). *Scientific Autobiography: And Other Papers*. Open Road Media.

Rothman, J.E. (2013). Rothman on breakthroughs in research. https://www.youtube.com/watch?v=uOcB-XjQc3M (accessed 19 March 2017).

Sanger, F. (2014). Transcript from an interview with Frederick Sanger, 1958 and 1980 Nobel laureate in chemistry. Nobelprize.org. Nobel Media AB 2014. http://www.nobelprize.org/nobel_prizes/chemistry/laureates/1980/sanger-interview-transcript.html (accessed 4 Mar 2017).

Scheffler, I. (2014). *Reason and Teaching (Routledge Revivals)*. London: Routledge.

Schekman, R.W. (2014). Nobel lecture: genetic and biochemical dissection of the secretory pathway. Nobelprize.org. Nobel Media AB 2014. http://www.nobelprize.org/nobel_prizes/medicine/laureates/2013/schekman-lecture.html (accessed 8 Mar 2017).

Siegel, H. (1980). Critical thinking as an educational ideal. *The Educational Forum* 45 (1): 7–23.

Simonton, D.K. (2013). After Einstein: scientific genius is extinct. *Nature* 493 (7434): 602–602.

Sinatra, R., Wang, D., Deville, P. et al. (2016). Quantifying the evolution of individual scientific impact. *Science*, 354 (6312): aaf5239. doi: 10.1126/science.aaf5239.

Tamir, D.I. and Mitchell, J.P. (2012). Disclosing information about the self is intrinsically rewarding. *Proceedings of the National Academy of Sciences* 109 (21): 8038–8043.

Tesla, N. (1919). My inventions Nikola Tesla's autobiography. *Electrical Experimenter Magazine*. http://www.teslasautobiography.com (accessed 22 June 2018).

Wallas, G. (1926). *The Art of Thought*. London: J. Cape.

Watson, J. (1970). Horace Freeland Judson: (RARE) Interview with James Watson and Francis Crick [Video file] https://www.youtube.com/watch?v=NGBDFq5Kaw0 (accessed 11 July 2016).

10

Learning from the Best

You cannot learn science from books, you have to learn science from other people, who give you the right imprint.

Carlo Rubbia (2015)[1]

Learning from the best is one of the most distinct ways to succeed. Identifying cutting-edge methodologies, best laboratories, trailblazing researchers, or leading institutions of the particular research area is the first step. Subsequently, the aspiring researcher needs to gather information, study practices, and, in some cases, seek employment or visiting opportunities.

In the information age, a large variety of resources is available to identify the best and learn about it. However, personal exposure and hands-on experience are the ultimate learning opportunities. One may be pleasantly surprised by the openness of many leading researchers in sharing experiences and offering advice. The real scientist, who is driven by curiosity, does not hide his or her thoughts but rather shares them and interested in your thoughts.

In modern times, mentor is the widely recognized English word describing someone who offers encouragement and shares knowledge with a less experienced colleague. Originally, Mentor was a mythological figure mentioned in Homer's Odyssey. While Odysseus was away on the expeditionary Trojan War, Mentor was in charge of staying home and advising the son of Odysseus. Today, mentoring is one of the most frequently highlighted ways of personal development, but learning from the best is a much broader opportunity that lasts for a lifetime.

Sometimes, status quo-loving people are asked to revisit their practices and learn from the best, but, in response, only the excuses proliferate (e.g. we are different, we do not want to be like them, and it is certainly not for us). In these

---

1 Rubbia, C. cited by Gatchell, M. (2015). Mentoring: the perspective of Nobel laureates. *Naturejobs | Naturejobs Blog* (28 July).

Innovative Research in Life Sciences: Pathways to Scientific Impact, Public Health Improvement, and Economic Progress, First Edition. E. Andrew Balas.

cases, it should be pointed out that if you dislike the best and do not even want to look at it, you are never going to be the best.

In spite of the obvious benefits of learning from the best, it remains one of the most challenging and most frequently neglected opportunities. It is quite surprising to find out how many researchers have never read a Nobel lecture or never watched video interviews with award-winning scientists. If the National Institutes of Health (NIH) measures its success based on the Nobel Prize winners supported, then today's researchers should know how they achieved their landmark discoveries.

Mentoring is always recommended and advantageous, but there are occasional successes without it. Some talented scientists have a trait of original thinking and intellectual curiosity from a very young age. For example, Mikhail Lomonosov was an accomplished Russian polymath scientist with a uniquely curious, hard-working, and eager-to-learn personality from a very young age (Crease and Shiltsev 2013). He went from Kholmogory to Moscow on foot to expand his studies when he was 19 (a more than 1000 km walk). Later, his scientific achievements included discovery of the law of mass conservation in chemical reactions and landmark contributions to the kinetic theory of gases.

The Benefits of Research Mentoring

In comparison with other educational experiences, *mentoring* has three distinguishing characteristics: dynamic and informal, individually tailored, and focused on mentee's interests and growth (Schunk and Mullen 2013). As opposed to apprenticeship, mentoring is a relationship that promotes the development of both the mentor and mentee.

Everything can be described in reports, but you can never learn everything about research only by reading books and articles. In scientific studies, there is much to learn about new methodologies, handling the many frustrations that come with failed experiments, rejection of articles, criticism of new ideas, and rejection of grant applications. Mentoring can model the human side of managing these challenges patiently and with a positive, can-do attitude.

In 2015, Nobel laureate Richard Roberts wrote an entertaining article on "simple rules" to win the prize. Among them are picking your family carefully, work in the laboratory of a previous prize winner, and work in the laboratory of a future prize winner. Apparently, 3 out of 10 rules are about being mentored for success. Roberts also highlighted that children of seven prize winners became prize winners themselves, and four married couples have jointly won the prize. For example, the father–son pairs of prize winners include Arthur Kornberg in medicine and Roger Kornberg in chemistry.

In looking at the history of significant scientific recognitions, it is readily apparent that several influential award-winning scientists mentored and generated many more great scientists. In Germany between 1920 and 1930, pioneering biochemist Otto Warburg mentored three future prize winners in his laboratory. Very succinctly, Warburg advised a PhD student in Philadelphia, "If you wish to become a scientist, you must ask a successful scientist to accept you in his laboratory, even if at the beginning you would only clean his test tubes" (Krebs and Schmid 1981).

Mentoring is more than a foremost instrument of career-relevant progress and performance development. Mentoring is a major contributor to personal growth, including learning skills, socialization, *relational competencies*, personal awareness, *emotional intelligence*, work–family balance, and others (Ragins and Kram 2007). Some people highlight that the list of beneficial non-work outcomes also includes wellness and physiologic outcomes. The benefits may come from better appreciation of non-work-related life balance as it plays a key role in devoting full attention to the quest for new knowledge.

Young researchers often talk about their need for mentoring. Mentors can do at least five important things to help the aspiring scientist:

- They can highlight promising research challenges worthy of study and probably solvable with available methodologies. Good research mentors know how to find essential questions and how to develop responsive methods to get an accurate answer.
- Mentors can teach sophisticated and cutting-edge methodologies not readily available to most other researchers. Frequent meetings with the mentor can become one of the most valuable sources of guidance and feedback.
- Mentors can be great sounding boards to discuss early-stage ideas and research plans before spending a lot of time and resources on them. Remember, talking through is one of the best ways to clarify ideas and forecast outcomes.
- More effective thinking is one of the great benefits of scientific mentoring. It includes both critical and creative aspects of reasoning. Such thinking helps in realizing when something really new surfaces in an observation, what can be derived from sometimes mundane observation, how a problem they encounter can be easily solved, or where the connections are with applicable conceptualizations.
- Mentors can be role models of focus, persistence, integrity, and many other essential ingredients of good science. Mentors can also be the source of encouragement and inspiration at a time when experiments fail, and digging through the data requires superhuman persistence.

Education has a tremendous power over the development of the young professional. As James Watson (2005) jokingly noted: "My father was raised to be an Episcopalian and republican but after one year in college he became an atheist and a democrat."

The Mentored Way to Success

Being mentored is a commitment that should start with a desire to succeed. It is an active process, a way of reaching out to attract interest and support (Figure 10.1). In the classic definition, mentoring is viewed as a means of fostering the mentee's acquisition of knowledge and skills to be used in the profession (Merriam 1983). In the mentoring relationship, the person mentored is sometimes called mentee or protégée.

Of course, the scientific mentor is not a nanny holding hands and leading unmotivated people to great science. Sitting around, waiting to be mentored, and finger pointing at the lack of mentoring are inappropriate and unlikely to lead to meaningful results. As one mentor expressed it, "My responsibility is to support your success, but your responsibility is to succeed."

The benefits of good mentoring have been recounted by many award-winning scientists. They recall the inspirational comments, exciting scientific experiences, and new techniques. At Yale University, the postdoc advisor of Venkatraman Ramakrishnan, 2009 Nobel laureate in chemistry, introduced him to work on the ribosome. "The ribosome turned out to be this long-term interesting problem – he essentially gave me a lifetime of work" (Ramakrishnan 2015).

Good mentors can serve as compass in the vast, disorienting jungle of scientific literature. Everything can be described in a report, but it is impossible to learn research only by reading books and articles. Prominent physicist Carlo Rubbia added that "you should never go where all of the other people go" (Rubbia 2015).

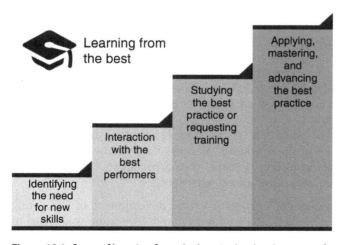

Figure 10.1 Steps of learning from the best in the development of research skills.

Most mentees become different people when they leave the mentoring lab. Their personality, work habits, ethics principles, confidence, responsibility, ability to function in a team, drive, and enjoyment of research should all go up. Ultimately, a happier and more effective scientist should transition to the chosen field of inquiry.

Mentoring partnership can last far beyond the postdoctoral years. Even award-winning scientists describe soliciting advice at a much later and more recognized phase of their career. *Mentor–mentee relationships* can last a lifetime with many significant benefits year after year.

It should benefit society when professionals with advanced science education take increasingly important and leadership positions in various companies and organizations. To prepare for informed choices, career discussions should occur not just at the end of PhD programs but also in the beginning. National data show that only 14% of PhDs in the biological sciences obtain tenure-track positions within a few years after graduation. Meanwhile, 43% were employed full time in nonacademic settings (Stephan 2012).

Mentoring can also help the aspiring researcher in becoming successful outside academia. At the University of California, San Francisco, a study pointed out that doctoral students of basic biomedical sciences are increasingly considering a broad range of career options and about one-third of them intends to pursue non-research pathways (Fuhrmann et al. 2011). Correspondingly, education and mentoring should put more emphasis on career planning and professional development to develop skills applicable in a variety of science-related but nonacademic occupations. Ultimately, students should be well prepared for success in a wide range of traditional and nontraditional science-related careers.

The National Institutes of Health has several grant programs designed to support mentored career development. Among them, the best known is the K award program designed to develop successful clinical researchers. The program has a considerable track record of success but eventually only about half of the funded recipients develop sustainable research programs (Fleming et al. 2012). The other major source of career development support is the *Clinical Translational Science Award* (CTSA) programs. The CTSA National Mentor Working Group prepared a series of white papers and recommendations for comprehensive mentoring programs.

Eric Kandel is an Austrian-born American physician and neuroscientist. During his influential years at New York University, he became interested in learning, memory, and the biological basis of the mind. His most significant studies were conducted as a faculty member of the Center for Neurobiology and Behavior at Columbia University.

In New York, Kandel had the opportunity to learn about pioneering methodologies of neurobiology. The laboratory of Harry Grundfest at Columbia

University showed him the use of the oscilloscope to measure action potential conduction of neurons (Kandel 2007). In moving forward, intracellular recordings of the electrical activity of small neurons of the vertebrate brain represented the major technical obstacle.

Fortuitously, he also learned about Stephen Kuffler's work on an experimentally more accessible system, the neurons isolated from marine invertebrates. After completing his medical school studies, he returned to the lab of Grundfest and learned how to make microelectrodes and obtain intracellular recordings from the crayfish giant axon. He also learned first insights into the universality of cellular processes. After completing his residency in psychiatry, Kandel visited the laboratory of Ladislav Tauc in Paris and learned about his studies on the marine mollusk *Aplysia californica*. Ganglia isolated from *Aplysia* allow the study of simple forms of learning such as sensitization, habituation, and conditioning.

With these experiences and mentoring, Kandel was on his way to developing a reductionist approach to learning and memory. As his Nobel lecture pointed out in 2000, "After an extensive search for a suitable experimental animal, I settled on the giant marine snail Aplysia, because it offers three important experimental advantages: its nervous system is made up of a small number of nerve cells; many of these are very large; and (as became evident to me later) many are uniquely identifiable" (Kandel 2001).

Ultimately, the experiments of Eric Kandel demonstrated that chemical signals change the structure of synaptic connections between cells. He also proved that short-term and long-term memories are formed by different signals. These fundamental findings of learning turn out to be generalizable for all animals from mollusks to humans. Kandel's Nobel lecture thanked Tom Cech and Gerry Rubin, whose "farsighted vision has encouraged Hughes investigators to take a long-term perspective so as to be able to tackle challenging problems."

Mentoring Minority Students and Researchers

When national statistics on the proportions of various minority groups are substantially different from the institutional ratios, questions can be raised regarding inclusivity of the environment and also the effectiveness of various initiatives in addressing issues of *underrepresentation*. In a result-oriented environment, issues of minority representation should have heightened significance.

According to 2016 US Census data, 13% of the US population is African-American, and 18% is Hispanic or Latino. However, in 2013, the representation of African-Americans was only 4.8% of all science and engineering doctorates. Similarly, Hispanics were only 7% of the Science, Technology, Engineering, and Mathematics (STEM) workforce in 2011. According to government data, women comprise 24% of STEM workers, whereas the representation of women in the workforce of the United States is 48%.

There are various and changing definitions of underrepresented minorities. Well-known and classic categories need to be continually revisited and updated based on improved understanding of minority status and challenges. Today, the list of minority groups is broad and diverse: ethnic minorities like African-Americans or Hispanics, religious minorities, women in the STEM fields, LGBT minorities, minorities coming from small and disadvantaged nations, and many others.

There are much statistical data and innumerable reports on the significant underrepresentation of women and minorities in science and engineering. In fact, below the average representation of minorities in higher education and scientific work is one of the quintessential components of marginalization.

In the STEM areas there have been many Nobel Prize-winning minority scientists, among them William Arthur Lewis (black), Ada E. Yonath (woman with children), César Milstein (Hispanic/Latino), Vladimir Prelog (Croatian in Bosnia), and others. The work of Alan Turing and others illustrates the significant accomplishments of LGBT scientists. While the successes of minority scientists are impressive, many of them also include stories of discrimination and marginalization.

Most importantly, these and other successes of scientists from various minority groups show the enormous societal loss when minorities are left out of research or directly marginalized. The loss to the creative environment and research productivity damages the majority and minority alike: research opportunities are squandered, and discoveries are missed.

In his study, Whittaker et al. (2015) presented stunning data on declining faculty diversity as faculty members are progressively promoted. Notably, the ratios of African-Americans as assistant professors, associate professors, and full professors are decreasing in the process of promotion according to data from the US Department of Education, National Center for Education Statistics, 2013 (6.3, 5.6, and 3.6%, respectively). Similarly, declining rate can be observed for Hispanic/Latino faculty members as well (4.3, 3.9, and 2.9%).

The *diversity of research workforce* is essential to offer role models of success, gain ideas from diverse sources, and support fairness and social justice. However, national statistics indicate that the racial and ethnic minorities are disproportionately underrepresented in the faculty, especially in STEM disciplines. In spite of significant efforts and funding that has been committed to increasing recruitment, retention rates are often below expectations.

To assess faculty characteristics and institutional culture, the National Initiative on Gender, Culture and Leadership in Medicine conducted a cross-sectional study of faculty from 26 representative academic health centers in the United States (Pololi et al. 2015). In their national sample, 43% of the respondents noted inadequate mentoring, and only 30% had a positive view of mentoring. Inadequate mentoring was associated with lower self-efficacy in career advancement and lower scores on the trust/relationship/inclusion scale.

Ramón y Cajal was the pioneer of brain exploration and founder of modern neuroscience. He worked at the Faculty of Medicine of Zaragoza, University of Valencia, University of Barcelona, and later Faculty of Medicine of Madrid. Shortly after becoming director of the Anatomical Museum of the University of Zaragoza, he went to visit the laboratory of Luis Simarro Lacabra, a neuropsychiatrist experimenting with the Golgi silver chromate staining method that greatly impressed his future research (Andres-Barquin 2002).

Y Cajal described the fundamental structure and organization of the nervous system. He laid the cornerstones of neuroscience with the development of the neuron doctrine and the law of functional polarization (i.e. the axipetal polarization and the dendrifugal conduction in nerve cells). The law of functional polarization remains a fundamental principle in neuroscience while recognizing some important exceptions to the usual rules of neuronal polarity (Urban and Castro 2010).

In addition to groundbreaking research, y Cajal was a dedicated teacher and passionate mentor for more than 50 years. Through involvement with several universities, he created a school that started the flourishing of neurology. The list of his very successful students and their accomplishments is exceptionally long. Y Cajal recalled learning from his father that "ignorance was the greatest of all misfortunes, and teaching the noblest of all duties" (y Cajal 1989).

Ramón y Cajal was also a talented artist who could illustrate anatomical observations with exceptional clarity and virtuosity. As he summed up, "The garden of neurology holds out to the investigator captivating spectacles and incomparable artistic emotions" (y Cajal 1989). His hundreds of legendary drawings about the delicate arborizations of brain cells are still used to teach neuroanatomy. In 1906, he was awarded the Nobel Prize in Physiology or Medicine together with Camillo Golgi, his scientific competitor.

The ratio of faculty who considered leaving their institution was highest among those who had inadequate mentoring (58%).

A survey of 153 of women working in STEM higher education suggested that about half of them had at least one child (Howe-Walsh et al. 2016). Perception of mentoring, organizational support, and network resources were significantly lower for women academics with children. Higher-ranking mentors should be able to share comparable personal experience in solving problems, give challenging assignments leading to new opportunities to learn, and serve as role models.

In a series of interviews with 37 African-American engineering and computer science major college students at two predominantly white public research universities, the responses suggested a desire to interact with same race faculty members (Newman 2015). Some respondents expressed concerns about negative interactions with a white faculty member and the potential implications of this experience.

Supporting minorities and providing inclusive environment has tremendous significance in the development of a healthy, vibrant, and innovative institutional culture. Other chapters talk about the intellectual significance and major advantages of cultural inclusiveness in creating the *environment of innovation*. This discussion focuses on minorities in the broadest sense of this term. Mentoring and role models have a special significance in attracting minorities and women to scientific careers.

In recent years, there have been many meritorious initiatives in the United States and elsewhere to increase diversity, but the rate of success is still below expectations. The patterns of marginalization have deep historical roots, evolved over decades or centuries, and long-standing biases urge correction.

Today, many institutions make the promotion of diversity part of their mission and strategic plans. Laws and regulation also set some expectations that need to be carefully observed. However, the reality in the academic offices and research laboratories often fail to reflect the framed, magnanimous statements on the walls.

To increase diversity in academic environments, the barriers need to be removed by addressing inequities in training, degree attainment, and recruitment and/or retention; changing established environmental cultures and traditions; reducing disparities in research grant support; integrations into academic communities and/or reducing isolation; and overall environmental support, changing negative stereotypes and implicit bias, leadership understanding, and will (Whittaker et al. 2015).

Some studies showed not only the positive impact of student–faculty interactions but also tremendous variations across minority groups. The study by Kim and Sax (2009) analyzed data on 58 281 student participants of the 2006 University of California Undergraduate Experience Survey (UCUES). Most remarkably, research-related student–faculty interaction predicts higher college GPAs for all groups. On the other hand, course-related student–faculty interactions are not as effective, especially in impacting the academic performance of minority in first-generation students. Asian-American students were more likely to participate in research-related interactions, while African-Americans and first-generation students were more active in course-related interactions.

Alan Turing was a pioneering mathematician, cryptanalyst, and theoretical biologist. He studied mathematics at the University of Cambridge. His mentor was Alonzo Church, pioneer of mathematical logic and computer science at Princeton. Turing received his PhD from Princeton University in 1938.

Early in his career, Turing's seminal paper titled "On computable numbers, with an application to the Entscheidungsproblem" focused on what mathematical statements are provable within a particular mathematical system (1937). It made a tremendous impact on the early evolution of computer science. Later, his so-called Turing machine concept was essentially a model for a general-purpose computer.

One of his most recognized achievements was the Turing test, which Turing developed while he worked at the University of Manchester (1950): a person should be called to judge whether electronic communication with another party suggests a human or machine. If it is indistinguishable then we can talk about machine intelligence. Turing led the design work on the Automatic Computing Engine and created a pioneering blueprint for store-program computers at the National Physical Laboratory in the United Kingdom. For these and other accomplishments, Turing is considered among the founders of modern computer science and artificial intelligence.

During World War II, Turing was part of the famous Bletchley Park code-breaking operation. Among others, his intellectual contributions and technical improvements helped to make decoding of intercepted German messages highly effective. By 1942, Bletchley Park was able to decipher two intercepted messages per minute, day and night giving tremendous strategic and operational information to the Allies. For his wartime work, Turing was appointed as Officer of the Order of the British Empire by King George VI in 1946.

In 1952 after a break-in, he called the police and admitted a homosexual relationship with the perpetrator, a gross indecency under British law of his time. To avoid imprisonment, he submitted to chemical castration through injections of synthetic estrogen. Two years later, he died, probably result of suicide. In 2009, British Prime Minister Gordon Brown and the government publicly apologized for Turing's utterly unfair treatment, and subsequently Queen Elizabeth II granted Turing a royal pardon.

Learn from Leaders to Become a Leader

Creative and elegant methodologies or at least substantial adaptations of existing methods are just as essential to scientific discoveries as the ability to identify crucial scientific questions (Figure 10.1). Elegant study means *elegant methodology*: responsive to the scientific question, creative in design, and yet simple enough to be practical.

Most doctoral students and postdocs attend educational courses on scientific methods. These courses are certainly worthwhile in reviewing the classic and most commonly used methodologies. However, believing that fruitful research can be pursued just by picking one of the widely taught off-the-shelf methods is entirely misleading. What is not in the established textbooks and not in the standard curriculum of doctoral education is mostly left to mentoring.

Besides the prevalent neglect of scientific modeling, ignorance of the need to design new methodologies responsive to new questions is probably among the most significant drawbacks of modern science. New questions often require the development of new methodologies and research tools. Having new tools can help to address not only the primary problem but also other phenomena for the first time. Ultimately, the new tools can lead to scientific discoveries and new products successful in improving everyday life. Consequently, learning about new methods and also acquiring the skills of developing new methodologies remain important reasons for good mentoring.

Howard Temin, the discoverer of the reverse transcriptase enzyme, got the idea of provirus after working with the Rous sarcoma virus in the laboratory of Professor Renato Dulbecco at the California Institute of Technology. Temin's doctoral thesis was on Rous sarcoma virus, and after graduation, he spent another year in the same lab (Drake and Crow 1996). According to the provirus hypothesis, the Rous sarcoma virus, an RNA virus, is replicated via a DNA intermediate. It came about when experiments demonstrated that actinomycin D, an inhibitor of DNA and RNA synthesis, also blocks the reproduction of the Rous sarcoma virus. The replication via DNA intermediate led to the landmark discovery of the reverse transcriptase. Dulbecco and Temin, mentor and mentee, were awarded the Nobel Prize in 1975.

Between 1926 and 1930, Hans Krebs worked as a research assistant in the laboratory of Otto Warburg at the Kaiser Wilhelm Institute for Biology in Berlin. Warburg had a small but very well-equipped laboratory. This was Krebs's first paid job, and he was jubilant. During this remarkable time, he learned the techniques of tissue slicing and manometric methods to measure the rate of respiration and glycolysis. These methodologies became profoundly influential in the research on the tricarboxylic acid (TCA) cycle or the Krebs cycle (Krebs and Johnson 1937).

Krebs also learned from Warburg integrity, scientific dedication, and rigorous *intellectual analysis*, which remained with him during his long and distinguished scientific life (Stubbs and Gibbons 2000). Later in his memoirs, Krebs recalled the high standards set by Warburg in science: "His dedication manifested itself in his long and regular working hours and his contempt for those who tried to further their careers by jockeying for position, by hobnobbing with and courting the influential, or by publishing trivia for publishing's sake. He was prepared to take infinite pains with every aspect of his work" (Krebs 1981). In addition to Krebs, Warburg also mentored two other Nobel Prize winners.

When you learn about a new, breakthrough methodology, you are launching research from an elevated level (Figure 10.1). Further improvements and methodology innovations will add to the already high level of sophistication and responsiveness in your research. Visiting laboratories that practice breakthrough methods, learning from senior scientists about developing methodologies that fit the questions, and discussing ideas for elegant methodologies with mentors are exceptionally important.

Learning and using the latest methods also raise the issue of copying. In the method sections, what level of duplication is acceptable based on today's standards of originality? In principle, if you use the same methodology then nearly identical wording should be acceptable, especially, when it comes from your earlier publication. Due to pertinent copyright rules, publishers can insist on original wording in every new publication. However, publishers often give permission for work to be reproduced as long as the reproduction is faithful to the original and accompanied by attribution.

If you want to do something at a high level, it is a good idea to look around for best methodologies and replicate the chosen ones in your laboratory. When you do it well, you can add new value, expand the use, and elevate the methodology to a new and higher level. This is a principle that works very well in many situations, ranging from productive scientific laboratories to high-tech economies.

It was 1947 when Rosalind Franklin got a "chercheur" position in the Laboratoire Central des Services Chimiques de l'Etat in Paris. There she learned X-ray diffraction techniques, a demanding but promising methodology, under the mentorship of Jacques Mering. Later, while working at the University of Cambridge with Raymond Gosling, a graduate student, she made numerous improvements to the technique. Ultimately, her mastery of X-ray diffraction produced the famous photo 51 that became the turning point in the discovery of the double helical structure of DNA. Her landmark paper on the molecular configuration in sodium thymonucleate came out in the same issue of *Nature* where Watson and Crick's discovery was published (Franklin and Gosling 1953). For their discovery of the structure of DNA, Watson, Crick, and Wilkins shared the Nobel Prize in Physiology or Medicine in 1962. In 1958, Franklin died of ovarian cancer at age 37, a great tragedy, and the Nobel Committee does not make posthumous nominations.

In summary, there are many desirable outcomes of learning from the best and benefiting from a successful mentor–mentee relationship (Pfund et al. 2016). The list includes research skills (e.g. disciplinary knowledge, technical skills, practice of ethical behavior and responsible conduct), interpersonal skills (e.g. active listening, building trusting and honest relationships), psychosocial skills (e.g. motivation, effective coping, developing science identity, developing a sense of belonging), cultural responsiveness (e.g. equity and inclusion, culturally responsiveness, reducing the impact of bias and stereotype threat), and skills of independence (e.g. promoting further professional development establishing professional networks, actively advocating). The emerging science identity portrays how an individual develops the professional personality in a particular scientific culture and forecasts future science-related progress.

Knowing what is the very best worldwide and learning from it are essential skills for research success. Obtaining relevant and pioneering scientific mentoring can put the talented beginner on the road leading to scientific discoveries.

References

Andres-Barquin, P.J. (2002). Santiago Ramón y Cajal and the Spanish school of neurology. *The Lancet Neurology* 1 (7): 445–452.

Crease, R.P. and Shiltsev, V. (2013). Pomor polymath: the upbringing of Mikhail Vasilyevich Lomonosov, 1711–1730. *Physics in Perspective* 15 (4): 391–414.

Drake, J.W. and Crow, J.F. (1996). Recollections of Howard Temin (1934–1994). *Genetics* 144 (1): 1.

Fleming, M., Burnham, E.L., and Huskins, W.C. (2012). Mentoring translational science investigators. *JAMA: the Journal of the American Medical Association* 308 (19): 1981–1982.

Franklin, R.E. and Gosling, R.G. (1953). Molecular configuration in sodium thymonucleate. *Nature* 171 (4356): 740–741.

Fuhrmann, C.N., Halme, D.G., O'Sullivan, P.S., and Lindstaedt, B. (2011). Improving graduate education to support a branching career pipeline: recommendations based on a survey of doctoral students in the basic biomedical sciences. *CBE-Life Sciences Education* 10 (3): 239–249.

Howe-Walsh, L., Turnbull, S., Papavasileiou, E., and Bozionelos, N. (2016). The influence of motherhood on STEM women academics' perceptions of organizational support, mentoring and networking. *Advancing Women in Leadership* 36: 54.

Kandel, E.R. (2001). Nobel lecture: the molecular biology of memory storage: a dialog between genes and synapses. *Bioscience Reports* 21 (5): 565–611.

Kandel, E.R. (2007). *In Search of Memory: The Emergence of a New Science of Mind*. New York: WW Norton & Company.

Kim, Y.K. and Sax, L.J. (2009). Student–faculty interaction in research universities: differences by student gender, race, social class, and first-generation status. *Research in Higher Education* 50 (5): 437–459.

Krebs, H. (1981). *Reminiscences and Reflections*. New York: Oxford University Press.

Krebs, H.A. and Johnson, W.A. (1937). The role of citric acid in intermediate metabolism in animal tissues. *Enzymologia* 4: 148–156.

Krebs, H. and Schmid, R. (1981). *Otto Warburg: Cell Physiologist, Biochemist and Eccentric*, 1–141. Oxford: Clarendon Press.

Merriam, S.B. (1983). Mentors and protégés: a critical review of the literature. *Adult Education Quarterly* 33: 161–173.

Newman, C.B. (2015). Rethinking race in student-faculty interactions and mentoring relationships with undergraduate African American engineering and computer science majors. *Journal of Women and Minorities in Science and Engineering* 21 (4): 323–346.

Pfund, C., Byars-Winston, A., Branchaw, J. et al. (2016). Defining attributes and metrics of effective research mentoring relationships. *AIDS and Behavior* 20 (2): 238–248.

Pololi, L.H., Evans, A.T., Civian, J.T. et al. (2015). Mentoring faculty: a US national survey of its adequacy and linkage to culture in academic health centers. *Journal of Continuing Education in the Health Professions* 35 (3): 176–184.

Ragins, B.R. and Kram, K.E. (2007). *The Handbook of Mentoring at Work: Theory, Research, and Practice*. Thousand Oaks, CA: SAGE.

Ramakrishnan, V. cited by Gatchell, M. (2015). Mentoring: before they were laureates. *Naturejobs | Naturejobs Blog* (4 August).

Roberts, R.J. (2015). Ten simple rules to win a Nobel Prize. *PLoS Computational Biology* 11 (4): e1004084.

Rubbia, C. cited by Gatchell, M. (2015). Mentoring: the perspective of Nobel laureates. *Naturejobs | Naturejobs Blog* (28 July).

Schunk, D.H. and Mullen, C.A. (2013). Toward a conceptual model of mentoring research: integration with self-regulated learning. *Educational Psychology Review* 25 (3): 361–389.

Stephan, P.E. (2012). *How Economics Shapes Science*. Cambridge, MA: Harvard University Press.

Stubbs, M. and Gibbons, G. (2000). Hans Adolf Krebs (1900–1981). His life and times. *IUBMB Life* 50 (3): 163–166.

Turing, A.M. (1937). On computable numbers, with an application to the Entscheidungsproblem. *Proceedings of the London Mathematical Society* 2 (1): 230–265.

Turing, A.M. (1950). Computing machinery and intelligence. *Mind* 59 (236): 433–460.

Urban, N.N. and Castro, J.B. (2010). Functional polarity in neurons: what can we learn from studying an exception? *Current Opinion in Neurobiology* 20 (5): 538–542.

Watson, J. (2005). How we discovered DNA. https://www.youtube.com/watch?v=2HgL5OFip-0 (accessed 27 December 2017).

Whittaker, J.A., Montgomery, B.L., and Acosta, V.G.M. (2015). Retention of underrepresented minority faculty: strategic initiatives for institutional value proposition based on perspectives from a range of academic institutions. *Journal of Undergraduate Neuroscience Education* 13 (3): A136.

Y Cajal, S.R. (1989). *Recollections of My Life*, vol. 8. Cambridge, MA: MIT Press.

11

Cracking Public Health Needs

Epidemiology provided hints for a successful search.
Harald zur Hausen (2008)[1]

Successful research requires an all-embracing collection of information from every imaginable, trustworthy source, as it has repeatedly been pointed out in this book. Epidemiology, the study of distribution and determinants of health conditions and diseases, is one of the most practical information sources of research in life sciences. In developing models, it is essential to get all information that is available to derive generalizable principles that fit the available facts.

The discipline of *epidemiology* started with the study of infectious diseases, but today the concepts and methodologies are much more widely applied to an almost infinite variety of health conditions. Descriptive epidemiology identifies the occurrence of diseases in terms of person, place, and time (i.e. who the affected people are in terms of their age, sex, occupation; where they were when got the disease; when people get the disease in terms of season or occurrence over time). The classic concepts of prevalence and incidence remain valuable tools, but the modern analyses are highly sophisticated and can be very revealing with the help of the latest methodologies.

Attempts to understand and modify the spread of diseases are inevitably applied scholarly endeavor. Epidemiology is the ultimate applied science as it is about what people have, what people need, and so much more. However, a significant number of biomedical research discoveries illustrate that epidemiology data can not only trigger pure basic science studies but also actually contribute to the progress of curiosity-driven research.

In reviewing epidemiologic data, the simpleminded approach is that let us look at the most frequent diseases and choose one of them for further research.

---

1 zur Hausen, H. (2008). The search for infectious causes of human cancers: where and why. Nobel Lecture (7 December).

Innovative Research in Life Sciences: Pathways to Scientific Impact, Public Health Improvement, and Economic Progress, First Edition. E. Andrew Balas.
© 2019 John Wiley & Sons, Inc. Published 2019 by John Wiley & Sons, Inc.

Nothing can be further from the truth. Such trivialization would miss tremendous basic science research opportunities that come from thoughtful analysis of population data.

Looking only at the frequency of certain diseases or the magnitude of societal problems misses the point that productive research also needs scientific opportunities. There are many examples of major diseases that did not have scientific opportunities to address them meaningfully for a long time (e.g. not smoking-related lung cancer offered minimal opportunities for progress for decades in the past).

The need for simultaneous consideration and balancing of the *societal burden* and *scientific opportunity* was eloquently explained in the report of the Institute of Medicine on the priority setting and public input processes of the National Institutes of Health (NIH 1998).

Based on epidemiologic studies, the broad variety of public health measures includes surveillance vector control, public education' appropriately targeted sanitation measures, improving nutritional status, and efficient delivery of health services for those in need, among others. Each of these components needs a good understanding of the problems, well-targeted applied research, and carefully designed response.

Meanwhile, the relationship between epidemiology and curiosity-driven basic research is also getting much stronger. Descriptive epidemiology offers a wealth of population-level data, which include vital statistics, aggregate data from registries, reportable conditions, health, surveys, and many other sources. Innovative researchers need the skills to recognize and access these valuable data and resources.

Epidemiology Triggers Basic Research

Epidemiology of diseases and conditions can raise the first suspicion of a causal relationship. Many emerging and reemerging infectious diseases illustrate how observations of classic epidemiology can trigger concentration of research efforts to diagnose, treat, and prevent major diseases. Identifying the causative agent, development of useful diagnostic tests and therapies are just parts of the solution.

The prompting role of epidemiology is illustrated by the classic studies of Ronald Ross who was first to identify the malaria parasite at a particular stage of life in the stomach of the mosquito (Ross 1902). The root of his search was the connection between malarial fever and stagnant water as recognized from epidemiologic observations (e.g. frequency of infection in warm moist climates, in the lower stories of the house). Ultimately, the discovery of the malarial parasite in the mosquito not only identified the vector but also laid the foundation for combating the disease with public health measures.

The HIV epidemic first recognized in 1981 illustrates the model of concentrated worldwide research effort and its impact (Folkers and Fauci 2001). Between 1981 and 2000, the NIH funding for HIV/AIDS research exceeded $20 billion. Other government and foundation sources plus industry funding were also significant.

In the research leading to the discovery of human immunodeficiency virus, Françoise Barré-Sinoussi extensively relied on epidemiologic information. She worked in the Institut Pasteur and received lymph node samples from one of the first AIDS patients in France. They halted all other research to meet the urgent challenge of the emerging epidemic. When the virus was isolated, a serological test was developed to perform seroepidemiological studies and demonstrate that the virus was actually responsible for the disease affecting AIDS patients (Barré-Sinoussi 2008).

Within three years after the detected outbreak, the responsible virus was identified, and sensitive diagnostic tests were developed. The blood supplies were screened for HIV for prevention. The relevant basic research identified the unique genetic organization and particular replication mechanisms of the human immunodeficiency virus.

At the 1995 height of the HIV epidemic, there were 50 610 deaths of persons with acquired immune deficiency syndrome in the United States. In 2014, the number of deaths was 6721, as a result of the intense, successful research and development efforts. In 1996 a 20-year-old person in the United States with AIDS had a life expectancy of three to five years and now expects to live to be 69 years.

Understandably, the applied nature of HIV research implied a focus on prevention and management of infection. However, there has always been an active basic research component with the advancement of studies on cellular factors of HIV penetration, viral reservoirs, immune response, and mechanisms of viral persistence (Stevenson 2012).

Confronting this infection became issues of domestic health and national security. Particularly in the beginning, there were many inaccurate public misconceptions about the spread of HIV, like handshakes and toilet seats. The need for facts and sound research has rarely been greater in the history of epidemics. On the other hand, there are not many examples of the comparable success of political advocacy in promoting the progress of science and public health.

Years later, stories spread about "Patient Zero" – a young flight attendant who presumably brought the infection to the United States for the first time. The more recent study by Worobey et al. (2016) serologically screened more than 2000 serum samples from the 1970s and developed an extremely sensitive methodology for recovering viral RNA from archival samples. It found that HIV started spreading a decade earlier and initially on the East Coast, contrary to the assumptions about the West Coast.

Rampant drug-resistant malaria of North Vietnamese soldiers during the war and similarly resistant malaria outbreaks in southern China triggered a massive search for effective treatment. Chinese President Mao Zedong contacted 50 laboratories and over 500 people to search for new and more effective treatment.

In the China Academy of Traditional Chinese Medicine, **Youyou Tu** was assigned to the secret drug discovery unit Project 523, named after its starting date of 23 May 1967. First, she went to the southern island of Hainan where she studied patients impacted by the disease.

Subsequently, Tu interviewed practitioners of traditional medicine nation-wide and also searched ancient Chinese traditional medicine texts for anti-fever and anti-malaria drugs. Over 380 herbal extracts were tested on mice.

In a book of traditional Chinese medicine written more than 1600 years ago, she found the plant sweet wormwood that was used for intermittent fevers, a hallmark of malaria. It was identified as an effective treatment for malaria. After several trials and refinements in methodology, she was also able to extract the essential ingredient, artemisinin.

Subsequent work led to the identification of the chemical structure and further enhancement in potency. Youyou Tu's discovery has saved millions of lives and earned the first Nobel Prize in Medicine or Physiology to a Chinese scientist in 2015.

There are many other examples of epidemiologic studies prompting basic research. The most influential environmental triggers of lung cancer have been first recognized in epidemiologic studies. Exquisite basic research followed these leads and made valuable discoveries about the mechanism of disease.

In several countries, case–control studies compared *H. pylori* seroprevalence status in gastric cancer patients with controls and showed somewhat inconsistent but overall positive association after matching on age and sex (IARC 1994). Today, the relationship is much clearer, especially in the case of gastric lymphoma, but basic science is still trying to figure out the exact nature of these carcinogenic interactions with mucosa.

The development of molecular and genetic epidemiology further elevated the beneficial partnership between basic and population sciences. This progress has been occasionally competitive and uneasy, but the understanding of human health and illnesses benefited tremendously. An example of such studies is the systematic synthesis and evidence grading of genetic association studies in schizophrenia (Allen et al. 2008). The researchers identified four significant genetic associations with "strong" epidemiological credibility.

Scientific Assumptions Meet the Brutal Truth of Epidemics

Even in cases when triggering is the primary role, epidemiologic information often adds further value in guiding research. One of the original goals of research and development of HIV vaccine has not yet been achieved, but the effectiveness of prevention and combination treatment brought about a dramatic decline in threat levels. The AIDS-related research was not only successful in addressing the targeted epidemic but also benefited many other infectious disease areas and the human genome project.

The discovery of the harmful effect of smoking illustrates how epidemiologic studies can lead to a multitude of basic science investigations. Initially recognized in the middle of the twentieth century, the relationship between smoking and lung cancer has been discovered and rediscovered repeatedly.

Patient chart reviews by German physician Franz Müller first identified a link between smoking and lung cancer in 1939. His dissertation represents the first controlled epidemiologic study of smoking and lung cancer (Proctor and Proctor 2000).

In the United Kingdom, Richard Doll and A. Bradford Hill (1950) rediscovered the ignored or forgotten link again through chart reviews. Both studies started with the (incorrect) hypothesis that asphalt fumes cause lung cancer. The landmark epidemiologic discovery was the basis of wide-ranging further studies and more importantly public health measures of tobacco control. Today, smoking is still recognized as the undisputed, most avoidable cause of all cancers in both men and women (Parkin 2011).

Considering the multitude of harms caused by smoking, it is not surprising that more and more basic science studies attempt to identify the exact relationship between this environmental exposure and changes in essential cell functions. Further research will likely uncover the full impact of *environmental exposure* on the somatic mutations and associated diseases.

As we age, the human body accumulates random somatic mutations in normal human tissues. Using massively parallel sequencing, a study of these somatic mutations pointed out that they spontaneously arise from endogenous and exogenous sources (e.g. DNA replication errors, environmental insults like smoking or sunlight) (Hoang et al. 2016). Mutation prevalence and range vary depending on carcinogen exposure, tissue type, DNA repair, and others.

Methods of laboratory research often have major limitations that cannot be surpassed without the help of epidemiology. For example, the Ames et al. (1973) test is the widely used screening method for potential carcinogens and other mutagens. It is a simple and sensitive test that uses rat or human liver homogenate for carcinogen activation (representing mammalian metabolism) and *Salmonella* histidine mutants for mutagen detection. Carcinogenic effect of many compounds becomes fully identified only in human epidemiologic studies (e.g. chlorinated solvent trichloroethylene in the environment).

Fibroblast growth factor 23 (FGF-23) is a hormone produced by osteocytes to target the kidney and regulate phosphate homeostasis. It plays a central role in maintaining normal serum phosphate levels in spite of variations in dietary intake. Through various endocrine signaling pathways, fibroblast growth factors (FGFs) are a group of growth factors involved in wound healing, angiogenesis, and embryonic development.

Among patients with chronic kidney disease, hyperphosphatemia and low 1,25-dihydroxyvitamin D levels are well known to be associated with increased mortality. The controlled clinical study led by Gutiérrez et al. (2008) demonstrated that higher FGF-23 levels are independently and additionally associated with higher mortality among patients starting hemodialysis renal replacement therapy. As the accompanying editorial correctly pointed out, this study is a classic case of physiology meeting epidemiology (Hsu 2008). Beyond the direct and predictive benefits of FGF clinical studies, basic scientists interested in this endocrine system would also take advantage of the many valuable observations of epidemiologic studies (Beenken and Mohammadi 2009).

One of the more distant impacts of epidemiology is the use of such data in theoretical research on social progress and economic development. To illustrate the far-reaching impact, the Nobel lecture of economist Robert William Fogel can be mentioned. It highlighted the relationships among economic growth, population theory, physiology, and the bearing of long-term processes on the development of economic policy.

Fogel's analysis of secular declines in morbidity and mortality examined the changes induced in physiological functioning since 1700. Between 1929 and 1939, there was a life expectancy increase by four years and a height increase of adults by 4 cm (Fogel 1993). As a lesson for the future, he pointed out that massive social investments made between 1870 and 1930 generated payoffs during years of the Great Depression.

Curiosity-driven Research Enlists Epidemiology

The discovery of hepatitis B virus illustrates how integrating epidemiologic, clinical, and serological observations can lead to breakthrough discoveries when the methods of virology alone could not produce results. In the studies of Baruch Blumberg, the original focus was that "patients who received large numbers of transfusions might develop antibodies against one or more of the polymorphic serum proteins (either known or unknown) which they had not inherited, but which the blood donors had" (Blumberg 1976).

Testing of transfusion-related antibody hypothesis produced the discovery of the Australia antigen (Au), originally from the blood of an Australian aborigine. Initially, Au (+) and Au (−) test results appeared to be characteristic to the individual. In 1966, the breakthrough came when a patient who was Au negative suddenly converted to Au positive and shortly afterward diagnosed with hepatitis. Further studies on the Au antigen confirmed the association of Au with hepatitis.

Particularly significant epidemiologic information was that donor blood that had Australia antigen was more likely to transmit hepatitis. Blood transfusion practices had to change immediately.

Florence Nightingale is widely recognized for being the founder of modern nursing, but her landmark biostatistics and public health achievements are equally remarkable. Contrary to the prevailing views of their times, Nightingale's father believed that women should be educated, and she had a chance to learn not only philosophy, Italian, Latin, Greek, and history but also mathematics.

During the Crimean War, she cared for gravely injured soldiers when the death rates were astoundingly high. A contemporary report in *The Times* newspaper described her as "the lady with the lamp" making rounds among the sick and trying to help them. The experience led her to the broader understanding that nonmedical interventions like improved nutrition, supplies, and working conditions are also needed to improve outcomes. When she came back from the Crimean War, her pioneering epidemiologic analyses and public health advocacy thrived.

After returning to Britain, she began collecting and reporting evidence to the Royal Commission on the Health of the Army. Her advocacy for improved sanitary living conditions led to significantly reduced peacetime deaths in the army. After 10 years of wide-ranging sanitary reform, mortality among the British soldiers in India had declined from 6.9 to 1.8%. As a particular focus on enforcement of hygienic requirements, her successful activism led to major improvements in the Public Health Acts of 1874 and 1875. The resulting improvements in drainage and decentralized enforcement were largely credited for increasing the average national life expectancy by 20 years between 1871 and the mid-1930s.

Nightingale was a passionate gatherer of facts and reporter of policy analyses (Kopf 1916). Among others, she compiled vast tables of mortality statistics and discovered that soldiers in peacetime died at twice the rate of civilians, a shocking discrepancy. Her classic "Diagram of the Causes of Mortality in the Army in the East" is an illustration of mastery in using statistical graphics. She pioneered the public health use of pie chart, polar area diagrams, and color coding.

Her groundbreaking public health reports and advocacy letters were very effective in changing policies and serving public health. Undoubtedly, she saved most lives with her statistics. Today, many people believe that it would be more fitting to remember her as the "lady with the data."

In many cases, initial suspicion based on incidental observations can be significantly reinforced by epidemiologic information, leading to further, much more focused studies in the laboratory. In other words, experiments of the basic scientist can become better targeted and more concentrated. Blumberg's research illustrated the intertwining of basic biomedical research in immunology and applied research methods of epidemiology.

In his lecture, Harald zur Hausen pointed out the tremendous contributions of epidemiology to the discovery of the viral origins of many cancers. Among the strongly *suggestive epidemiologic factors* were geographic coincidence, regional clustering of cases, dependence on sexual contacts, and observations of cancers arising under immunosuppression (zur Hausen 2008). They also showed that HPV16 and HPV18 were high-risk viruses and major risk factors for the development of cervical cancer.

Furthering these studies, zur Hausen also published a comprehensive book on the subject of infections causing human cancer (zur Hausen 2006). About 20% of cancers can be linked to infectious agents, including viruses, bacteria, and parasites. Epidemiologic research has been invaluable in exposing these and other linkages.

Sometimes laboratory findings are surprising and deviate from common perceptions. In such cases, concurrent epidemiologic analyses can show the public health feasibility of the contentious experimental evidence. In a somewhat indirect way, epidemiologic evidence can corroborate the experimental findings.

Working with Robin Warren on the role of *Helicobacter pylori* infection in subsequent peptic ulcer disease, Barry Marshall synthesized epidemiologic information from a wide variety of sources: "I then re-read papers I had discovered which described epidemics of gastritis in laboratory volunteers. There too achlorhydria had been observed. Suddenly the whole process became clear" (Marshall 2005). Corroborating their experimental findings, he demonstrated that varying patterns of infection could explain variations in ulcer incidence around the world.

Not infrequently, basic and theoretical scientists effectively use epidemiology to highlight societal responsibilities and opportunities for public health action. Occasionally, the synergistic message of laboratory experiments and public health observations can send a particularly powerful message.

On the road to his discoveries concerning the interaction between tumor viruses and the genetic material of the cell, Renato Dulbecco pointed out that epidemiologic evidence can raise suspicion of causal relationships but often not compelling enough to convince the general public and the government about the need to take important public health measures (e.g. expansion of tobacco control). Given the high correlation between mutagenicity and carcinogenicity, there is a case for testing pro-mutagens in the environment as a public health routine (Dulbecco 1975).

Some of the most fundamental basic science discoveries, like the CRISPR/Cas9 system, have a surprisingly immediate interaction with epidemiologic considerations. With this new technology, gene editing is becoming a therapeutic promise for diseases like sickle cell trait or cystic fibrosis.

In a larger epidemiologic context, we have to recognize that carrying one copy of the sickle cell gene was a historic advantage in preventing the worse symptoms of malaria. Similarly, carrying one gene for cystic fibrosis appears to be protective against tuberculosis, causing the high cystic fibrosis prevalence among peoples of European ancestry (Mowat 2017). In other cases, the complex interplay between genes and environment is even more influential. For example, in the border area between Russian Karelia and Finland, autoimmune disorders like celiac disease and type 1 diabetes are six times more common on the Finnish side than in Russia, while the genetic inheritance is very similar (Kondrashova et al. 2013).

Therefore, one may have to be careful in editing genes arbitrarily. What appears to be a disadvantage today may become an advantage in surviving the next epidemic of an emerging infectious disease. Recently, an expert group of the National Academy of Sciences and the National Academy of Medicine concluded that genome editing for research purposes and clinical trials of genome editing in treatment or prevention of somatic cell diseases should continue within the existing ethics and regulatory framework. However, genome editing for enhancement of physical traits and capacities and particularly germline editing that creates inheritable changes should proceed only with great caution and only in cases meeting strict criteria (National Academies of Sciences 2017).

The Unparalleled Value of Retrospective Studies

Real-life data can tremendously benefit use-inspired researchers and increase understanding of biological systems and functions. As a result, epidemiologists can be excellent translational science partners in curiosity-driven research. The books by Buck et al. (1998) and Holland et al. (2007) offer far-reaching information about the development and methodologies of epidemiology.

Notably, electronic health records can offer a lot of information about disease processes. The electronic health record is also the accurate reflection of actual processes of care and its documentary basis for later clinical assessments. In other words, big data from electronic health records reflect the reality of care documentation in daily practice. Consequently, identification of most efficient or best practices can use such records time and again.

On the other hand, researchers using *retrospective health records* should be cautious as they have little control over the quality of what already has been collected. Electronic health records often do not provide complete information about disease characteristics and processes. For example, a study in the field of ophthalmology documented significant discrepancies between patient-reported symptoms and information recorded in electronic health records (Valikodath et al. 2017). Overall, information in the records proved to be narrower, and many important details were missing.

Retrospective analyses, epidemiologic studies, and research on electronic health data not only have enormous scientific potential but also have many vocal critics. The vast amount of electronic health data, often called big data, should not be viewed as fully accurate source of information about biology and pathophysiology. The data are not clean, there are many errors of omissions and commissions, and they have never been collected with the rigor expected in prospectively planned scientific experiments.

At the same time, we also need to recognize several foremost benefits of real world data and epidemiology. There are at least seven principal reasons for using retrospective research, including epidemiologic studies, for the purposes of advancing fundamental understanding:

1) In general terms, if you are interested in understanding reality, you have to study it. There is no experiment that can fully replicate what is usually going on in real life.
2) In human studies, harmful health risks cannot be studied prospectively, only retrospectively. For obvious reasons, harms like smoking cannot be tested in prospective clinical trials, and, therefore, retrospective epidemiologic studies are the only viable option for study.
3) Exploration of potential health risks needs scanning a large number of conditions and factors as opposed to narrowly or specifically targeted investigations.
4) Similarly, life expectancy can only be calculated through large population data. Such analyses are very much needed to answer the first question of every cancer patient immediately after learning their diagnosis.
5) Health care is known for enormous practice variations, and correspondingly outcomes are also highly variable. Identifying best practices based on superior results remain highly effective in support of meaningful quality improvement in health care.
6) One of the generators of non-reproducible research is reporting based on isolated and singular findings. Good access to archived data is an excellent opportunity to strengthen experimental results with already recorded real-life data.
7) Wide-ranging cause–effect and nonrandom coincidence studies can play a significant role in developing an understanding of functional relationships and scientific models. Historically, epidemiological modeling of diseases effectively helped to understand the origins, risks, transmission, and mechanism of many infectious diseases.

Bill Foege is an accomplished public health scholar and former director of the Centers for Disease Control (CDC). Born in Iowa, Dr. Foege attended medical school at the University of Washington where his lifelong commitment to public health started.

Among others, he was the author of the trailblazing study on actual causes of death in the United States in 1990 (McGinnis and Foege 1993). According to these estimates, the leading external, non-genetic factors that contributed to death included tobacco, diet and activity, alcohol, microbial agents, toxic agents, firearms, sexual behavior, motor vehicles, and illicit drug use. The listed numbers of deaths can be considered avoidable and represent targets for prevention by public health measures.

One of his creative initiatives was using a surveillance/containment strategy in the fight against smallpox epidemics. He came up with the idea in Nigeria where vaccines were in short supply. Based on informal surveillance of the outbreaks, Foege focused on inoculating market villages and places where relatives of victims were living (i.e. containment of the disease). As a result, the disease was stopped with less than 50% rate of inoculation, while traditional approaches would have required 80–100% inoculation rate.

His excellent book titled *House on Fire: The Fight to Eradicate Smallpox* (Foege 2011) makes the point that "We were an optimistic group... I tell students there is a place for cynicism and a place for pessimism and whenever you need it, contract for it but don't get those people on your payroll. They will ruin your day."

Bill Foege's work illustrates the tremendous practical benefits of applied epidemiology. His wide-ranging scholarly and leadership contributions to public health were recognized among others by the Mary Woodard Lasker Award for Public Service in 2001 and the Presidential Medal of Freedom in 2012.

Population-based and epidemiologic studies can highlight physiologic mechanisms in unexpected ways. Vitamin A has been known to play a significant role in vision, gene transcription, hematopoiesis, and many other physiologic functions. As mentioned, the pioneering clinical trial by Alfred Sommer documented that children with vitamin A deficiencies are also at increased risk of preventable death, especially in developing countries (Sommer 2013; Sommer et al. 1986).

Interestingly, the history of scientific awards reveals that scientists who produced landmark epidemiologic studies have never been recognized with the Nobel Prize. There are various explanations for this shortage of recognition. One of them is that scientific disciplines with little opportunity for controlled experimentation have a hard time to meet the criterion for groundbreaking discovery (Adami 2009). It also takes more than one confirmatory study to document a cause–effect relationship, and therefore attribution to no more than three people can be challenging. Finally, there is a prevalent lack of understanding and appreciation of epidemiology among biomedical researchers. People often point to the tendency of epidemiology to detect only associations but not causality.

Epidemiology used to be the study of large numbers, while biochemistry traditionally relied on small numbers of hypothesis-driven experiments. With high-throughput technologies, the dynamics of research is rapidly changing. Today, it is possible to conduct large-scale, hypothesis-free study of biochemical changes and their prevalence in sprawling populations. The resulting data on the population distribution of biochemical and physiologic characteristics opens up previously unimaginable opportunities to study major diseases, their predictors, and susceptibility to treatment.

All cancers are caused by somatic mutations. Investigation of the underlying triggers and also potential infectious origins can benefit from epidemiology. Alexandrov et al. (2013) analyzed 4 938 362 mutations from 7042 cancers and extracted more than 20 distinct mutational signatures. These signatures appear to be associated with age of the patient, specific mutagenic exposures, and defects in DNA maintenance, but many have an unknown origin. There is an apparent diversity of mutational processes underlying the development of cancer. Further studies need to clarify the relationship between cancer epidemiology and the distribution of mutational changes.

The study of Ganna et al. (2014) performed a mass spectrometry-based non-targeted metabolomics study to explore possible associations with coronary heart disease events in two groups of 1028 and 1670 individuals, respectively. The results identified four lipid-related metabolites with evidence for a causal role in coronary heart disease development and clinical utility as well. In large prospective epidemiological studies, analyses of circulating metabolites should lead to better understanding and improved treatment of coronary heart and many other diseases.

In summary, everybody knows the most frequent diseases. However, only the best researchers know how to use the fine details of community health information and epidemiologic data to decipher fundamental mechanisms in biology, understand human needs, and develop innovative models. By seeking and using epidemiologic information, the collaborative, astute scientist gets a chance to benefit from a much broader range of information and accelerate research discoveries.

References

Adami, H.O. (2009). Epidemiology and the elusive Nobel Prize. *Epidemiology* 20 (5): 635–637.

Alexandrov, L.B., Nik-Zainal, S., Wedge, D.C. et al. (2013). Signatures of mutational processes in human cancer. *Nature* 500 (7463): 415–421.

Allen, N.C., Bagade, S., McQueen, M.B. et al. (2008). Systematic meta-analyses and field synopsis of genetic association studies in schizophrenia: the SzGene database. *Nature Genetics* 40 (7): 827–834.

Ames, B.N., Durston, W.E., Yamasaki, E., and Lee, F.D. (1973). Carcinogens are mutagens: a simple test system combining liver homogenates for activation and bacteria for detection. *Proceedings of the National Academy of Sciences* 70 (8): 2281–2285.

Barré-Sinoussi, F. (2008). HIV: a discovery opening the road to novel scientific knowledge and global health improvement. *Nobel Lecture* (7 December).

Beenken, A. and Mohammadi, M. (2009). The FGF family: biology, pathophysiology, and therapy. *Nature Reviews Drug Discovery* 8 (3): 235–253.

Blumberg, B.S. (1976). Australia antigen and the biology of hepatitis B. *Nobel Lecture* (13 December).

Buck, C., Llopis, A., Nájera, E., and Terris, M. (1998). *The Challenge of Epidemiology: Issues and Selected Readings*. PAHO Scientific Publication No. 505. Washington, DC: Pan American Health Organization.

Doll, R. and Hill, A.B. (1950). Smoking and carcinoma of the lung. *British Medical Journal* 2 (4682): 739.

Dulbecco, R. (1975). From the molecular biology of oncogenic DNA viruses to cancer. *Nobel Lecture* (12 December).

Foege, W.H. (2011). *House on Fire: The Fight to Eradicate Smallpox*, vol. 21. Los Angeles: University of California Press.

Fogel, R.W. (1993). Economic growth, population theory, and physiology: the bearing of long-term processes on the making of economic policy. *Nobel Lecture* (9 December).

Folkers, G.K. and Fauci, A.S. (2001). The AIDS research model: implications for other infectious diseases of global health importance. *JAMA: the Journal of the American Medical Association* 286 (4): 458–461.

Ganna, A., Salihovic, S., Sundström, J. et al. (2014). Large-scale metabolomic profiling identifies novel biomarkers for incident coronary heart disease. *PLoS Genetics* 10 (12): e1004801.

Gutiérrez, O.M., Mannstadt, M., Isakova, T. et al. (2008). Fibroblast growth factor 23 and mortality among patients undergoing hemodialysis. *New England Journal of Medicine* 359 (6): 584–592.

Hoang, M.L., Kinde, I., Tomasetti, C. et al. (2016). Genome-wide quantification of rare somatic mutations in normal human tissues using massively parallel sequencing. *Proceedings of the National Academy of Sciences* 113 (35): 9846–9851.

Holland, W.W., Olsen, J., and Florey, C.D.V. (eds.) (2007). *The Development of Modern Epidemiology: Personal Reports from Those Who Were There*. Oxford: Oxford University Press.

Hsu, C.Y. (2008). FGF-23 and outcomes research – when physiology meets epidemiology. *New England Journal of Medicine* 359 (6): 640–641.

IARC (International Agency for Research on Cancer) (1994). *Schistosomes, Liver Flukes and Helicobacter pylori*, IARC Monographs on the Evaluation of Carcinogenic Risks to Humans, vol. 61. Lyon: IARC Press.

Institute of Medicine Committee on the NIH Research Priority-Setting Process (1998). *Scientific Opportunities and Public Needs: Improving Priority Setting and Public Input at the National Institutes of Health*. Washington, DC: National Academies Press.

Kondrashova, A., Seiskari, T., Ilonen, J. et al. (2013). The "hygiene hypothesis" and the sharp gradient in the incidence of autoimmune and allergic diseases between Russian Karelia and Finland. *APMIS* 121 (6): 478–493.

Kopf, E.W. (1916). Florence Nightingale as a statistician. *Quarterly Publications of the American Statistical Association* 15 (116): 388–404.

Marshall, B.J. (2005). Helicobacter connections. *Nobel Lecture* (8 December).

McGinnis, J.M. and Foege, W.H. (1993). Actual causes of death in the United States. *JAMA: the Journal of the American Medical Association* 270 (18): 2207–2212.

Mowat, A. (2017). Why does cystic fibrosis display the prevalence and distribution observed in human populations?. *Current Pediatric Research* (30 January).

National Academies of Sciences, Engineering, and Medicine (2017). *Human Genome Editing: Science, Ethics, and Governance*. Washington, DC: National Academies Press.

Parkin, D.M. (2011). The fraction of cancer attributable to lifestyle and environmental factors in the UK in 2010. *British Journal of Cancer* 105: S2–S5.

Proctor, R.N. and Proctor, R. (2000). *The Nazi War on Cancer*. Princeton: Princeton University Press.

Ross, R. (1902). Researches on malaria. *Nobel Lecture* (12 December).

Sommer, A. (2013). *Ten Lessons in Public Health: Inspiration for Tomorrow's Leaders*. Baltimore, MD: John Hopkins University Press.

Sommer, A., Djunaedi, E., Loeden, A.A. et al. (1986). Impact of vitamin A supplementation on childhood mortality: a randomized controlled community trial. *The Lancet* 327 (8491): 1169–1173.

Stevenson, M. (2012). Review of basic science advances in HIV. *Topics in Antiviral Medicine* 20 (2): 26–29.

Valikodath, N.G., Newman-Casey, P.A., Lee, P.P. et al. (2017). Agreement of ocular symptom reporting between patient-reported outcomes and medical records. *JAMA Ophthalmology* 135 (3): 225–231. doi: 10.1001/jamaophthalmol.2016.5551.

Worobey, M., Watts, T.D., McKay, R.A. et al. (2016). The 1970s and "patient 0" HIV-1 genomes illuminate early HIV/AIDS history in North America. *Nature* 539 (7627): 98–101.

zur Hausen, H. (2006). *Infections Causing Human Cancer. 1*. Weinheim: Wiley-VCH.

zur Hausen, H. (2008). The search for infectious causes of human cancers: where and why. *Nobel Lecture* (7 December).

12

Engaged Research

Science and everyday life cannot and should not be separated. Science, for me, gives a partial explanation of life.

Rosalind Franklin (1940)[1]

Researchers always have been taught rigor and accuracy. They must rely on quality scientific literature and accurate methodologies. Painstaking cleaning of scientific data, careful validation of measurements, and preference for controlled experimental data are of particular importance. According to these principles, only the most precise scientific sources and data deserve the attention of the research community. In scientific communications, any hints of considering less than scientific information sources is a vulnerability and invites criticism by peers.

In recent years, second thoughts are surfacing regarding the completeness of the purist approach to scientific investigation. Based on reading a wide range of the scientific literature and discussing research projects with colleagues at scientific conferences, significant effort goes into research that is precise, but has little connection with actual life or the potential for real scientific or practical impact. According to a review of research publications in clinical epilepsy, 214 out of 300 papers (71%) were categorized as having no enduring value because they were inherently unimportant (55.6%), were not novel (38.8%), or had significant methodological flaws (22.0%) (Gregoris and Shorvon 2013).

In response to these challenges, the engaged researcher should explore every possible source of information that might shed light on how nature or the particular principle in focus works. Many great scientific discoveries illustrate the power of integrating information from diverse, unconventional sources and gaining a broader understanding of how nature works.

---

1 Franklin, R. (1940). Letter to her father, Ellis Franklin (undated, summer 1940 while she was an undergraduate at Cambridge). Excerpted in Brenda Maddox, *The Dark Lady of DNA* (2002), 60–61.

Innovative Research in Life Sciences: Pathways to Scientific Impact, Public Health Improvement, and Economic Progress, First Edition. E. Andrew Balas.
© 2019 John Wiley & Sons, Inc. Published 2019 by John Wiley & Sons, Inc.

Science Confronts the Bubble Bias

One of the dangers of scientific work is being overly immersed in the familiar and close professional groups. Such approach misses the opportunity of gathering information from the widest possible range of sources to stimulate creative ideas and answer important scientific questions. There are significant limitations of talking only to the same scientific colleagues and reading only the peer-reviewed scientific papers of a highly specialized profession.

At the same time, it should be acknowledged that meticulous activity of numerous specialized researchers studying what might appear to be mundane issues but when interpreted in light of all other studies/results has led to remarkable discoveries and treatment paradigms that we all enjoy today. We all want to cure every disease, but unfortunately that is not how science works. Sometimes serendipity prevails, but without a focused study in the first place, there is no chance for serendipity.

Bubble bias is a particular danger of the Internet age when researchers can direct their entire information seeking toward sources that they like or prefer. Relying only on a limited set of sources can become inhibiting factor and lose sight of the humanistic mission. Sometimes, researchers prefer news only from people with similar views. Such communities of mutually supportive views are like bubbles or echo chambers with a propensity to validate member views. Overly committed participation in bubbles of like-minded people can become a dangerous source of bias.

Not infrequently, the status quo becomes a bubble. Insiders react to outside information, contrarian views, or simply changes by explaining that it cannot work, it should not work, and so it does not work. Harald zur Hausen mentioned the Florida conference on cervical cancer research where everybody was preoccupied with the presumed role of herpes virus and suggestions of the role of papillomavirus were unwelcome (zur Hausen 2014). One may wonder how much distraction and waste were generated by the herpes virus bubble.

Research conducted in a bubble or disengaged research is also at risk of becoming misguided by the observations, perceptions, or illusions of a single person. Perhaps in some cases, time, dedication, and passion may make a difference, but often the chance of success is diminished, and examples of productivity are rare.

Appropriately, the peer-review process should and does provide strong criticism of work, often requiring additional studies and better analysis. Such feedback can improve the work and also help to prevent only one opinion or bias from dominating the discussion.

Not just applied researchers but also basic scientists should establish collaborations and access information outside traditional means. Expansion of horizons and development of more innovative and translational research ventures promise advancing science instead of moving it laterally.

Resources and Indispensable Benefits of Engaged Research

In trying to understand nature or societal phenomena, researchers should never limit themselves to input from the peer-reviewed scientific literature. Every source of information should be considered to gain a full understanding of natural phenomena and their use in solving problems of people and societies (Figure 12.1). In many ways, engaged research goes beyond traditional scientific communications with multiple sources of information and wide-ranging networks of interactions.

The work of innumerable leading scientists illustrates the power of engaged research. The suitable resources of engaged research in three particular benefit categories are identified in Table 12.1.

A deeper understanding can be best served by checking and using a much wider range of information sources, far beyond just reliance on scientific publications and conferences. Some of the tools are relevant data sources, while others are opportunities to check the validity of assumptions. The efficient use of earlier published case studies was illustrated when the causative role of papillomavirus in cervical cancer was correctly hypothesized.

Figure 12.1 Network of partners in engaged life sciences research.

Table 12.1 Resources of engaged research beyond scientific circles.

Deepening understanding	Pointers to potential information	Learning about human experience
National statistical databases and registries	Science section of news media	Answering questions of laypersons
Patent and intellectual property databases	Internet searches, Google, Wikipedia	Meeting patients with relevant diseases
Law and regulation, reimbursement rules	News about emerging technologies, corporate strategies	Conversations with clinicians
Published case studies, factual databases	History of current understanding	Patient advocacy and community organizations
Industry practices	Studying the work of competing laboratories	Solving practical problems for immediate use
Debating partnership	Conversations with people from other professions	Literary and visual art descriptions

Solving practical problems always has been the beneficial side effect of good science. At the end of the nineteenth century, the spectacular Adriatic resort island of Brioni was virtually depopulated due to malaria. To identify and eliminate sources of this infection, Robert Koch, already famous for his studies in bacteriology, was invited to the island (Mülder 2001). He successfully treated the infected population and eliminated the mosquito breeding areas on the island. Within one year, a dramatic drop occurred in the number of infections to just one case.

According to the studies of Contopoulos-Ioannidis et al. (2003) on the translation of highly promising basic science research into clinical applications, less than 5% of those research results become clinically successful. However, some form of industry involvement in the study and publication is associated with improvement in chances of clinical success. With very few exceptions, scientists need to connect with nontraditional stakeholders like industry to avoid being detached from reality.

Reading patents and descriptions of other intellectual property has noteworthy practical significance for researchers. Knowing what has or has not been protected can guide to avoid useless directions and facilitate turn toward promising or uncovered areas. Some researchers go further and create their specialized collection of IP documents, patents, and other by-products of research.

Professor John D. Lambris, PhD, a prolific research innovator at Penn State, created a free open-source database for patents on the complement system. At the time of writing this book, the database has about 1200 records. The complement database includes US applications, US granted patents, WIPO documents, European documents, and Japanese documents (Yang et al. 2013).

Over the past decade, specialized research databases of the NIH National Library of Medicine became an invaluable resource for researchers. Outside the long-established bibliographic databases of the peer-reviewed literature, the newer research resources include BLAST genetic sequence databases, Chemical Carcinogenesis Research Information System (CCRIS), ClinicalTrials. gov, database of Genotype and Phenotype (dbGaP), Dietary Supplement Label Database, and many others.

Pointers to sources of information are another major benefit of engaged research. The scientific literature is vast and increasingly complex. The surging number of scientific publications makes it impossible to keep up with the latest developments even in a narrowly defined specialty area. Studies show that expert searchers can find only a fraction of highly relevant articles in reference databases of peer-reviewed literature like Google scholar or the National Library of Medicine's PubMed. As much as possible, using multiple different channels to find relevant information makes a lot of sense and increases the likelihood of new findings.

Targeted searches can elicit a reasonable percentage of relevant publications. However, many commonly used search techniques will never retrieve all important discoveries that might be relevant, but have not been considered by the researcher (Boeker et al. 2013). This discrepancy is one of the reasons why the science sections of newspapers and websites can actually serve as useful pointers to trustworthy scientific sources.

Among others, general Google searches and Wikipedia searches are not replacements but helpful additions to high-quality reference databases. Due to their enormous variability and high vulnerability to errors, general searches should almost never be used as references in scientific publications. However, they can be very helpful in providing a quick introduction to various subjects and offering pointers to high-quality scientific sources. Most pointers will not work, but some will prove to be very useful. The general rule can be summed up by enlisting "always use but never cite" sources.

Studying the work of competing laboratories and conversations with people from other professions represent other important sources of valuable pointers. Engaging these sources has the potential for broader and higher effects as discussed in the chapter on multicultural convergence.

Understanding real life is not surprisingly the quintessential precondition and ultimate test of *scientific validity*. Engaging sources that describe the other side of the story by those who experience them tends to be exceptionally valuable (Figure 12.1). Real-life sources can increase understanding, motivate research in new and unexpected directions, shape the message of science, and facilitate implementation of scientific discoveries.

In the past, engaged research has often been simplified to a partnership with community organizations. In such a limiting interpretation, community organizations are the primary partners in implementation, and the challenge is implementing what has already been discovered. Obviously, the opportunities, tools, and benefits of engaged research are much broader, and the potential impact on the process of research itself can be much greater.

Discussing scientific concepts and directions with laypeople like family, friends, or patient advocacy organizations can not only spread the message but also inform the research process by highlighting needs, misunderstandings, and expectations. Enthusiastic researchers often create concepts and vocabulary that are largely incomprehensible to outsiders. Conversations with laypeople have the power to shape and benefit the research.

Forecasting societal reaction can be particularly valuable for applied research. Scientists make discoveries about nature, but it is society's role to decide what and how it will be used. Therefore, scientists interested in benefiting society with their discoveries should be well informed regarding societal perceptions relevant to their area of research. To facilitate successful implementation of their results, scientists need to gain some understanding not only of nature but also how society reacts to their discoveries.

Scientists also are advised to stay in contact with society to avoid being disconnected, as discoveries move to potential applications. Many issues generate intense public interest, and sometimes this interest can be surprising or distorting. No clearer illustration of this challenge exists than the *in vitro* fertilization research of Nobel laureate Robert Edwards, an intensely emotional and political issue at the time of development. In 1974, James Watson attacked the idea of IVF by warning a congressional committee that "All hell will break loose, politically and morally, all over the world" (Rorvik 1974).

In Europe, the highly politicized societal debate about genetically modified food is another example of disregard for evidence on the part of the general public and many politicians. Assuming that only the genetically modified food has genes is obviously incorrect. All the foods that we eat contain billions of genes. Innumerable studies have focused on the hypothesized health risks of genetically modified food; however, no conclusive evidence has emerged regarding those presumed harms. As an EU commissioners also acknowledged, until now nobody has ever died because of eating genetically modified food.

Prolific Partnerships of Arguers

The powerful medium of research partnerships features prominently in the history of many significant scientific discoveries. Thought-provoking collaborations between two researchers frequently become uniquely potent sources of discoveries (Figure 12.1). Continuously challenging each other and criticizing each other's interpretations can build a robust foundation for productive research.

One of the most famously stimulating scientific partnerships was between Watson and Crick, co-discoverers of the double helix structure of the DNA. Francis Crick started out as a physicist. James Watson was a microbiologist and zoologist and committed enthusiast of the role of DNA as the genetic material. The two met in the Wilkins laboratory of the University of Cambridge in 1951, what was later called an instantaneous meeting of minds. Always debating, constantly arguing, and often seeking input from others, they developed the revolutionary model of the double helix structure.

In his book, Watson noted that Francis Crick, "who many thought of as an arrogant man who talked too much, and whose brilliance was appreciated by few," had a tendency to be annoyingly candid (Watson 1968/2001). They endlessly argued with each other over their different theories and the possibility of determining a fitting model of the DNA structure. In an interview, Watson reiterated: "We were both readers. We liked facts. We wanted to find out which facts were correct… We want the answer more than proving that we were right yesterday" (Judson 2015).

Rosalyn Sussman Yalow was born in New York and got her PhD in physics at the University of Illinois. At the VA Hospital in the Bronx, the eminently productive collaboration of Rosalyn Yalow and Solomon A. Berson led to the discovery of pioneering techniques for the early detection of diseases, including radioimmunoassay. With her strong physics and chemistry background, Yalow was an ideal collaborative research partner of the internist Berson. Their joint investigations focused on insulin metabolism in patients with diabetes.

To get a better understanding, they needed accurate methods for the measurement of insulin levels. The opportunity came by attaching an atom of a radioactive isotope of iodine, iodine-125, to the insulin molecule. In a generalization of this approach, adding hormone molecules tagged with isotopes and an antibody against the hormone to the patient's blood can be used to measure very small amounts of a hormone in the specimen.

The discovery of radioimmunoassay earned Yalow the Nobel Prize in 1977, the Lasker Award, and the National Medal of Science by President Ronald Reagan. Unfortunately, Solomon A. Berson died a few years earlier and therefore could not be named as corecipient. In her autobiography for the Nobel Foundation, Rosalyn Yalow explained the powerful ingredient of their scientific partnership: "We learned from and disciplined each other and were probably each other's severest critic" (Odelberg 1978).

In 1969, Harold Varmus joined Michael Bishop's laboratory as a postdoctoral fellow at the University of California, San Francisco. Within a short period, he and the leader of the laboratory became exceptionally productive partners in research. For many years, the two researchers worked together, leading to the discovery of the cellular origin of retroviral oncogenes.

In his Nobel lecture, Harold Varmus recognized Mike Bishop as someone "who has been a generous and challenging colleague for nearly two decades" (Varmus 1993). Much of their work has been the result of this collaboration as they shared facilities, personnel, and funds. Mike Bishop later recalled their collaboration as a relationship of "coequals, and the result was surely greater than the sum of the two parts."

In many ways, these and other productive partnerships illustrate the potential of debating research partners in facilitating scientific understanding and development of innovative ideas. The desire to seek such opportunities and the skills to develop such close, productive partnerships represent productive methods of engaged research.

Talking Through to Achieve Understanding

"People learn while they teach," wrote Seneca the Younger in the Roman times (Seneca and Campbell 1969). Throughout history, an enduring alliance has grown between the pursuit of science and the practice of teaching. In recent decades, the concept of learning by teaching (German: Lernen durch Lehren, short LdL) gained further attention (Martin and Kelchner 1998).

Researchers cannot easily develop sophisticated models without intense conversation, consultation, and interactions. Nobel laureate Paul Nurse (2014) made the point: "I recommend that you talk a lot with your colleagues…it's by talking that you can make sense whether it's sensible or not… Talk about it. Don't keep it to yourself. Talk about it."

Various aspects of complex models need to be vetted with trusted advisors, laypeople, or consumers and, occasionally, with fierce critics. *Talking through* to reach an understanding is at the heart of success in partnerships of arguers. The chapter on bioentrepreneurship already highlights the significance of outreach, which is a great way to talk through as well.

Every week should be an outreach week for the researcher. Reaching people whom you have never contacted before has tremendous research value in soliciting collaborations, getting advice, offering support, and looking for opportunities. Every week, the researcher can send an email, phone, arrange a meeting, or ask to be introduced to someone new. These actions may elicit valuable input from others and have impact beyond current limits.

Conversations with other professionals and laypeople can help to identify salient points as well as develop a vocabulary that is clear and understandable to the intended users of the results. Many researchers are lost in the acronyms of their narrow field, and the scientific gobbledygook becomes disconnected from the ultimate mission of science: intellectual contributions, public health improvement, and economic development.

Particularly, when the purpose of research is to help a group of patients, interacting with the actual priority group and understanding their perceptions and needs is essential. Conversations with others also force researchers to articulate the most practical aspects of the research for implementation.

The world's most famous theorist, Albert Einstein not only had an intense focus on the complex study of mass and energy but also had 18 patented inventions and repeatedly explained his research to laypeople and nonexperts in newspaper articles. He was clearly a theorist but also an engaged researcher. The following quote, attributed to Albert Einstein, captures the extraordinary value of modeling and communication in conceptualizing scientific ideas: "everything should be as simple as it can be but not simpler" (Sessions 1950).

Implementation of research results often requires an understanding and active participation of people in the community. In such cases, the researcher reaches out to the community, develops the research with community input, and, after getting the results, works with the community on implementation. These interactions rely on skills of effective and persuasive communication.

Academic Engagement with the Community

By focusing more on the diversity of interactions, academic engagement is defined as a knowledge-related collaboration by academic researchers with nonacademic organizations (e.g. collaborative research, contract research, consulting, ad hoc advice, or networking with practitioners) (Perkmann et al. 2013). Interpretations of the concept of *community engagement* often center on desirable interactions with organizations of underserved communities.

Academic engagement tends to be pursued by well-established and well-connected scientists who are more senior and have wider social networks, more publications, and more government grants (Hicks et al. 2012; Wallerstein and Duran 2010). According to several studies, a positive correlation appears to exist between engagement and government grants and engagement and scientific productivity. Overall, academic engagement is often considered valuable similarly to the benefits of the better-known intellectual property commercialization.

In 2006, NIH launched the Clinical and Translational Science Awards (CTSA) program to facilitate the translation of basic science research discoveries into clinical and practical applications. Currently, over 60 academic health centers and other institutions participate in the CTSA program (Liverman et al. 2013). An essential component of the program is promoting beneficial collaborations with patient groups, partnerships with patients, family members, health-care providers, community organizations, private foundations, industry, government agencies, and regulatory bodies.

In the NIH Clinical Translational Science Award (CTSA) program, the most frequent interpretation of engaged research refers to community-engaged projects. Such engagement is often viewed as the key to reaching minority and underserved communities. In 2013, the Institute of Medicine of the National Academies assembled a committee to review the CTSA program (Liverman et al. 2013). The committee concluded that "community engagement is critical in all phases of clinical and translational research from basic research to clinical practice and public health research." Therefore, the committee report recommended that partnerships with the community need to be preserved, nurtured, and expanded in the CTSA program.

In another major federal research support initiative, the Patient-Centered Outcomes Research Institute (PCORI) aims to fund projects of clinical effectiveness research. Consistent with the requirements of academic engagement, the 2016 scientific methodology standards of PCORI require that the researchers engage "people representing the population of interest and other relevant stakeholders in ways that are appropriate and necessary in a given research context" (PCORI 2016). While community interaction often is highlighted, genuinely engaged research is far more complex and inclusive, ranging from lay sources to industry collaboration and others.

Corporate Competence in Life Sciences

Among the many opportunities for engaged research, *corporate competence* appears to be an essential skill, especially in the applied health sciences area (Figure 12.1). The list of award-winning researchers who were engaged in fruitful and mutually beneficial industry collaborations is very long, while the loners are relatively few.

According to several studies, a synergistic relationship exists between industry partnerships and research productivity. A wide-ranging study suggested that life sciences research faculty receiving industrial support had more papers published, five more patent applications, and participated in more academic administrative activities than faculty members who did not receive such support (Blumenthal et al. 1986). According to Brown et al. (2006), research projects funded by private industry had a higher average methodology score than studies funded by traditional academic sources.

Historically, the long list of eminent theoretical and applied scientists illustrates that working with industry and considering business implications are relatively lesser, but very natural, considerations in the pursuit of breakthrough scientific discoveries. Interactions with business interests can be easily recognized in the work of many great scientists: during his most productive years, Albert Einstein worked as a patent officer in Geneva. Louis Pasteur, the founder of modern microbiology, developed pasteurization, a fundamental food industry process. The Nobel Prize-winning Paul Ehrlich immunologist discovered Salvarsan for the treatment of syphilis and worked with Hoechst AG on development and marketing.

Learning the ins and outs of collaboration with corporations is an essential competence of the engaged researcher. To make the introduction and experience more positive, researchers need to learn basic communication skills. There are significant benefits of corporate competence and information exchange with industry.

Only drug companies make drugs. Therefore, collaboration with such companies is essential for the successful development of new, effective pharmaceutical treatments. A similar statement can be made about other industries, including but not limited to medical devices, prosthetic manufacturers, information system vendors, and others. This type of insight can be enormously helpful in fine-tuning research questions and making the results readily available as a product or service.

Be Ready to Take. For obvious reasons, one of the excellent sources of technical information about the latest technologies, prevailing practices, and production processes is corporate knowledge. Various businesses, especially larger companies, routinely collect such information and use it for product and service development, new marketing initiatives, and strategic plans. Conversations with companies also can be very helpful in assessing the viability and sustainability of the research discovery and its potential impact on day-to-day practice.

Commercial companies have unmatched *market competence* and *daily contact with the market* and what people need. Companies develop and deliver products and services that are responsive to customers' needs. Correspondingly, corporate understanding of customer preferences has tremendous intellectual value, not just for the company but also for researchers with relevant scientific questions and interest in realizing societal benefits. Applied science researchers will particularly appreciate industry's insight into market needs and public priorities.

Keeping in contact with commercial companies also may help researchers understand the *intellectual property and patenting landscape.* Notably, applied health sciences researchers can benefit greatly from access to corporate information. In any research area, recognizing what is covered by existing knowledge and where gaps exist is not easy or trivial.

The interaction with industry also might provide *insight on what competitors are doing in the field*, including researchers from commercial companies who are less visible in the published literature and those who are identified by industry as important sources of information. The traditional wisdom is that competitors make us better. Certainly, the drive to compete and watching competitors are demonstrated by the thinking of many award-winning scientists. In his book, DNA double helix discoverer Watson describes their intense competition with Linus Pauling and its stimulating effect on the pace and intensity of the work (Watson 1968/2001). In fact, Pauling announced that he had discovered the structure of DNA before Watson, but it was an incorrect model.

In the age of non-repeatable research, commercial and industry interest in *replicable results and functioning prototypes* might be particularly beneficial. Therefore, industry interactions also have the potential to promote more trustworthy research results. As described in the modeling chapter, the reliably functioning prototype is one of strongest ways to prove reproducibility (e.g. drug, device, procedure, technology).

Not collaborating with commercial companies in developing new drugs or devices can become an insurmountable obstacle to practical relevance, implementation, and dissemination. A surprisingly large ratio of basic science projects has corporate relevance due to their inherent potential for commercialization (e.g. new scientific measurement technologies, various production processes of cells and tissues, vaccine research).

And Give. Preparing for corporate discussions requires advance study of the general principles and publicly available specifics of the corporation, as well as discussing preparations with academics who routinely interact with corporations (e.g. well-networked researcher colleagues, technology transfer office of the university). Corporate competence should not be confused with entrepreneurship that focuses on launching a new product or startup. Obviously, the most successful collaborations are based on matching the expertise of the researcher with the needs of a company. Research and business collaborations require a certain sense of curiosity, affability, and responsiveness on both sides.

Detail what you expect to achieve and what you have already published. Do not sell out everything for nothing. Ask questions that show your interest, competence, and acumen. Without selling out, make the conversation useful for them as well. They also want to make a living like you do. Asking about something that turns out to be corporate secret (e.g. the recipe for Coke) is pardonable, but aggressively going after it is a blunder. A successful applied researcher with major industry partnership experience summed it up: if you want to build industry collaboration and get corporate funding, go to the company, ask them what they need, and then do it. The "ask and do it" model can work very well in a wide range of applied research collaborations.

Academic–industry collaborations tend to be focused on applied research in the interests of product development or marketing. Commercial companies rarely, if ever, award unrestricted grants to basic scientists who do not produce intellectual property. Typically, commercial companies cannot afford funding curious minds and exploratory projects of basic sciences. The quarterly pressures of producing business results push most companies into targeted activities. The simple "just give me a research grant to fund my project" request is unlikely to initiate productive partnerships or work effectively in industry collaboration.

For obvious reasons, corporations are very selective in sharing essential information, and one may need a high level of trust in the relationship before access. At the same time, researchers must be careful in disclosing details about their latest projects to avoid unwise sellouts or interference with future commercialization. Nobody wants to end up in a situation where the company learns all the insight of someone's research results without making any commitment or compensation in return.

Sound and ethical research that is supported by industry collaboration or funding is essential not only for the progress of life sciences but also for *economic development*. Tierney et al. (2016) pointed out that pharmaceutical and device manufacturers fund more than half of the medical research in the United States and such support has increased over the past 20 years, while federal funding has declined.

Guided by Principles of Humanism. Partnering with industry has major potential benefits, as well as significant risks. The clear majority of industry collaborations are honest and ethical – very much consistent with good science. While creative industry partnerships should be encouraged, principles of ethics can never be compromised, and accuracy of observations should never be negotiated. Researchers need to recognize dangers that include distortion of scientific results for financial gains and undesirable distraction from seeking answers to major scientific questions.

When biomedical researchers partner with industry to receive funding, accelerate innovation, and develop intellectual property, ethical concerns are frequently raised. They range from the potentially biased influence of industry preferences on the objectivity of research to fears that the desire to gain and retain industry funding will lead to unpublishable research results that are contrary to societal interests. Marcia Angell, former editor of the *New England Journal of Medicine*, provides a passionate overview of the threats to the impartiality of research in her book (Angell 2005).

Indeed, there have been many well-publicized cases of research bias introduced by industry interests. For example, concerns have been raised in connection with research funding from the sugar industry that de-emphasized the role of sugar consumption in obesity and cardiovascular mortality (Kearns et al. 2016). Not surprisingly, similar but greater concerns have been raised by research on smoking that is funded by the tobacco industry.

In the unfortunate, but infrequent, situation when the business partner may press for not disclosing, distorting, or hyping research results, the researcher should simply and unambiguously say NO. The ethical concerns and long-term adverse effects of research improprieties are so severe that it is not just anxiety producing, but not worth compromising on basic principles of scientific professionalism.

Being attentive to concerns of ethics and conflict of interest has paramount importance, but it should not obstruct and paralyze collaboration with industry. The most significant ethical concerns of commercialization and industry partnership are further explored in the chapters on humanism and engaged research.

Team Science and Collaborative Research

In modern science, a solo researcher can achieve almost nothing alone. Successful progress of virtually all research projects demands networking and collaborative engagement and *team science*. Being able to knock down barriers and work with many other professions makes science exciting and productive.

The engaged researcher effectively interacts with professional colleagues to test ideas and debate various concepts and possible explanations. Such interactions often provide tremendous opportunities for learning from the expertise of others.

Most interestingly, competition can alternate with collaboration in certain phases of work where the partners can clearly benefit from each other. Again, Watson's book provides excellent examples from the history of double helix discovery. A more recent example of competition turned collaboration is the joint work of Jennifer Doudna and Emmanuelle Charpentier in the discovery of the CRISPR/Cas DNA engineering system.

Willem Einthoven was born in Semarang in Java, now Indonesia. After moving to Holland, he studied at the University of Utrecht as a medical student. After graduation and qualifying as a general practitioner, he became professor of physiology at the University of Leiden (Einthoven 1965).

Very early, his research interests turned toward interactions between physics and physiology. He investigated various geometric–optical illusions, form and magnitude of the electric response of the eye to stimulation by light, registering heart sounds, using a capillary electrometer, and mathematical correction of the errors in the photographically recorded results. Einthoven finally devised the string galvanometer, the most reliable tool in electrocardiography, that earned great recognition. His laboratory in Leiden became a site of pilgrimage by scientists from all over the world.

In his Nobel lecture, Einthoven not only duly recognized the work of other scientists that influenced his investigations but also highlighted the happy circumstance of engaged research: "A new chapter has been opened in the study of heart diseases, not by the work of a single investigator, but by that of many talented men, who have not been influenced in their work by political boundaries and, distributed over the whole surface of the earth, have devoted their powers to an ideal purpose, the advance of knowledge by which, finally, suffering mankind is helped" (Einthoven 1925).

Based on analysis of tens of thousands of scientific publications in premier peer-review journals, specific co-authorship and collaborative patterns were identified by Nature Index for a variety of institutions worldwide. With the help of such authorship analyses, one can identify clusters of institutions that collaborate frequently (Adams and Loach 2015). For example, the French National Center for Scientific Research (CNRS), Max Planck Group, Cambridge University, Imperial College, and Oxford University provide strong regional focal points. The US average for international co-authorship is about one-third, but, for example, Harvard has an international co-author of about half of its publications.

Randy Schekman's research illustrates the enormous power of cooperation and competition. His engaged approach was singularly focused that brought him outstanding recognition: discovering the machinery that regulates vehicle traffic, a major transport system in our cells. His Nobel lecture is an impressive list of mobilized talents – graduate students, fellows, and friendly or competing laboratories (Schekman 2013).

At Emory University, Dr. Schinazi is one of the most prolific serial innovators in life sciences. Among his many initiatives is the annual DART (Design of AntiRetroviral Therapy) conference series. It is not your usual "contemporary issues," or "let's discuss anything submitted" conference. The sharpshooting focus on killing the retrovirus makes DART uniquely useful. The organizers and devoted attendees of the conference can have an exceptionally good understanding of the latest opportunities in antiretroviral therapy research. Among the many successful engaged research initiatives of Emory University is the Vaccine Dinner Club (VDC), which is designed to facilitate networking and collaboration among researchers, clinicians, policymakers, and historians/journalists. The club's mission is to advance the practice of vaccine science by stimulating intellectual potential and research productivity of the vaccine research community.

Bouncing Ideas in the Real World

Successful research almost invariably involves multiple sources of information that are far beyond narrowly defined peer-reviewed publications of the particular field, laboratory experiments, and scientific conferences. Limiting someone's thinking to a narrow base is often labeled as the academic ivory tower.

In summary, the engaged researcher is interested in all information that might be available about the natural phenomena and might shed further light on its internal workings. The history of award-winning discoveries has innumerable examples of benefiting from engaged research, developing fruitful collaborations, and learning from diverse sources of knowledge beyond the scientific literature.

Bouncing off ideas in the real world can help to reestablish the connection between reality and scientific pursuits. Obviously, none of the tools of engaged research can replace the established scientific methodologies and the pursuit of rigor in scientific investigations. Methods of engaged research add to scientific investigations by strengthening its foot and improving preparations for practical use of study results.

References

Adams, J. and Loach, T. (2015). Comment: a well-connected world. *Nature* 527 (7577): S58–S59.

Angell, M. (2005). *The Truth About the Drug Companies: How They Deceive Us and What to Do About It*. New York: Random House Trade Paperbacks.

Blumenthal, D., Gluck, M., Louis, K.S. et al. (1986). University-industry research relationships in biotechnology: implications for the university. *Science* 232: 1361–1366.

Boeker, M., Vach, W., and Motschall, E. (2013). Google Scholar as a replacement for systematic literature searches: good relative recall and precision are not enough. *BMC Medical Research Methodology* 13 (1): 131.

Brown, A., Kraft, D., Schmitz, S.M. et al. (2006). Association of industry sponsorship to published outcomes in gastrointestinal clinical research. *Clinical Gastroenterology and Hepatology* 4 (12): 1445–1451.

Contopoulos-Ioannidis, D.G., Ntzani, E.E., and Ioannidis, J.P. (2003). Translation of highly promising basic science research into clinical applications. *The American Journal of Medicine* 114 (6): 477–484.

Einthoven, W. (1925). The string galvanometer and the measurement of the action currents of the heart. *Nobel Lecture* (11 December).

Einthoven, W. (1965). The string galvanometer and the measurement of the action currents of the heart. In: *Nobel Lectures, Physiology or Medicine 1922–1941*. Amsterdam: Elsevier Publishing Company.

Franklin, R. (1940). Letter to her father, Ellis Franklin (undated, summer 1940 while she was an undergraduate at Cambridge). Excerpted in Brenda Maddox, *The Dark Lady of DNA* (2002), 60–61.

Gregoris, N. and Shorvon, S. (2013). What is the enduring value of research publications in clinical epilepsy? An assessment of papers published in 1981, 1991, and 2001. *Epilepsy & Behavior* 28 (3): 522–529.

Hicks, S., Duran, B., Wallerstein, N. et al. (2012). Evaluating community-based participatory research to improve community-partnered science and community health. *Progress in Community Health Partnership* 6: 289–299.

Judson, H.F. (2015). (RARE) Interview with James Watson and Francis Crick. Published on 16 December 2015 [Video file]. https://www.youtube.com/watch?v=NGBDFq5Kaw0 (accessed 11 July 2016).

Kearns, C.E., Schmidt, L.A., and Glantz, S.A. (2016). Sugar industry and coronary heart disease research: a historical analysis of internal industry documents. *JAMA Internal Medicine* 176 (11): 1680–1685.

Liverman, C.T., Schultz, A.M., Terry, S.F., and Leshner, A.I. (eds.) (2013). *The CTSA Program at NIH: Opportunities for Advancing Clinical and Translational Research*. Washington, DC: National Academies Press.

Martin, J.P. and Kelchner, R. (1998). *Lernen durch lehren* (ed. J.P. Timm), 211–219. http://www.lernen-durch-lehren.de/Material/Publikationen/timm.pdf (accessed 22 June 2018).

Mülder, C. (2001). Malaria. Robert Koch auf Brioni. *Deutsches Ärzteblatt* 98 (48).

Nurse, P. (2014). You can have many more good ideas than you can possibly pursue. Paul Nurse, Nobel Laureate. https://www.youtube.com/watch?v=-StnPpyoqFc (accessed 11 August 2017).

Odelberg, W. (ed.) (1978). *Les Prix Nobel. The Nobel Prizes 1977*. Stockholm: Nobel Foundation.

PCORI Methodology Standards (2016). http://www.pcori.org/research-results/research-methodology/pcori-methodology-standards (accessed 19 October 2016).

Perkmann, M., Tartari, V., McKelvey, M. et al. (2013). Academic engagement and commercialisation: a review of the literature on university–industry relations. *Research Policy* 42 (2): 423–442.

Rorvik, D. (1974). The winner will be a brave new baby conceived in a test-tube and then planted in a womb. *New York Times* (15 September).

Schekman, R. (2013). Genes and proteins that control the secretory pathway. *Nobel Lecture* (7 December).

Seneca, L.A. and Campbell, R. (1969). *Epistulae morales ad Lucilium*, vol. 210, c. 65 AD. Harmondsworth, UK: Penguin.

Sessions, R. (1950). How a "difficult" composer gets that way. *New York Times* 89.

Tierney, W.M., Meslin, E.M., and Kroenke, K. (2016). Industry support of medical research: important opportunity or treacherous pitfall? *Journal of General Internal Medicine* 31 (2): 228–233.

Varmus, H. (1993). *From Nobel Lectures, Physiology or Medicine 1981–1990* (Editor-in-Charge T. Frängsmyr, ed. J. Lindsten). Singapore: World Scientific Publishing Co.

Wallerstein, N. and Duran, B. (2010). Community-based participatory research contributions to intervention research: the intersection of science and practice to improve health equity. *American Journal of Public Health* 100 (Suppl 1): S40–S46.

Watson, J.D. (1968/2001). *The Double Helix: A Personal Account of the Discovery of the Structure of DNA*. New York: Touchstone.

Yang, K., Deangelis, R.A., Reed, J.E. et al. (2013). Complement in action: an analysis of patent trends from 1976 through 2011. *Advances in Experimental Medicine and Biology* 734: 301–313.

zur Hausen, H. (2014). Transcript of the telephone interview with Harald zur Hausen immediately following the announcement of the 2008 Nobel Prize in Physiology or Medicine. Nobelprize.org. Nobel Media AB 2014. http://www.nobelprize.org/nobel_prizes/medicine/laureates/2008/hausen-telephone.html (accessed 4 February 2017).

13

Cross-cultural Convergence

If you're doing research in both chemistry and biology, you are involved in a lot of excitement.

Gobind Khorana (2011)[1]

Many researchers do not fully realize the tremendous opportunities of cross-cultural collaboration. They read the journals of their profession, go to the same professional meetings annually, and often use the established tools and models of their particular specialty. By working in such specialty environment, some of the most essential ingredients of successful research are missing, particularly the inspiring convergence of multiple disciplines and engagement with the diverse cultures of the world.

Meanwhile, many amazing discoveries have been generated when scientists from diverse disciplines came together. One of the classic examples of joining forces is the interaction between chemistry and biology. The concept was pioneered by Otto Warburg in the first half of the twentieth century (Krebs and Schmid 1981). The excitement continues as expressed eloquently by Nobel laureate Har Gobind Khorana, the scientist who played a key role in understanding the role of genetic code in protein synthesis (2011). Khorana crossed not just disciplinary boundaries but also international borders. He was born in India and reached his breakthrough discoveries at the University of Wisconsin–Madison. Other examples include computer science and sequencing of the human genome, bioengineering, and rehabilitation, or nanotechnology and pharmacology.

In describing the exceptional value of diversity, Mario Capecchi, the Italian-born American molecular geneticist pointed out: "Yes, I seek diversity. I have people joining our laboratory from medicine, molecular biology, neurobiology and developmental biology... There is a constant flux of people coming and

---

1 Khorana, H.G. (2011). Interview. UW-Madison's Nobel Prize Winners 2011; Nobelist Khorana nears new achievement – the gene. *Medical Tribune.* https://www.youtube.com/watch?v=g-BidjlCnHs (28 May 2017).

Innovative Research in Life Sciences: Pathways to Scientific Impact, Public Health Improvement, and Economic Progress, First Edition. E. Andrew Balas.

leaving, so I always bring in people from different disciplines... If you have a group of people from different backgrounds, they will look at problems from very different vantage points, and I think that enriches the whole lab" (Kain 2008).

When Paul Lauterbur was awarded the Nobel Prize, he noted that in that particular year a chemist and a physicist received the Nobel Prize in Physiology or Medicine. Such recognition shows how the various seemingly very disparate branches of science can and indeed come together.

The *interdisciplinary* character of life sciences methodology development is further highlighted by the many physics and chemistry awards given for methods and studies with high relevance to biomedical research and life sciences in general.

The chemistry Nobel prizes for life sciences discoveries include mechanistic studies of DNA repair (Chemistry, 2015); studies of the structure and function of the ribosome (Chemistry, 2009); studies of the molecular basis of eukaryotic transcription (Chemistry, 2006); discoveries concerning channels in cell membranes (Chemistry, 2003); development of crystallographic electron microscopy and structural elucidation of biologically important nucleic acid–protein complexes (Chemistry, 1982); fundamental studies of the biochemistry of nucleic acids, with particular regard to recombinant DNA (Chemistry, 1980); and work on the structure of proteins, especially insulin (Chemistry, 1958).

The illustrative physics Nobel prizes with direct relevance to life sciences include discovery and interpretation of the Cherenkov effect (Physics, 1958); fundamental work in electron optics and the design of the first electron microscope (Physics, 1986); demonstration of the phase contrast method, especially invention of the phase contrast microscope (Physics, 1953); and discovery of the remarkable rays subsequently named after Röntgen (Physics, 1901), among many others. Indeed, many of the most innovative tools of modern medicine were developed not by life scientists but by physicists who created technologies such as X-rays, nuclear, magnetic resonance, ultrasound, particle accelerators, and radioisotope tagging. These diagnostic and therapeutic advances have made an enormous difference in the quality and outcomes of health care.

Being able to knock down barriers and working with many other people and professions make science exciting and productive. The fusion of best ideas from different cultures and disciplines promises solution to intractable problems of our times. This chapter explores the scientific and societal value of international collaborations, mobility, disciplinary convergence, team science, and minority thoughts.

International Collaborations and Mobility

The tremendous value of cross-cultural and international collaborations has long been recognized. Throughout history, there have been many shining examples of countries, institutions, and scientists benefiting from the latest and most relevant knowledge and technologies all over the world. Conversely, isolationism has long been viewed as limiting and disadvantaging.

Successfully gathering scientific discoveries, technological ideas, and talents from all over the known world was the hallmark of the Royal Library of Alexandria in antiquities or the House of Wisdom in the medieval Baghdad. At the turn of the first millennium in Europe, King Stephen also advised future rulers: "guests and newcomers generate great benefits" (Békefi 1901). Today, many productive research universities have an extremely diversified scientific workforce as they focus a great deal of effort on attracting and supporting the best minds from all over the world.

The Royal Library of Alexandria was one of the most famous hubs of scientists in antiquity. It was much more than just a warehouse of science. The chief librarian of Alexandria, Eratosthenes, founded the discipline of geography, and he was first in figuring out the circumference of the Earth. Probably, Archimedes also studied in Alexandria. Founder of geometry Euclid also lived and worked there. The Royal Library of Alexandria was a center of enlightenment, knowledge, and learning. In many ways, the Royal Library of Alexandria was the world's first genuine research center.

Not surprisingly, there are many historical references to the role of immigrants in bringing new ideas and creative energies to many countries, and particularly to the United States. Probably, the best known among them is Albert Einstein who arrived as a refugee from Germany.

Many consider the renowned ninth-century Persian mathematician **Al-Khwarizmi** the founding father of algebra. Algebra is a mathematical discipline in which letters and symbols are used to represent quantities in formulas and based on given axioms; rules are applied for manipulating these symbols. He was probably the most famous scientist among those who worked at the House of Wisdom (Lyons 2011).

In the medieval Baghdad, the House of Wisdom was an almost unparalleled fusion of scientific achievements and talents from many diverse cultures. Starting as a translation institute and library, the House of Wisdom gathered scientific and technological discoveries from all over the world, including Greek learning from Europe and Alexandria; scientific discoveries from Persia, India, and Sumer; production of paper from China; and many others. Christian, Jewish, and Islamic translators and scientists gathered in the House of Wisdom.

The House of Wisdom was more than translations of scientific work from earlier researchers and scientists from other countries. It was also verification of their results and improvement of their methodologies. The merger and expansion of many intellectual traditions promoted unprecedented scientific productivity and led to not only major scientific discoveries but also the launching of new scientific disciplines.

Al-Khwarizmi's famous *Book of Restoring and Balancing* (820/1831) gave us the concept and the term "algebra." He introduced the methods of reduction and balance in dealing with equations and provided a comprehensive explanation of solving quadratic equations. These and other achievements were built on the accomplishments of the earlier work of the Greek mathematician Diophantus and others. Al-Khwarizmi also played a pivotal role in spreading the Hindu–Arabic numeral system. Later the decimal point was introduced by one of his collaborators.

Other influential scientists of the House of Wisdom include Ibn al-Haytham, creator of the modern scientific method and accomplished scientist of optics and vision. Al-Jahiz wrote the *Book of Animals* based on observations of the social organization and communication among animals and the effects of diet and environment. Reportedly, the first to make astrolabe a breakthrough navigational instrument was the eighth-century mathematician Muhammad al-Fazari, another scientist affiliated with the House of Wisdom.

Other examples include the Manhattan Project, the major nuclear weapons development program during World War II. It was effectively initiated by the Hungarian-American Leo Szilard and his persistent advocacy that reached the highest level of the government. When the project was finally launched by President Roosevelt, the number of contributing immigrant scientists was astounding. Among them, five were born at the beginning of the twentieth century in Budapest (Theodore von Kármán, Leo Szilard, Eugene Wigner, John von Neumann, and Edward Teller). As mentioned, they were called the Martians due to their strange language and mysterious origins in the faraway Budapest (Hargittai 2008).

A recent study found that immigrants launched more than half (44 of 87) of startup companies valued at \$1 billion or more in the United States (Anderson 2016). They also serve as key members of management and development teams in over 70% of these companies. Another study estimated that more than 40% of America's Fortune 500 companies were founded by an immigrant or a child of an immigrant (Partnership 2012). Many observers credit the success of Silicon Valley partly to creative and hardworking immigrants or a process sometimes referred to as brain drain.

Emmanuelle Charpentier, one of the codiscoverers of CRISPR/Cas system, summarized her cross-cultural approach as she repeatedly moved: "I define myself as a mobile researcher...series of moving around for 25 years in 5 different countries, 10 different cities and 10 different labs...this has been very worthy for me personally to develop my personality as a scientist...because I got to have to deal with different types of cultures, different ways of working and also the idea that nothing is impossible" (Charpentier 2016).

According to studies of publication impact, higher international co-authorship rates correlate with higher citation rates of articles as reflected by the citation per paper index (Khor and Yu 2016). It is well known that various scientific fields have very different average citation rates and old institutions have a citation advantage in the published research. However, some general overarching trends are clearly emerging. One of the most important among them is that international co-authorship has been on the rise and it has a significant positive contribution to the field-weighted citation impact of the institution.

Understandably, some university ranking systems consider the ratio of international faculty an important academic quality indicator (e.g. QS World Ranking). Meanwhile, other university ranking systems use the number of articles with international collaboration as an essential indicator of research quality (e.g. *Times*; SCImago; University Ranking by Academic Performance [URAP]; U.S. News & World Report – Global Ranking; EU Multirank; Webometrics).

Convergence of Disciplines

The concept of *convergence* highlights the importance of collaboration of disciplines that have been traditionally viewed as separate and disjointed (Sharp et al. 2011). Biomedical research and ultimately patient care have much to gain from such merger of life, physical, social, and engineering sciences. It is very powerful when diverse groups of expertise and different thinkers come together.

In recent years, the concept of convergence gained further prominence in some leading institutions. It has been particularly actively promoted by the Massachusetts Institute of Technology (MIT). In 2011, 12 leading MIT researchers released a major report at the meeting of the American Association for the Advancement of Science (Kaiser 2011). Much of scientific innovation comes from the convergence of ideas, expertise, disciplines, and fields. The MIT report called for policies to support crosscutting studies.

Accordingly, MIT launched several major initiatives to promote convergence through hiring practices, creating physical space, and supporting interdisciplinary research, among others. These efforts make sure that people of different professions and specialties come together to solve the same problem. Such approach is creating unique opportunities for cross-fertilization. Nobel laureate Susumu Tonegawa spoke eloquently about the power of convergence: "There is a gold mine in interdisciplinary science. Science became so specialized and often scientists don't talk to each other. Wonderful opportunities become unrealized" (Tonegawa 2007).

Many universities and research institutions put great emphasis on promoting team science initiatives and collaborative approaches. According to Stokols et al. (2008), the following levels of cross-disciplinary team science collaborations can be distinguished:

- *Multidisciplinarity* is a sequential process where researchers from different disciplines work independently with a goal of eventually combining efforts.
- *Interdisciplinarity* is an interactive process in which researchers from different disciplines work jointly to address a common research problem.

- *Transdisciplinarity* is an integrative process in which researchers develop and use a shared conceptual framework and create new models to address a common research problem.

In 2014, the National Research Council released a report on convergence, facilitating the *transdisciplinary* integration of life sciences, physical sciences, engineering, and beyond. The report enumerates the special needs of launching and sustaining convergent research: establishing effective organizational cultures, structures, and governance; addressing faculty development and promotion needs; creating education and training programs; forming stakeholder partnerships; obtaining sustainable funding; designing facilities and workspaces for convergent research; and establishing partnership arrangements across institutions and others.

In research collaborations, the opposites attract each other and ignite creativity. Simply working with people from the same specialty is rarely inspiring or

---

Gregor Mendel, the founder of modern genetics, was a passionate, cross-cultural natural scientist and an Augustinian friar in the nineteenth-century Brünn (today Brno, Czech Republic). Over a period of eight years, Mendel's pea plant experiments established the laws of Mendelian inheritance that are still widely benefiting research and clinical practice as well (Mendel 1866).

Heredity of traits has always been an intriguing question, and farmers have long been familiar with the benefits of crossbreeding. However, the more exact rules of heredity had not been established before Mendel's work, and many speculative theories of the time apparently failed to explain how nature works.

At the conclusion of his long series of experiments, Mendel's landmark paper titled "Experiments in Plant Hybridization" was published in 1866. It remains one of the most exemplary classic research reports in the history of life sciences. The accuracy of the observations of nature, elegant use of mathematics, and matter-of-fact reporting style are just as astounding today as it was 150 years ago.

Mendel's theory of the separate inheritance of characteristics also ran against the prevailing assumption of his time, the joint and total inheritance of biological characteristics. In the resistance against Mendel's work, scientific authority played a key role as well-established botanists of the time were harshly critical of the "provincial work" of a monk.

Mendel's studies represented an unprecedented cross-cultural interaction of two different scientific fields, botany and mathematics, that led to launching the new scientific discipline of genetics. Historians point out the anti-mathematical conceptualization of botany during Mendel's time. The outdated isolationism was primarily responsible for the great delay in appreciating Mendel's landmark discoveries.

informative. The simplistic invitation for unidisciplinary collaboration sounds like this: "your university has environmental health research, and their university has the same, so you should collaborate." Such invitations often sound hollow and marginally useful. Collaborations are much more exciting when different professions come together for a common purpose.

One of the most highly cited papers that revolutionized interaction with technology is "Man-Computer Symbiosis" by J. C. Licklider (1960). The pioneering concept put emphasis on facilitating thinking and cooperating in decision-making with technology. Through his visionary papers, Licklider became the pioneer of interactive computing, the Internet, and graphical user interface. Characteristically, he received triple bachelor's degree in physics, mathematics, and psychology, effectively personifying the concept of convergence. In recent years, joint degrees and double majors gained unprecedented popularity in higher education, and many ambitious students seek such university degrees.

In his Kyoto Prize commemorative lecture, Leonard Herzenberg, developer of the fluorescence-activated cell sorter (FACS), explained: "Much of my work in science, in fact, has involved collaborations with colleagues whose interests, sometimes quite different from mine, have opened the way to studies that neither they nor I could have done independently" (Herzenberg 2006).

Convergence is the preferred term in promoting cross-disciplinary interactions that include not only dedicated team science collaborations but also numerous, ad hoc, and brief interactions among researchers from various disciplines. Using team science terminology, one may say that convergence is built on interdisciplinarity and transdisciplinarity. Convergence is a particularly favored concept in the basic science and technology development world.

At research universities, the concept of convergence can be promoted and facilitated through a variety of measures:

- Establishing interdisciplinary research centers and institutes.
- Recruitment of research faculty from diverse areas to work on a common set of problems.
- Special grants and other incentives for interdisciplinary teams of scientists.
- Building that houses research from diverse disciplines together and offers meeting places.
- Organizing scientific events and discussions that purposefully attract scientist from diverse fields.

When directing their philanthropic effort, Zuckerberg and Chan decided to invest in the establishment of a "BioHub" that connects engineering and life sciences research from multiple institutions. The new hub can draw talents from the University of California at Berkeley and San Francisco, Stanford University, and also Silicon Valley. The entire effort is focused on increasing interface among different disciplines and developing convergence for productive innovation (Baltimore 2016).

Evolution of Team Science

Scholars of management science have repeatedly pointed out that the biggest challenge of innovation is internal and one of the factors is teamwork or more precisely the lack of it. As pointed out in the previous chapter, team science is also an essential component of engaged research. Nobel laureate Tim Hunt complained about silos in research: "Today everything is compartmentalized. You don't even know what is going on over there. Once everything was under all under one roof, but now it's under many, many roofs" (Hunt 2014).

Frequently, the term team science is used as a synonym of coordinated cross-disciplinary collaborations. Team science initiatives are designed to promote collaborative – occasionally unidisciplinary but most frequently cross-disciplinary– approaches to analyzing research questions and finding practical solutions.

"In these days, science is not a solo enterprise; it is the work of a team" – as Nobel Prize winner James Rothman pointed out. In leading research laboratories, everyone contributes to the successes, and everyone shares the failures in search for new scientific knowledge (Rothman 2013). Yet, researchers are incentivized and recognized only as individuals at many universities.

Charles Kelman, the pioneer of cataract surgery, was an ophthalmologist working at the Manhattan Eye, Ear, and Throat Hospital in New York. In the 1950s, cataract surgery was a painful procedure, requiring a large incision and prolonged hospitalization.

Notably, removal of the cataract lens proved to be an intractable problem. In searching for cataract lens removal method, Kelman spent $250 000 in grant money but failed to find a solution. The breakthrough came when he had a dentist appointment to have his teeth cleaned with an ultrasonic probe. This was a fortuitous convergence of expertise and technologies in addressing a major public health need. Kelman quickly realized the opportunity of ultrasound to emulsify the cataract lens and aspirated through a much smaller opening. The vibrations of the ultrasonic probe break and liquefy the cataract lens, a convergence of ophthalmology and ultrasound technology.

The nearly noninvasive procedure is called phacoemulsification, and it has become the standard of care under local anesthesia (Kelman 1967). Ultimately, the method reduced cataract surgery recovery from a 10-day hospital stay to outpatient surgery worldwide. The ophthalmologists were shocked when Kelman started discharging his patients on the same day of surgery and let them return to full activity on the second postoperative day.

In recognition of his landmark discovery, Kelman was recognized with the Albert Lasker Clinical Medical Research Award in 2004. In 1992, he was awarded the National Medal of Technology, and in 1994, his peers named him "Ophthalmologist of the Century." Today, at least three million people benefit from the quick, outpatient procedure in the United States annually.

The successful development of the HPV vaccine is considered one of the finest examples of team science. It was a nested analysis of the efficacy of a bivalent HPV 16/18 vaccine against anal HPV 16/18 infection among young women in the Costa Rica Vaccine Trial. Collaborators included the National Cancer Institute, NIH, Bethesda, Maryland, USA; Proyecto Epidemiológico Guanacaste, Fundación INCIENSA, Costa Rica; DDL Diagnostic Laboratory, Voorburg, The Netherlands; Information Management Systems, Rockville, Maryland, USA; and the International Agency for Research on Cancer, Lyon, France (Kreimer et al. 2011). The study found that the ASO4-adjuvanted vaccine provides effective prevention of anal HPV, particularly among women more likely to be HPV naïve at vaccination.

The goals of team science initiatives can be wide ranging (e.g. facilitating scientific discovery; training the new generation of researchers; and addressing practical, clinical, translational, public health, or policy development problems) (Stokols et al. 2008). Critics of cross-disciplinary and team science initiatives contend that it takes away valuable resources from necessary discipline-based research.

The concept of *team science* tends to come up with particular frequency in applied and public health sciences as an attempt to find answers to major practical questions. For example, it has been advocated for exploring the joint influence of social, behavioral, and biogenetic factors on cancer etiology and prevention (Hiatt and Breen 2008).

There have been many more great accomplishments coming out of team science. For example, the World Health Organization (WHO) faced the crisis of the spreading SARS (severe acute respiratory syndrome) pandemic in 2002. To identify the pathogen responsible for SARS deaths, the WHO brought together 11 researchers from 9 countries. The effort was built on the Global Outbreak Alert and Response Network that pools human and technical resources for rapid identification and response to outbreaks. The project adopted several fundamental principles of effective teams: shared commitment to a common goal and frequent communication about data and next actions. Within a few months after the outbreak, the team identified a previously unrecognized coronavirus as the causative agent of SARS.

The National Institutes of Health recognized the significance of team science by releasing a field guide (Bennett et al. 2010). It covers major aspects of team building in research endeavors: planning and preparing for team science, building a research team, fostering trust, developing a shared vision, communicating about science, sharing recognition and credit, handling conflict, strengthening team dynamics, and navigating and leveraging networks and systems. Scientists need to recognize that collaboration can happen at various levels: independent investigators primarily only reading about each other's results, research collaborations that involve data sharing or brainstorming together, and integrated research team that shares leadership responsibility, decision-making authority, data, and credit.

The American Association for Cancer Research, for example, has created the Team Science Award that recognizes and highlights the role of interdisciplinary teams in the understanding of cancer and the translation of research discoveries into clinical cancer applications. For example, in 2012, the Award went to the Institute of Cancer Research (ICR) and Royal Marsden Hospital: Cancer Research UK Cancer Therapeutics Unit and Drug Development Units. This interdisciplinary team was behind the discovery of 16 cancer drug development candidates over the previous six years. The team includes experts in cancer biology, pharmacology, medicinal chemistry, medical oncology, and also various other academic and industrial partners.

In 2015, the National Research Council published its report on enhancing the effectiveness of team science. This landmark report not only summarized the results of research on groups and teams but also highlighted opportunities for improving the team and group effectiveness.

Several key factors have been identified. Task-relevant diversity is certainly critical ingredient of success. Professional development interventions are also effective in improving performance (e.g. knowledge development training to increase sharing of individual knowledge and improve problem-solving). There are additional benefits in providing professional development for leaders of science teams and larger groups. Finally, appropriate technology support for virtual collaborations is essential to facilitate effectiveness in geographically dispersed team science.

University policies, including but not limited to *promotion and tenure* considerations, and initiatives can play a vital role in initiating, sustaining, and advancing team science. Promotion of team science can be greatly facilitated by revising tenure and promotion guidelines (Mazumdar et al. 2015). Such revision should create standards as to ways in which team science, intellectual property, and entrepreneurship should be documented and valued. Certainly, funders of research can also set some expectations of team science and may require demonstration of effective collaborative competencies. There are some calls that major scientific awards should recognize not only individuals but also team science.

Recently, the dean for graduate education and professor of materials science and engineering at the Massachusetts Institute of Technology, Christine Ortiz decided to leave her job to build a new type of institution centered on *project-based learning*. For her research she received the National Science Foundation Presidential Early Career Award for Scientists and Engineers. Certainly, the project-based approach has the promise of bringing the benefits of team science to higher education. Her teaching and project philosophy was summed up in an interview with the *Chronicle of Higher Education* (Young 2016): "We were studying the exoskeletons of animals. We were looking at that to create bio-inspired armor for the military — human armor, exoskeletons. ... I've had many students from different disciplines in my research group... It really changed the game for the entire group... It stimulated creativity in all of the students."

If You Have Never Been in Minority, You Have Never Said Anything Original

Beyond diversity of disciplines and international cultural diversity, there appears to be special merit to the concept of minority status in advancing scientific creativity. The reasons for synergies between innovation and *minority thought* are often visibly apparent. Obviously, the essence of innovation is often thinking and saying entirely different statements that run contrary to the perception of the majority. Not infrequently, minority experience appears to support the diversification of general human understanding and development of above average resilience that leads to success.

Examples of minority successes are the many significant scientists who came from a minority background. There are innumerable inspiring examples of great scientists coming from very small and sometimes marginalized minorities. In the history of science, such examples are almost endless: Nikola Tesla, a Serbian from Croatia in the Austrian Empire, Marie Skłodowska Curie,a Polish scientist from the Russian Empire, the African-American Charles R. Drew who developed the first large-scale blood banks, and many others.

There were four Nobel Prize winners from the very small Italian-Jewish minority (less than 50000 people): Levi-Montalcini, Emilio Segrè, Salvador Luria, and Franco Modigliani. In reflecting on her achievements and many challenges, Rita Levi-Montalcini, the pioneering discoverer of growth factors, famously stated: "my only merits have been commitment and optimism."

Inclusivity and involvement of minorities is an important resource for responsive research. It is a matter of fairness, social justice, and also attention to all segments of the population in identifying needs and developing solutions. Furthermore, various minorities represent a tremendous pool of talent and opportunities. It is a great loss to scientific progress when minorities are marginalized. The more effective effort of many academic institutions to increase involvement of minorities in STEM education is long overdue and has a great promise for everyone.

Diversity and minority representation have become essential not only in the success of scientific teams but also in the educational environment that develops future professionals. In 2013, the Supreme Court issued its decision in Fisher v. University of Texas at Austin. The Court recognized that colleges and universities have a compelling interest in ensuring student body diversity and they can consider diversity in their admissions program.

While various explanations can be more easily identified for the societal resistance to new discoveries, the resistance by scientists to scientific discoveries is a more challenging subject. In 1961, Professor Barber of Columbia University wrote a remarkable essay about resistance in the scientific community. He pointed to cultural blinders as a constant source of resistance to innovation. Established culture tends to define the situation usually helpfully, but at the same time, it can harmfully blind to other ways of conceiving that situation.

In a graduation address to medical students, Lister warned against blindness to new ideas in science, ignorance such as he had experienced when he developed the theory of antisepsis (Barber 1961). The eminent physicist Lord Kelvin called the announcement of Wilhelm Röntgen's discovery of X-rays a hoax. Nobel laureate Lauterbur made a sarcastic comment in his Nobel lecture: "You don't know that you have a good idea until at least three Nobel Prize winners tell you that you are wrong."

Not infrequently, scientists have to ask themselves what is more important, being right or fitting in. Particularly, prejudice against mathematics was a major source of resistance to innovation in biology for a long time. Silos and *overspecialization* can also slow down acceptance of innovation as illustrated by the opposition of the medical profession against the nonphysician Pasteur's germ theory. Obviously, being right and serving as a trusted source of knowledge should be the driving principle of good science, often in the face of adversities.

Opposition to new scientific ideas is rarely pleasant but often useful as it represents probing the reproducibility of research results and can lead to better understanding of natural phenomena. Decades after Mendel's death, Ronald Fisher, the famous statistician, reconstructed his landmark inheritance studies and questioned their validity (Franklin et al. 2008). The resulting controversy lasted for several years but ended up with the exoneration of Mendel's work.

At the other extreme, marginalizing and shunning innovative scientists can be tough to handle and harmful to the progress of science. The cocoon of established scientific authorities can be challenging to penetrate. It may take time, effort, and courage of the great scientist to stand up for reproducible results that are contrary to prevailing views. Ultimately, it should be very reassuring to know that, in debates of scientific authorities, nature always wins on the long run.

Multicultural convergence has many benefits. Instead of long-established tracks of thoughts, great research discoveries tend to come from diverse environments and the spark of unprecedented interactions among various professions, different expertise, dissimilar cultures, and conflicting assumptions.

References

Anderson, S. (2016). *Immigrants and Billion Dollar Startups*. NFAP Policy Brief (March). Arlington, VA: NFAP.

Baltimore, D. (2016). The boldness of philanthropists. *Science* 1473–1473.

Barber, B. (1961). Resistance by Scientists to Scientific Discovery. *Science* 134 (3479): 596–602.

Békefi, R. (1901). István király Intelmei. *Száz* 35: 922–990.

Bennett, L.M., Levine-Finley, S., and Gadlin, H. (2010). *Collaboration & Team Science: A Field Guide*. Bethesda, MD: NIH. http://teamscience.nih.gov (accessed 23 June 2018).

Charpentier, E. (2016). L'Oréal Foundation. Pr. Emmanuelle Charpentier – L'Oréal-UNESCO Laureate 2016 – Germany (14 March) [Video file]. https://www.youtube.com/watch?v=xldariJBojY (accessed 23 June 2018).

Franklin, A., Edwards, A.W.F., Fairbanks, D.J., and Hartl, D.L. (eds.) (2008). *Ending the Mendel-Fisher Controversy*. Pittsburgh, PA: University of Pittsburgh Press.

Hargittai, I. (2008). *Martians of Science: Five Physicists Who Changed the Twentieth Century*. Oxford: Oxford University Press.

Herzenberg, L.A. (2006). The more we learn. Kyoto Prize Lecture.

Hiatt, R.A. and Breen, N. (2008). The social determinants of cancer: a challenge for transdisciplinary science. *American Journal of Preventive Medicine* 35 (2): S141–S150.

Hunt, T. (2014). Interview. Nobelprize.org. Nobel Media AB 2014. http://www. nobelprize.org/nobel_prizes/medicine/laureates/2001/hunt-interview.html (accessed 20 May 2017).

Kain, K. (2008). The first transgenic mice: an interview with Mario Capecchi. *Disease Models & Mechanisms* 1 (4–5): 197–201.

Kaiser, J. (2011). MIT calls for more "Convergence" in research. *Science* (4 January).

Kelman, C.D. (1967). Phacoemulsification and aspiration: a new technique of cataract removal. *American Journal of Ophthalmology* 64: 23–25.

Khor, K.A. and Yu, L.G. (2016). Influence of international co-authorship on the research citation impact of young universities. *Scientometrics* 107 (3): 1095–1110.

Khorana, H.G. (2011). Interview. UW-Madison's Nobel Prize winners 2011; Nobelist Khorana nears new achievement – the gene. *Medical Tribune*. https://www.youtube.com/watch?v=g-BidjlCnHs (accessed 28 May 2017).

Krebs, H. and Schmid, R. (1981). *Otto Warburg: Cell Physiologist, Biochemist and Eccentric*, 1–141. Oxford: Clarendon Press.

Kreimer, A.R., González, P., Katki, H.A. et al. (2011). Efficacy of a bivalent HPV 16/18 vaccine against anal HPV 16/18 infection among young women: a nested analysis within the Costa Rica vaccine trial. *The Lancet Oncology* 12 (9): 862–870.

Licklider, J.C. (1960). Man-computer symbiosis. *IRE Transactions on Human Factors in Electronics* (1): 4–11.

Lyons, J. (2011). *The House of Wisdom: How the Arabs Transformed Western Civilization*. New York: Bloomsbury Publishing.

Mazumdar, M., Messinger, S., Finkelstein, D.M. et al. (2015). Evaluating academic scientists collaborating in team-based research: a proposed framework. *Academic Medicine* 90 (10): 1302–1308.

Mendel, G. (1866). *Experiments in Plant Hybridization (1865)*. Im Verlage des Vereines Brünn. © 1996, Electronic Scholarly Publishing Project, http://old.esp. org/foundations/genetics/classical/gm-65-a.pdf (accessed 28 May 2017).

Al-Khwarizmi, Muhammad ibn Musa (820/1831). The compendious book on calculation by completion and balancing. In: *The Algebra of Mohammed Ben Musa*, vol. 17 (ed. F. A. Rosen). London: Murray.

National Research Council (2014). *Convergence: Facilitating Transdisciplinary Integration of Life Sciences, Physical Sciences, Engineering, and Beyond.* Washington, DC: National Academies Press.

National Research Council (2015). *Enhancing the Effectiveness of Team Science.* Washington, DC: National Academies Press.

Partnership for a New American Economy and the Partnership for New York City (2012). Not Coming to America: Why the US Is Falling Behind in the Global Race for Talent (May). http://www.renewoureconomy.org/sites/all/themes/pnae/not-coming-to-america.pdf (accessed 23 June 2018).

Rothman, J.E. (2013). *Rothman on breakthroughs in research.* https://www.youtube.com/watch?v=uOcB-XjQc3M (accessed 20 May 2017).

Sharp, P.A., Cooney, C.L., Kastner, M.A. et al. (2011). *The Third Revolution: The Convergence of the Life Sciences, Physical Sciences, and Engineering.* Cambridge, MA: Massachusetts Institute of Technology.

Stokols, D., Hall, K.L., Taylor, B.K., and Moser, R.P. (2008). The science of team science: overview of the field and introduction to the supplement. *American Journal of Preventive Medicine* 35 (2): S77–S89.

Tonegawa, S. (2007). Advice from a Nobel Laureate. https://www.youtube.com/watch?v=yRpp21lXDzI (accessed 28 May 2017).

Young, J. (2016). Why this MIT dean is leaving her job to start a new kind of university. *Chronicle of Higher Education* (20 April).

14

Targeting and Repurposing

Research is to see what everybody has seen and think what nobody has thought.

Albert Szent-Györgyi (1957)[1]

While the inspiring mission of scientific research is rightly universal, choosing an area of scientific interest and subsequently picking promising targets for the next research project are deeply personal decisions of the researcher. Choice of interest area and targets has colossal importance.

Such target selection has major consequences regarding hopes invested, the amount of time devoted, resources spent, and ultimately results produced in the chosen direction. The statement "I am interested in" sounds casual but can easily have multimillion dollar consequences in addition to significant scientific discoveries achieved or missed.

Before starting research projects, the design work tends to be preoccupied with the motivating factors, either curiosity or desire to solve a practical problem. Both considerations are important, but they are primarily focused on the past and the present. Another way to look at research planning is considering anticipated outcomes.

The scientific literature has plenty of marginal, inconsequential, and non-repeatable research results. Science as a profession needs to do better. Universities teach future researchers how to answer questions accurately, but they rarely teach them how to ask the right questions. What could make research consequential? How can we get there?

There are many ways to choose research targets. This chapter will summarize some of the methods of prioritization based on the experiences of award-winning scientists and successes of research universities: consequential research, targeting, repurposing, and embracing the unexpected.

---

1 Szent-Györgyi, A. (1957). *Bioenergetics*, 64–73. New York: Academic Press.

Innovative Research in Life Sciences: Pathways to Scientific Impact, Public Health Improvement, and Economic Progress, First Edition. E. Andrew Balas.

Consequential Research Generates Value to Others

In educational psychology, Samuel Messick (1993) was a pioneering advocate of checking consequential validity in testing. Later, Shepard (1997) argued that the consequential aspect includes interpretation (value implications) as well as test use (social consequences). Correspondingly, designers of new tests are advised to investigate both positive/negative, intended/unintended, and actual/potential consequences.

In environmental sciences, consequential life cycle assessment is recommended to understand the environmental impact of changes in policy, technology, market, and consumer behaviors (Ekvall and Weidema 2004). Consequential life cycle assessment informs about the effects of changing the level of consumption and disposal of a product, both inside and outside the product life cycle. A variety of models have been proposed for such assessments, and one may need to consider multiple scenarios in interpreting projected harms and benefits (Plevin et al. 2014).

Lack of consequential value has also been noted in connection with the latest life sciences studies. Darnovsky (2008) cautioned that germline (i.e. inheritable) modification is the most socially consequential and ethically dubious application of human biotechnology. There may be scientific and therapeutic benefits, but risks appear to outweigh beneficial results.

Consequential research is designed to produce reproducible and generalizable results that are useful for further scientific studies or solve practical problems. In moving forward, emerging definitions can be made consistent with the terminology used in other scientific fields and also sufficiently actionable to distinguish consequential and inconsequential studies to assist better research project planning.

Consequential research is designed to generate value that benefits other people, among them scientists, practitioners, or fellow human beings. Such research is based on consideration of what comes after the project is completed. In other words, the value of research should be defined not only by reasons for conducting it but also by the anticipated scientific and practical impact of the results.

As described before, consequential research can be conceptualized in life sciences by the dual set of criteria for good science and inventions (see Chapter 1). Correspondingly, criteria for consequential research include the following: (i) novel (repeating already achieved results may have marginal value), (ii) non-obvious (no new study should produce trivial results), (iii) reproducible (high-quality execution of the research is the basis of trusted results), (iv) generalizable (scientific results should reflect general principle), and (v) useful (reported results respond to questions of the next scientists or practitioners).

Business theorist Argyris (1996) pointed out that "knowledge produced by empirical research can have external validity, which means it is relevant to the everyday world…the concept of causality that underlies much rigorous empirical research makes it difficult to transform knowledge with high external validity into actionable knowledge." If used correctly, consequential knowledge from empirical research should lead to predicted effects and not to something that is contrary to the prediction.

Serious ethics concerns can be raised regarding the participation of patients in *inconsequential research*. It could be fair to say that no preclinical research results should enter into human trial without successful and stated self-replication of preclinical study results plus replication by others. In listening to recollections of award-winning scientists, including Nobel Prize winners, you can hear frequent references to the time and effort they invest in cross-checking their experiments and replicating their results.

The challenge and importance of target selection are illustrated by the use and limitations of the concept of *druggability*. In the development of new pharmaceuticals, druggability is the likelihood of being able to modulate a target with a small-molecule drug (Owens 2007). Currently, only a small portion of the human genome represents druggable targets.

Predicting druggability is especially difficult but very important exercise to improve success rates and minimize losses. One of the most useful guidelines in this area is Lipinski's rule of five to prioritize compounds with a high likelihood of oral absorption in drug discovery (Lipinski et al. 1997). Lipinski's rule of five evaluates the molecular properties of new compounds and the likelihood of becoming orally active drug in humans. These chemical properties are major influencers of absorption, distribution, metabolism, and excretion (e.g. hydrogen bond donors, acceptors, molecular mass, and octanol–water partition). With the progress in pharmacology, drug development is already expanding beyond Lipinski's rules (Doak et al. 2014).

Concerns with inconsequential clinical studies were overviewed by Deborah Zarin at the NLM-FNLM Consequential Clinical Research Conference in 2017. Examples include studies not supported by systematic review of available data, not based on substantial question, overly restrictive eligibility criteria, inappropriate comparators, inadequate power, questionable design like single arm or small numbers, questionable outcome measures, non-standardized outcome measures, problems with accrual, study terminated without results, non-reporting of results, unacknowledged protocol deviations like changes to outcome measures, and selective outcome measure/adverse effect reporting.

Stanford researcher John Ioannidis has been in the forefront of studying the validity of published research findings (Ioannidis 2005). He highlighted that not only most findings are false, but also most of the true findings are not useful in contemporary clinical research (Ioannidis 2016). He suggested the following questions to consider in assessing the usefulness of clinical research: Was there a systematic assessment of prior evidence to show the need for new studies? Is there a health problem that is big/important enough to be addressed by research? Are methods, data, and analyses verifiable, unbiased, and reproducible? Does the research reflect real life or, if deviates, does it matter? Is the proposed study large and long enough to be informative? Is the research worth the money?

Targeting Mechanisms Goes Beyond Observation

Research without underlying model assumptions and multiple probing of key aspects of the modeled mechanism can be called observational study or documenting research. Such research limits itself to inspecting symptoms and documenting signs. In many areas of applied sciences, the observational approach is the only option, at least initially (e.g. various clinical research projects, prevalence, and incidence of diseases).

In *observational research*, when the researcher inspects a limited number of "things" without an underlying model, usually there is only one measurement for each "thing." The tested multiple "things" do not necessarily have clear, logical connections with each other or an underlying model relating them to each other. For obvious reasons, the studies are particularly vulnerable to investigative bias.

Consequently, such measurement of "things" is entirely at the mercy of accuracy and statistical analysis, confidence intervals, and tests of significance. The slightest deviation or miscalculations can cause irreparable harm to reproducibility without warning signs. There is no good way to verify the results, except conducting the same measurements in a different sample and perhaps by a different laboratory.

The research focused on a generalized model of research translation tries to understand an underlying principle, *law of nature*, or societal process. In such cases, the same model can be tested from a variety of directions cumulatively supporting or rejecting the model. The method of testing the same thing from different directions is called triangulation. When a comprehensive model is developed, there are opportunities to test it from multiple directions.

In spite of the serious drawbacks, many journal reviewers still prefer to see just one but presumably accurate testing as opposed to multiple corroborating measurements. For example, the study by Balas and Boren (2000) presented a stepwise model of translating clinical trial evidence into day-to-day patient care. The study used two independent measurement methods for estimating the time needed for translation from clinical trials to practical application. According to one estimate, the average transfer time turned out to be 17 years. The other measurement estimated the same period as 15.3 years needed to reach 50% utilization in patient care nationwide.

Obviously, the two estimates were not identical but quite close to each other. Together, they increase confidence in the transfer time approximations.

Ultimately, the 17-year estimate went into the scientific folklore and became widely cited by the research community and also by President Barack Obama in his Chicago address at the AMA meeting.

Focusing on the grand challenges is not for the fainthearted as it requires broad vision, knowledge of trends in science, and understanding of societal issues. In the records of science and technology, there have been many successful examples of such bold and audacious approach that eventually led to unexpected but important discoveries.

Nobel laurate Otto Warburg was quite vocal about the need to focus on the greatest unsolved mysteries of our time. Correspondingly, his research concentrated on the energy cycles of cells, oxygen consumption, and respiration in general. Later his attention turned to cancer but again with an emphasis on energy production and consumption. The so-called Warburg effect is the characteristic of many tumor cells that exhibit the high activity of anaerobic glycolysis even in the presence of sufficient oxygen. Warburg's studies on metabolism led to many new discoveries and profoundly changed our understanding of fundamental mechanisms of life.

Without being limited to preconceived notions, the researcher has to be open-minded and seek to learn unexpected new functional relationships and mechanisms. It is always easier to recognize what we already know as opposed to recognizing and understanding something entirely new.

Repurposing What Is Already Known

In managing the unexpected, *repurposing* is a particularly important skill. When surprising but reproducible results come up, the researcher is challenged to recognize the potential value, either scientific or practical application. It may sound trivial but frequently major challenge as the deviating results come from research that was initially designed to achieve something else.

In the history of pharmacology, several drugs developed for a particular disease started to exhibit unexpected side effects that became "accidental" opportunities to treat an entirely different condition. Some of the examples include minoxidil, an antihypertensive vasodilator that was found to promote hair regrowth; sildenafil, developed for the treatment of pulmonary arterial hypertension but turned out to be useful to treat erectile dysfunction under the trade name Viagra; and chlorpromazine, initially tested as a surgical anesthetic but later turned out to be an effective antipsychotic medication used particularly in the treatment of schizophrenia. Originally, AZT was developed with great hopes for cancer treatment but turned out to be not very useful. A few years later, it was repurposed as the first safe and effective drug against HIV.

A very effective way to make accidental discoveries is looking for unusual findings and, after finding them, aggressively focusing on investigating what is behind them. For example, chemistry Nobel laureate Harry Kroto conducted studies of carbon molecules and discovered the C60 molecule, highly stable structure similar to a geodesic dome, the so-called buckminsterfullerene. His discovery, the fullerenes, has been used as a carrier for gene and drug delivery systems and serum protein profiling and biomarker discovery (Bakry et al. 2007). One of his colleagues noted that "whenever he saw anything a little bit unusual, he would really key in on it."

Stephanie Kwolek was a pioneering chemist and the discoverer of Kevlar. She was born in Pennsylvania, the daughter of immigrants from Poland. Kwolek graduated from Carnegie Institution of Technology with a chemistry degree (Selle 2004). After graduation, she was hired by DuPont.

She started working on specialty fibers in Wilmington. Kwolek specialized in developing low-temperature processes for finding innovative petroleum-based synthetic fibers. DuPont was looking for fibers stronger and lighter than steel to improve the performance of automobile tires.

She turned her attention to poly-*p*-benzamide and poly-*p*-phenylene terephthalamide, but they were difficult to dissolve. Thank her creativity and perseverance, she found a solvent to dissolve long-chain polymers into a solution in 1965. In comparison with other polymer solutions, this one was thinner and more watery.

She convinced an apprehensive colleague to put the solution into a spinneret that turns liquid polymers into fibers. Once spun, the polymer's molecules lined up in parallel (Kwolek and Yang 1993). The resulting fiber showed extraordinary stiffness and tensile strength, unlike anything they had made before, five times stronger by weight than steel.

Later, Kwolek recalled, "I was very excited, as was the whole laboratory excited, and management was excited because we were looking for something new, something different, and this was it" (Kwolek 2013). In 1971, poly-*p*-phenylene terephthalamide was released under the trade name Kevlar commercially.

Kwolek received the National Medal of Technology and was elected member of the National Academy of Engineering. Over the years, Kevlar became an essential ingredient of oven gloves, tennis rackets, skis, boat hulls, car tires, airplanes, ropes, fire fighter boots, cables, hockey sticks, and most importantly lightweight helmets and bullet-resistant vests saving countless lives. In an interview, Stephanie Kwolek summed it up: "I think the role of science is to improve human life in general and improve the world as well" (Kwolek 2013).

Eureka Moment of Discovering the Unexpected

A further generalization of the concept of repurposing is readiness to recognize the unexpected in research studies. In purposeful targeting, the first approach is looking into the future and examining the most pressing challenges for science and humanity. It is about asking what the greatest shortcomings in our understanding are and which questions should be answered by appropriately focused scientific research. Discovery occurs when the prioritized, systematic investigation encounters unforeseen results.

Prevention and treatment of cancer have been the ultimate motivator and eminent target for many successful life scientists. The research of Jan Vilček also started with a focus on cancer and the tumor necrosis factor. However, the course of his studies on the immune system led him to major discoveries and effective treatment of inflammatory autoimmune diseases. In other words, research on cancer led to many landmark discoveries that ultimately had little to do with cancer but still resulted in profound discoveries that changed millions of lives.

As a postdoctoral student, Randy Schekman had the chance to learn and work in the laboratory of Arthur Kornberg, discoverer of the DNA polymerase. After a few years of very productive research, Schekman came to a divergent conclusion as later described in his Nobel lecture (Schekman 2013): "I resolved to strike off in a new direction where I might have the chance to establish my own identity and not be dependent on or overshadowed by Kornberg's reputation."

Schekman attended a meeting of the American Society for Cell Biology in San Diego, and he came away with the feeling that "cell biology had yet to enter the molecular world of biochemical mechanism. Here was an enormously complex pathway of membrane transformation in the secretory pathway and yet not a single protein had been ascribed a specific role in this essential process." This observation became the starting point of his successful research for years to come and ultimately his Nobel Prize-winning discoveries.

Elon Musk, prominent technology entrepreneur and one of the most extraordinary readers of cutting-edge science, pursued the grand challenges approach. Specifically, he identifies the following far-reaching questions that humanity needs to answer: sustainable energy, making life multi-planetary, artificial intelligence, and genetics. Indeed, his high-tech entrepreneurship is mostly and, so far, quite effectively focused on the first three challenges (electric car, solar technologies, open artificial intelligence, and reusable space transportation).

The origin of the word *Eureka* goes back to antiquity when Archimedes was presumably taking a bath and all of a sudden realized that the volume of water displaced must equal to the volume of the submerged part of his body. This recognition was the key to answering the king's question and determining whether the crown was pure gold. He shouted Eureka happily while running out to the street of Syracuse. There are many questions regarding the fidelity of this story. However, one thing is widely accepted that Eureka means "I have found it" in Greek. Today, Eureka is the state motto of California and also the frequent reference point in discussions about science.

Eureka identifies the pinnacle of scientific problem solving: when the researcher suddenly understands something that was previously incomprehensible. It is an objective achievement but also immensely satisfying personal experience. Sometimes, the time elapsed between recognition of the problem and the emergence of solution can be quite lengthy. Unsurprisingly, Eureka or sometimes called "aha" moment or insight has also been the subject of considerable scientific investigations.

The study by Jung-Beeman et al. (2004) observed two objective neural correlates of insight. Functional magnetic resonance imaging suggested increased activity in the anterior superior temporal gyrus of the right hemisphere. Meanwhile, scalp electroencephalogram recordings indicated a sudden burst of high-frequency neural activity in the same area a few seconds before the realization of the insight solutions. The sudden flash of insight occurs when neural and cognitive processes recognize connections that were previously incomprehensible.

Many excellent scientists recall the role of withdrawal and thinking in making their discoveries. Gobind Khorana was nowhere to be found when the Nobel committee wanted to contact him for the announcement of the Prize. "Well, I find it very difficult sometimes to get enough time to think. I need really just to be myself. I get away, and I use a place in the country. Get away from the phone, from the school obligations" (Khorana 2011). Eventually, his wife drove out to him and conveyed the amazing message.

According to the study of Wagner et al. (2004), pivotal insights can be gained through sleep, consistently with anecdotal reports on scientific discovery. Apparently sleep, by restructuring new memory representations, facilitates extraction of explicit knowledge and gaining insight. One may say that sleeping can be a facilitator of Eureka and an obstacle of serendipity.

Theodore Kármán, the recipient of the Daniel Guggenheim Medal and the National Medal of Science, came up with the concept of the Kármán line, the edge of the outer space, a physical boundary that lies at an altitude of 100 km, where aerodynamics stops and astronautics begins. Below the Kármán line airspace belongs to each country, and above it is free, international space. Kármán wrote, "the finest creative thought comes not out of organized teams but out of the quiet of one's own world" (Von Kármán and Edson 1967).

Wilhelm Röntgen was a German engineer and physicist. He is credited with the discovery of X-rays or Röntgen rays in 1895. Röntgen was born in the Netherlands but studied mechanical engineering in Switzerland. Eventually, he got his PhD at the University of Zürich. After graduation, he became a faculty member at the University of Strasburg and leader at the University of Würzburg (Glasser 1934).

Röntgen was a very talented student, astute scientific observer, and creative, innovative thinker. He quickly rose in rank, and various universities started to compete for his talent. Röntgen's most famous experiments were conducted at the University of Würzburg. After his discovery of X-rays, he was awarded an honorary doctorate by the University, and a few years later he was elected rector or president of the university.

At that time, many scientists experimented with cathode rays, but their results were conflicting. Röntgen conducted a series of experiments with the Crookes vacuum tubes. When electrical discharge passed the tube, he noticed that some rays were coming out, in spite of the tube being fully covered to block penetration of light. Röntgen also noticed that a nearby fluorescent screen showed the contour of various objects at a distance that was greater than the distance traveled by cathode rays. Soon afterward, he observed the contour of his bones on the florescent screen. On the next day of this discovery and after further experimentation, Röntgen created the world's first X-ray picture of the hand of his wife.

At the time of these experiments, Röntgen already had a tremendous scientific reputation. With the discovery of X-rays, he became an instant scientific sensation. In 1901, he received the first Nobel Prize in Physics.

From Serendipity to Practical Applications

Since the early days of science, stories of accidental discoveries have been plenty. From antiquity to modern days, the sudden, unanticipated discovery always attracted people's fascination (Kohn 1989). In interpreting the accidental discoveries, two extreme views also emerged.

One believes that the majority of breakthrough discoveries are accidental. People just should do good research and accidents will happen here and there. Obviously, such fatalistic view of science runs contrary to several observations: many Nobel Prize winners trained a disproportionately large number of future Nobel Prize winners in their laboratory. Serial research innovators produce hundreds of scientific articles and patents that are licensed, hardly fruits of serial accidents. If accident is the only driver of success, then many other factors known to accelerate good science are supposed to be useless (e.g. convergence, engaged research, cutting-edge technology, and others). Obviously, crossing your fingers is not a credible research proposal.

If accidental discoveries are exceedingly rare and marginal, only systematic research effort could produce meaningful results. The voluntarist approach disregards the possibility of accidental discoveries and limits itself to pre-planned results. Consequently, the long list of valuable but accidental discoveries would be disregarded. Understandably, such rigid and deterministic view of scientific progress would narrow down research discoveries to what is already expected and ignore the unexpected. It is noteworthy that accidental discoveries are most frequently lucky, practically useful findings but rarely profound recognitions of nature's law.

Between these extremes lies the most likely answer: well-planned research that is firmly rooted in the results of past observations and focused on developing new models is likely to produce the best results, either by achieving its targeted model or by finding something different accidentally but certainly happily. Chance favors only the prepared mind and the hardworking researcher who can recognize the scientific or practical value of unexpected results. As it was stated by Joseph Henry, a professor at Princeton and later the first secretary of the Smithsonian Institution, "The seeds of great discoveries are constantly floating around us, but they only take root in minds well prepared to receive them" (Henry 1886).

Ostensibly, the word *serendipity* comes from Horace Walpole who discovered the Persian fairy tale *The Three Princes of Serendip* in 1754. These were heroes who always made accidental discoveries. Apparently, Serendip is an old name for Sri Lanka. Today, serendipity is synonymous with pleasant surprises and fortunate happenstance (Editorial 1965). The list of accidental discoveries is very long with many practical solutions among them.

Serendipity and accidental discoveries are frequently mentioned in the discussion about major scientific accomplishments (Meyers 2007). Seemingly, the stories appear to justify such interpretation. Fleming accidentally left the Petri dish in the laboratory before going for vacation and unexpectedly discovered the inhibiting effect of the Penicillin mold afterward. Wilhelm Röntgen experimented with the mysterious X-ray and unexpectedly noticed the change on the screen. When Howard Temin studied RNA tumor virus replication, he all of a sudden realized that it could only be possible if it involves a DNA intermediate through reverse transcription.

These and other narratives may portray science in a fairytale world. They can also give a convenient excuse to anyone who has been doing research for a long time but has never bumped into any such happy accidents. Of course, the reality of research is quite different from the fairytale.

Just looking at the above-listed stories can convince anyone that an extraordinarily prepared mind, keen observational skills, and original interpretation of the facts were behind most of them. For example, the discovery of nonselective (broad-spectrum) active transporters was the result of lengthy, meticulous work. The serendipitous discovery of the permeability glycoprotein was the first among them, but without the preceding studies, it would have never happened.

Furthermore, many of the accidental observations were actually not new but something that was observed by others many times before, but those observers came down to some trivial and everyday interpretations without realizing the opportunity for discovery.

In many ways, Röntgen repeated an experiment that was conducted by others as well, but others never realized the full implications (Walden 1991). It should be noted that the Crookes tubes were invented around 1870, and subsequently many scientists experimented with them. Undoubtedly several of them could see effects of X-ray without realizing what was behind the phenomena. Table 14.1 presents a side-by-side comparison of the experimentation by two talented scientists, Arthur Goodspeed in the United States and Wilhelm Röntgen in Germany, one leading to discarded results and the other one making the discovery and receiving the Nobel Prize (Walden 1991).

Pioneering investigators should also be appropriately well prepared and cautious in their enthusiasm. A year after the discovery by Röntgen, Professor

Table 14.1 The benefit of tracking down unusual findings: comparison of experiments by Arthur Goodspeed in 1890 and Wilhelm Röntgen in 1895.

Researcher	Arthur Willis Goodspeed	Wilhelm Conrad Röntgen
University	Assistant professor, University of Pennsylvania	Chair of physics, University of Würzburg
Year of experiment	1890 (at age 30)	1895 (at age 50)
Experiment	Crookes tubes, partially evacuated glass bulb with two metal electrodes, cathode and anode, at the end. Applying high voltage between the electrodes projects electrons in straight lines from the cathode-causing fluorescence	Same experiments with Crookes tubes
Observation	Tubes produce contours of objects on covered photographic plates	Covered tubes produce contours of objects and bones on a screen or photographic plates
Interpretation	Several fogged pictures, one with the shadow of a mysterious disc, unexplained, disregarded	Faint shimmering on the screen, the discovery of a new kind of ray, called X-ray
Recognition	Photographic plates put aside, only to be rediscovered years later, after Röntgen's discovery	Nobel Prize in physics

John Daniels wanted to make a pioneering skull X-ray in the United States. He exposed the head of Charles Dudley dean of the Medical School at Vanderbilt University to X-ray radiation for 60 minutes. The skull X-ray experiment ended up in complete failure (Daniels 1896). More visibly, Dean Dudley lost his hair three weeks later.

In the history of science, it was not the first time that the penicillin mold landed on an agar plate filled with *Staphylococcus aureus*, but Fleming was the first with an intense interest in the inhibiting effect. It should also be added that Fleming missed the opportunity of exploring the possibility of clinical application. Similarly, shortly after the emergence of Temin's theory of reverse transcriptase, other scientists came to similar experimental conclusions as well.

Satoshi Ōmura tested thousands of soil samples before finding ivermectin. When the pioneering scientists test many compounds or samples and achieve breakthrough, then not only the "accidental" discovery should be noticed, but the clever targeting of screening, backbreaking work of repetitive evaluations, and ingenious interpretation should also be recognized.

Choosing the right target for research is of course a question of tremendous implications. In our quest for knowledge, we tend to idealize the averages and ignore the outliers. Looking at the history of science, surprisingly large number of discoveries came from the careful investigation of outliers. Good scientists need both skills, Eureka and serendipity, for targeting research and repurposing unexpected discoveries.

To streamline the process, successful research needs clearly set targets that are well grounded in past observations and set audacious but realistic milestones. In the case of unexpected discoveries, the creative researcher also finds ways to previously unimagined theoretical progress and practical applications.

References

Argyris, C. (1996). Actionable knowledge: design causality in the service of consequential theory. *The Journal of Applied Behavioral Science* 32 (4): 390–406.

Bakry, R., Vallant, R.M., Najam-ul-Haq, M. et al. (2007). Medicinal applications of fullerenes. *International Journal of Nanomedicine* 2 (4): 639.

Balas, E.A. and Boren, S.A. (2000). Managing clinical knowledge for health care improvement. In: *Yearbook of Medical Informatics 2000: Patient-Centered Systems* (ed. J. Bemmel and A.T. McCray), 65–70. Stuttgart, Germany: Schattauer Verlagsgesellschaft mbH.

Daniels, J. (1896). The x rays. *Science* 3: 562–563.

Darnovsky, M. (2008). Germline modification carries risk of major social harm. *Nature* 453 (7196): 720–720.

Doak, B.C., Over, B., Giordanetto, F., and Kihlberg, J. (2014). Oral druggable space beyond the rule of 5: insights from drugs and clinical candidates. *Chemistry & Biology* 21 (9): 1115–1142.

Editorial (1965). Serendipity and the three princes. *The New England Journal of Medicine* 273: 51–52.

Ekvall, T. and Weidema, B.P. (2004). System boundaries and input data in consequential life cycle inventory analysis. *The International Journal of Life Cycle Assessment* 9 (3): 161–171.

Glasser, O. (1934). *Wilhelm Conrad Röntgen and the Early History of the Röntgen Rays*. Springfield, IL: Charles C. Thomas.

Henry, J. (1886). The method of scientific investigation, and its application to some abnormal phenomena of sound. *Bulletin of the philosophical Society of Washington*, 1877, volume 11, pp. 162–174. In: *Scientific Writings of Joseph Henry*, vol. 550 (ed. J. Henry). Washington, DC: Smithsonian Institution.

Ioannidis, J.P. (2005). Why most published research findings are false. *PLoS Medicine* 2 (8): e124.

Ioannidis, J.P. (2016). Why most clinical research is not useful. *PLoS Medicine* 13 (6): e1002049.

Jung-Beeman, M., Bowden, E.M., Haberman, J. et al. (2004). Neural activity when people solve verbal problems with insight. *PLoS Biology* 2 (4): e97.

Khorana, G. (2011). UW-Madison's Nobel Prize winners. https://www.youtube.com/watch?v=g-BidjlCnHs (accessed 4 July 2017).

Kohn, A. (1989). *Fortune or Failure: Missed Opportunities and Chance Discoveries*. Oxford: Blackwell.

Kwolek, S. (2013). Museum of science Stephanie Kwolek. https://www.youtube.com/watch?v=UISIOB9tnRw (accessed 3 July 2017).

Kwolek, S.L. and Yang, H.H. (1993). History of aramid fibers. In: *Manmade Fibers: Their Origin and Development*, 315–336. Barking, UK: Elsevier Applied Science Publishers Ltd.

Lipinski, C.A., Lombardo, F., Dominy, B.W., and Feeney, P.J. (1997). Experimental and computational approaches to estimate solubility and permeability in drug discovery and development settings. *Advanced Drug Delivery Reviews* 23 (1–3): 3–25.

Messick, S. (1993). Foundations of validity: meaning and consequences in psychological assessment. *ETS Research Report Series* 1993 (2).

Meyers, M.A. (2007). *Happy Accidents: Serendipity in Modern Medical Breakthroughs*. New York: Arcade Publishing.

Owens, J. (2007). Determining druggability. *Nature Reviews Drug Discovery* 6 (3): 187–187.

Plevin, R.J., Delucchi, M.A., and Creutzig, F. (2014). Using attributional life cycle assessment to estimate climate-change mitigation benefits misleads policy makers. *Journal of Industrial Ecology* 18 (1): 73–83.

Schekman, R. (2013). Genes and proteins that control the secretory pathway. *Nobel Lecture* (7 December).

Selle, R.R. (2004). The woman who created kevlar. *The World & I* 19 (3): 44.

Shepard, L.A. (1997). The centrality of test use and consequences for test validity. *Educational Measurement: Issues and Practice* 16: 5–24.

Von Kármán, T. and Edson, L. (1967). *The Wind and Beyond: Theodore von Karman, Pioneer in Aviation and Pathfinder in Space*. Boston: Little, Brown.

Wagner, U., Gais, S., Haider, H. et al. (2004). Sleep inspires insight. *Nature* 427 (6972): 352–355.

Walden, T.L. Jr. (1991). The first radiation accident in America: a centennial account of the x-ray photograph made in 1890. *Radiology* 181 (3): 635–639.

15

Trailblazing Technologies

> *Biology should drive the choice of the technology developed and the technology should lift barriers to the deciphering of important biological information.*
>
> Leroy Hood (2002)[1]

Innovative research characteristically uses some ingenious and straightforward, otherwise called elegant, methodologies designed by the lead researcher or perhaps collaborators. Answering new questions typically requires new methodologies. This need is not surprising as established methodologies have already been used by many people and major discoveries are less likely with repeated use. Moreover, entirely new questions predictably need new ways to answer them. The examples of success are countless.

New methodologies also have a tendency to attract a greater number of citations by other researchers. The journal *Nature* published an eye-opening ranking of the top 100 papers published during 1900–2014 based on the number of citations attracted (Van Noorden et al. 2014). It noted that none of the well-known landmark discoveries, like the DNA double helix and others, got even close to the group of most cited papers. The vast majority of best-cited papers described experimental methods or software that proved to be essential tools for subsequent research (Table 15.1).

A similar but more focused analysis of the most-cited articles published in the *Proceedings of the National Academy of Sciences* also suggested comparable advantages for methodology papers (Gerow et al. 2018). Further examination reveals that particularly methodologies for protein studies fared well in subsequent use and citations, ahead of DNA, RNA, or lipid studies.

---

1 Hood, L. (2002). My life and adventures integrating biology and technology. A commemorative lecture for the 2002 Kyoto Prize in Advanced Technologies.

Innovative Research in Life Sciences: Pathways to Scientific Impact, Public Health Improvement, and Economic Progress, First Edition. E. Andrew Balas.
© 2019 John Wiley & Sons, Inc. Published 2019 by John Wiley & Sons, Inc.

Table 15.1 The top 10 cited scientific papers according to Web of Science.

Paper	Subject	Citation
Lowry et al. (1951)	Protein measurement with the Folin phenol reagent	305 148
Laemmli (1970)	Cleavage of structural proteins during the assembly of the head of bacteriophage T4	213 005
Bradford (1976)	A rapid and sensitive method for the quantitation of microgram quantities of protein utilizing the principle of protein–dye binding	155 530
Sanger et al. (1977)	DNA sequencing with chain-terminating inhibitors	65 335
Chomczynski and Sacchi (1987)	Single-step method of RNA isolation by acid guanidinium thiocyanate-phenol-chloroform extraction	60 397
Towbin et al. (1979)	Electrophoretic transfer of proteins from polyacrylamide gels to nitrocellulose sheets: procedure and some applications	53 349
Lee et al. (1988)	Development of the Colle-Salvetti correlation-energy formula into a functional of the electron density	46 702
Becke (1993)	Density-functional thermochemistry. III. The role of exact exchange	46 145
Folch et al. (1957)	A simple method for the isolation and purification of total lipids from animal tissues	45 131
Thompson et al. (1994)	Clustal W: improving the sensitivity of progressive multiple sequence alignment through sequence weighting, position-specific gap penalties, and weight matrix choice	40 289

Source: Data from Van Noorden et al. (2014).

Behind Every Great Discovery, There Is an Elegant Methodology

On the road to scientific discoveries, novel methodologies can provide the much-needed insight and new information that can precisely answer the research question. In examining the history of significant discoveries, finding creative use and benefits of trailblazing scientific methodologies is almost universal.

Successfully learning a new methodology and applying it in practice can greatly increase the confidence of the researcher in reproducibility of results and also provide essential insight into the functioning of nature. Many award-winning scientific discoveries not only uncovered novel concepts but

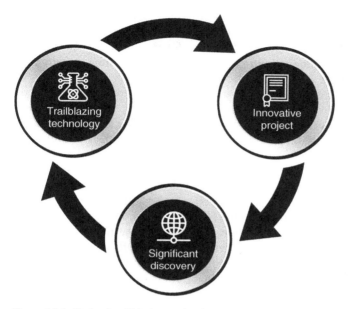

Figure 15.1 Circle of trailblazing technology innovation.

also used trailblazing measurement techniques and their unprecedented results (Figure 15.1).

One of the best examples of the remarkable insight provided by a novel methodology is the discovery of the DNA double helix. Several years earlier, Rosalind Franklin learned X-ray crystallography under the direction of Jacques Mering in Paris. In the focused, analytical work of Watson and Crick, the new X-ray diffraction image of DNA taken by Raymond Gosling, a PhD student of Rosalind Franklin provided an essential clue (i.e. the famous Photograph 51).

In X-ray crystallography, first an adequate crystalline specimen is obtained, and a beam of X-rays is shined through it. In the crystal, electron clouds of the atoms scatter the beam that can then be observed on a screen behind the crystal. Based on the intensities, an experienced crystallographer can create a three-dimensional picture of the electron density, and from it, the mean positions of the atoms within a crystal can be determined. Subsequently, the crystal must be rotated stepwise through various angles to get hundreds of photos and eventually a three-dimensional structure. At this point, the structure of the molecule can be considered "solved."

Often, application of the new methodology can be very challenging, and simply reading the methods section of a manuscript may not be sufficient to reach proficiency. Correspondingly, many researchers benefited from not only discussions with the designer of new methodology but also spending some time in their laboratory to practice. The breakthrough DNA visualization in the Franklin laboratory came from the proficient use of X-ray crystallography. With this new methodology, the opportunities for mistakes are endless, and only well-trained crystallographers can succeed. Such learning experiences represent another great way of scientific mentoring.

Development and installation of new methodologies are at the center of every ambitious research project. Constant and systematic search for *innovative methodologies* relevant to someone's scientific objectives is one of the most important activities of developing a new major research project. Monitoring competitors, scanning the literature, and discussing the need for measuring new phenomena are all part of this systematic search. Consulting other researchers on how to measure certain, essential characteristics can be very helpful in generating ideas for new and effective sources of information.

When Marshall Nirenberg received the Nobel Prize for interpretation of the genetic code and its function in protein synthesis, several new technologies were instrumental in making the achievement possible (Office of NIH History 2017). The French press, designed by Stacey French, was an instrument to rupture cells. The sample is placed in a chamber, then the pressure is raised, and subsequently the cells are forced into another chamber with a different pressure level. The rapid change in pressure ruptures the cells. This method was much more efficient than the earlier one used originally by Nirenberg (i.e. mortar and pestle to grind up cells). At NIH, Philip Leder and Charles Byrne designed another ingenious technology, the multiple millipore filtration instrument, named the "multi-plater" that tests 45 samples at once. In Nirenberg's studies, it separated radioactively labeled proteins from the solution in which they were suspended. Again, this method was far superior to the earlier precipitation and differential centrifugation methodology.

Continuous monitoring of activities in competing laboratories also makes good sense for learning and advancing. On the other hand, there is no need to over-observe and being driven by what others are doing as the community of researchers can be broadly misguided by illusions (see the chapter on engaged research about bubble bias). Scientists should be driven by the problem they try to solve and not by prevailing views, politics, or competition.

The breakthrough Warburg manometer, spectrophotometry, and tissue slicing methodologies played key roles in studies of aerobic and anaerobic perspiration. In 1931, Warburg received the Nobel Prize "for his discovery of the nature and mode of action of the respiratory enzyme."

The Warburg manometer was designed to measure aerobic respiration rates in tissue slices and small organisms. The basis for the measurement was that the amount of CO_2 produced is equivalent to O_2 uptake by a tissue sample placed in a flask. The carbon dioxide released from respiring tissue inside a flask is absorbed by potassium hydroxide. In the closed system, the absorption of carbon dioxide results in a drop in pressure that is measured by a manometer. As a result, the quantity of gas generated or absorbed is measured by the Warburg manometer.

The *Warburg apparatus* consists of 12–14 flasks that are hermetically connected to manometers by stopcocks. The precise measurement is very sensitive to temperature, which is controlled by immersing the flasks in a temperature-controlled water bath equipped with a heat-regulating device and a rocking mechanism. Changes in atmospheric pressure do not influence results in the closed-circuit system. The very thin slices of tissue are in fact small colonies of cells that survive and can be observed for the rate of perspiration for a few hours (Krebs 1981).

Beyond Warburg's Nobel Prize-winning work, the significance of his manometer was further illustrated when one of his research trainees, Hans Krebs, was forced to leave Germany in 1933. The Nazis allowed him to take only very few personal belongings but, surprisingly, let him take laboratory equipment, including 30 Warburg manometers. This equipment proved to be enormously valuable in jumpstarting the research of Hans Krebs in England.

In studying the problem of perspiration and putting the Warburg manometers to good use, Krebs generated innumerable hypothetical, pencil and paper, pathways, and most of them were wrong. Ultimately, the systematic studies and many experimental measurements offered valuable insight and valuable findings. His studies discovered the series of chemical reactions

---

Otto Warburg is a physician, physiologist, Nobel laureate, and one of the most influential scientific personalities of the twentieth century. He was a meticulous, unrelenting, and technically brilliant methodologist. Warburg was primarily concerned with the biological energy transformation. He noticed that oxygen consumption increases sixfold after fertilization. He was among the first to appreciate that processes of living matter follow the laws of physics and chemistry, a trivial observation today but not accepted during his time.

Warburg was a master of developing innovative research methodologies that were also very precise. Examples include his manometry method to measure gas exchange, spectrophotometry for measuring the reaction rates and quantities of metabolites, and a tissue slice technique based on his mathematical model for optimal thickness. Warburg published experimental papers at the rate of about five per year.

Three Nobel laureates benefited from mentoring in his laboratory: Meyerhof in 1922, Krebs in 1953, and Theorell in 1955. Warburg was first to create purified and crystallized enzymes and in effect created a new branch of the chemical manufacturing industry. An executive of enzyme manufacturer Boehringer Mannheim later acknowledged that "we consider ourselves as the executors of Warburg's pioneer work." (Krebs 1981).

used by all aerobic organisms to generate energy through the oxidation of acetyl-CoA. In 1953, Hans Krebs received the Nobel Prize "for his discovery of the citric acid cycle."

Sought-after Technique Becomes the Goal of Research

In science, the novel technology is often not just an instrument of inquiry but more importantly a development effort to meet major needs. Creation of the instrument or technology becomes the purpose of the research project. Making new technology the research goal resonates with the observation that the functioning prototype is one of the most convincing demonstrations of reproducible research results. In some cases, the resulting technology becomes a valuable tool for further research, while in other cases it becomes used by the general public to improve public health.

Development of a new measurement technique is often more than just technology development and involves proper **conceptualization of a phenomenon** that should be measured (Figure 15.1). No one should trivialize development of new measurement techniques by assuming that the concept of what should be measured is straightforward and clear from the beginning. Science has many examples of *conceptual evolution*. Time, temperature, or visual acuity appears to be simple and trivial concepts today, but in fact, they went through a series of changes in interpretation and also measurement.

The history of conceptual understanding of blood pressure and development of practical tools to measure it illustrates how groundbreaking methodologies evolve technically and conceptually over time and across many countries. Today, this measurement makes possible the recognition and treatment of one of the most devastating conditions, the essential and secondary hypertension diseases, a direct contribution to better public health. Beyond everyday clinical applications, it is also one of the most basic measurements in animal and human research experiments alike.

Multiple discoveries led to our current understanding and daily use of the measurement of systolic and diastolic blood pressure. In England 1616, William Harvey recognized that there was a finite amount of blood circulating one direction in the body. The first invasive blood-pressure recording was made by Stephen Hales by inserting one end of a brass pipe into the artery of a horse in 1733. In the 1800s, the French Poiseuille introduced the mercury hydrodynometer and the mmHg unit of measurement. In 1896, the Italian Scipione Riva-Rocci developed the modern mercury sphygmomanometer but could only measure systolic pressure (Roguin 2006). In Russia 1905, Nikolai Korotkoff noticed the sounds made by the constriction of the artery, and this made measurement of the diastolic blood pressure practical for widespread use today.

When Rosalyn Yalow and Solomon Berson worked on diabetes studies, they investigated the Mirsky hypothesis that maturity-onset diabetes is a consequence of abnormally rapid degradation of insulin as opposed to a deficiency in insulin secretion. They studied the metabolism of 131I-labeled insulin following intravenous administration to nondiabetic and diabetic subjects and observed that such insulin disappeared more slowly from the plasma of patients who had received insulin previously. They hypothesized that this was due to binding of labeled insulin to antibodies generated in response to previous administration of exogenous insulin. This significant observation led to the development of a new methodology, the radioimmunoassay (RIA) that revolutionized the measurement of peptide hormones in the blood and many other areas of scientific and clinical testing.

Not infrequently, a **major discovery forms the basis of a methodology** for further scientific studies. Recognition of a natural phenomenon becomes a unique opportunity for the development of a research methodology that yields multiple subsequent benefits. Rous sarcoma retrovirus (RSV) was the first oncovirus discovered by Peyton Rous at Rockefeller University in 1911. He observed that sarcoma growing on a domestic chicken could be transferred to a healthy bird by a cell-free filtrate.

In the beginning, the idea of viral transfer of cancer was received with great skepticism. Over time, the *Rous sarcoma* virus became an overwhelmingly effective and influential instrument of discovery in cancer research. While the discovery of the Rous sarcoma was a significant milestone in the understanding of cancer, the practical impact came from the use of this experimental model in further scientific investigations.

First created by Nobel laureate Mario R. Capecchi, *knockout mice* provide a particularly effective gene-**targeting method** for the large-scale study of the relationship between genotype and phenotype. The knockout mouse model is a uniquely valuable research method in the study of human diseases. It is developed when a gene is deleted or made inactive (i.e. knocked out). Such animal models offer unparalleled opportunity to study the missing effect of specific genes.

The knockout mouse model development greatly benefited from the study of Wigler et al. (1977) in which they showed that added DNA was randomly incorporated into the host cell genome. To increase efficiency, Capecchi made tiny hypodermic needles and stuffed the DNA directly into the nucleus itself. The result was a success in one of every three cells, instead of one in a million cells. Another fortunate breakthrough came when Martin Evans successfully isolated mouse embryonic stem (ES) cells (Evans and Kaufman 1981). The combination of technologies made the large-scale, targeted production of knockout mice possible (Capecchi 1994).

Every year, millions of knockout mice are used to study diseases ranging from cardiovascular diseases, metabolic conditions, various cancers, and

In the early years of his research, **LeRoy Hood**, a successful biologist and serial innovator, focused on the genetic foundation of antigen variability, a previously unresolved inconsistency between DNA stability and highly variable, adaptive immune response. He found that random genetic changes add to the variability of antibody genes. For his discoveries in immunology, Hood received the Albert Lasker Basic Medical Research Award in 1987 (Pollack 2001).

Working at Caltech and later at the University of Washington, he not only made many extraordinary discoveries but also designed and realized major new measurement technologies that revolutionized the field. In his development of innovative methodologies, Hood created high-throughput data accumulation through automation and parallelization of the protein and DNA analyses. His innovative instruments include the protein sequencer, protein synthesizer, DNA synthesizer, automated DNA sequencer, and ink-jet DNA synthesis technology. For his landmark designs, he received the Kyoto Prize in Advanced Technology in 2002.

Particularly, his automated DNA sequencer played a pivotal role by making the human genome project feasible. Francis Collins, director of the National Human Genome Research Institute, readily acknowledged, "We would not be here today if not for the innovation in technology" (Pollack 2001). Among others, Hood also helped to launch several major biotech companies, including Amgen and Applied Biosystems.

Most recently, Hood became an advocate of predictive, preventive, personalized, and participatory medicine: "The medicine of the past worked on population averages, the medicine of the future will focus on the individual" (Hood 2011).

others. The two methods used to create knockout mouse are gene targeting and gene trapping, i.e. specifically manipulate a gene or use a random process. Most frequently, the targeted gene is knocked out by a similarly engineered but dysfunctional DNA that is incorporated into the mouse DNA. In both cases, artificial DNA is entered into embryonic mouse stem cell. Commercial production and availability of knockout mice led to the development of thousands of strains, often named after the gene that was knocked out.

What is called *bioinformatics* is a rapidly growing set of multiple methodologies based on algorithms and mathematical formulas. The primary focus of bioinformaticians is to create tools that make research on complex biological data easier and more effective. The growing array of bioinformatics tools illustrates that the application of complex mathematics has far greater opportunities to support life sciences than the currently widely used but limited statistical estimation and hypothesis-testing methods.

One of the classic bioinformatics success stories was the development of the Basic Local Alignment Search Tool (BLAST). The design team of bioinformaticians at NIH was led by David Lipman, a biologist and the director of the National Center for Biotechnology Information (NCBI) since 1989. The other influential member of the team was Stephen Altschul, a mathematician at the Computational Biology Branch of NCBI. BLAST can be used for a variety of tasks including searching a nucleotide database using a nucleotide query, protein database using a protein query, protein database using a translated nucleotide query, translated nucleotide database using a protein query, or translated nucleotide database using a translated nucleotide query (Lipman and Pearson 1985; Altschul et al. 1997). Over time, BLAST became one of the most widely used search tools to find similarities with earlier described protein or nucleotide sequences.

Data science goes further by processing inference. It is about the machine doing the analysis for us because there is just too much data. As people say, data is the new oil that makes everything move better. Analytics and research are becoming much more intertwined, and scientific progress needs data analytics more than ever. Augmented reality means that you see the person or object, but the system also shows a large amount of relevant, insightful information about it. With the increased emphasis on big data, the tools of data science are expected to play a rapidly growing role in life sciences research.

The Technical Opportunity Presents Itself to the Researcher

The discovery of insulin by Frederick Grant Banting in Canada illustrates how a scientific breakthrough can quickly turn into technology development and industrial product. Banting was an orthopedic surgeon who was also interested in teaching and research.

When Banting started working on the problem of diabetes, it was already known that the metabolism of sugar is regulated by a protein hormone produced by the islands of the pancreas. However, innumerable efforts to make extracts from the pancreas to treat diabetes were unsuccessful. Banting got the idea of experimenting with the obstruction of the pancreatic duct while preparing for a medical school lecture. At this time, an article by Moses Baron described that ligation of the pancreatic duct leads to the degeneration of the digestive enzyme-producing cells.

Banting's hypothesis was that the harsh proteolytic enzymes of the pancreas destroy the product that prevents diabetes. If he could prevent this from happening, he would have clean insulin for treatments. To test his hypothesis, he worked with Macleod, a chemist in the neighboring institution. After several trials and errors, he was able to purify insulin and even demonstrated the benefits of insulin therapy on one patient. The results were almost miraculous. Frederick Grant Banting and John Macleod received the Nobel Prize for the discovery of insulin in 1932.

As soon as he convincingly demonstrated the beneficial impact of insulin, Banting started *turning discovery into technology* development and large-scale availability with Eli Lilly and Company. He sold the license of insulin to Eli Lilly for one dollar. On the effect of the resulting technology, animal insulin has been used with great results for many decades. The inevitably deadly disease became a manageable condition with the help of insulin available to millions of patients.

With long-term treatment using animal insulin, eventually rejection reactions limit effectiveness. To solve this problem, new discoveries of life sciences and resulting technologies for the production of human insulin came into the forefront. Today, millions of people with diabetes are treated with insulin that is produced by genetically modified bacteria based on human DNA. The production of insulin by using recombinant DNA technology is one of the best examples of life sciences achievements, leading to major improvements in clinical outcomes and also public health.

Many research projects focus on the study of a natural phenomenon, like cellular mechanism, which not only is interesting but also gives a sense of potential relevance to manufacturing processes in biotechnology. One may reasonably assume that the intense focus of a researcher on such mechanisms is further and strongly encouraged by the high potential for practical application.

Pharmaceutical production of human insulin repeatedly benefited from basic research that followed the scent of technology in its pursuit of understanding nature. This was a much-needed progress at a time when the lack of purity and resistance to animal-extracted insulin became increasingly recognized obstacles in the treatment of diabetes.

As a result of the Chemistry Nobel Prize-winning discovery of Paul Berg, the advent of recombinant DNA technology provided an unprecedented opportunity of synthesizing insulin of human sequence. For the first time, recombinant human insulin production was designed by a collaboration of scientists from City of Hope and Genentech in 1978 (Walsh 2005). Instead of cloning human genes, the researchers designed from scratch the genetic code on paper (Riggs 1981). The resulting codes were in two genes that contained a sequence of nucleotide for insulin production in the chosen bacteria *E. coli.*

In seeking further understanding of the *production system*, Günter Blobel hypothesized that a coding system guides the newly synthesized protein to a specific location in the cell and also provides access beyond otherwise impermeable membranes. These codes are in the form of signal peptide sequences attached to each protein. Indeed, his research found that the protein, being too bulky to fit through the narrow channel opening, unfolds from its highly complex structure and passes through the membrane as a long chain of peptides.

This research was greatly advanced by the replication of some highly relevant and newly published research results of Leder and Milstein. Replicating cutting edge results placed Blobel on an elevated platform in launching his experiments. Ultimately, Blobel was awarded the 1999 Nobel Prize in Physiology or Medicine for the discovery of signal peptides, a cellular mechanism to direct newly synthesized molecules to their proper location within the protein molecule. He also received the Lasker Award in 1993, the King Faisal International Prize in 1996, the Louisa Gross Horwitz Prize in 1989, and the Gairdner Foundation International Award in 1982.

In an interview, he summed up the practical impact of the discovery on the production of insulin: "let's take insulin, which was formerly gotten from slaughterhouses and was extracted, can now be made in bacteria and you can use a zip code to get the insulin out of the bacteria and then can separate it from bacteria much easier" (Blobel 1999).

The research of Randy W. Schekman, a Nobel Prize winner and cell biologist at the University of California, Berkeley, also illustrates the intense scientific interest in various production systems of the cell. On average, about 10% of proteins are packaged and shipped out of the cell. His research discovered the export mechanisms of proteins after intracellular synthesis in baker's yeast, which is exactly how it works in humans as well. His work was recognized for the discoveries of machinery regulating vesicle traffic, a major transport system in our cells. Currently, about one-third of human insulin is produced by yeast in the biotechnology industry.

When the Engineer Decides to Serve Public Health

Focus on direct service to public health is one of the most strenuous and ambitious goals that a researcher/developer can put forward. From historical experience, it is clear that such endeavors can take many decades and require unusual persistence. It is also obvious that choice of the technology goal is the enormously important turning point that needs to consider not only the public health need but also the scientific opportunity to create such technology.

The Austro-Hungarian **Béla Barényi**, inventor and *industrial designer*, is credited for a broad range of unprecedented car safety innovations. After completing engineering studies and working in the automotive industry for a few years, he moved to Germany with his family. Barényi was not only the original designer of the Volkswagen Beetle concept but also the remarkable pioneer of car safety.

In 1939, he applied for a job at Mercedes-Benz but was turned down. However, he applied again and got a one-on-one interview with chairman Haspel. In the interview, Barényi explained to him that everything done in the design department was wrong. Haspel was impressed with his criticism, vision, and perfectionism and promptly hired him. Subsequently, he remained with Mercedes until retirement in 1972.

Barényi recognized the significance of safety long before others paid much attention to it. Throughout his design career, Béla Barényi developed the concept of passive safety, crumple zone (his famous patent DBP 854.157), non-deformable passenger cell, protection against side impact, collapsible steering column, hidden windshield wiper, and many others. His biographers note that at a time when safety was not a priority, "Barényi followed up his ideas with incredible determination" (Niemann 1994).

Over his lifetime, Barényi produced 2500 patented innovations, twice as much as Edison, although some of his patents were registered in different countries. Correctly, Barényi is credited for developing the concept of the Volkswagen Beetle as a young engineer. When the concept of the famous people's car became attributed to Ferdinand Porsche, Barényi sued, demanded recognition plus one (=1) German Mark compensation. Ultimately, the court recognized his claim and awarded the compensation.

One of the most inspiring examples of making a particular, desirable technology the central goal of someone's research was the development of *in vitro* fertilization technology. The entire research work and the many diverse studies of Robert Edwards centered on the creation of a viable and functional *in vitro* fertilization methodology, an ultimately successful aspiration.

When the creation of an effective and sustainable service to patients is the goal, an early but compelling vision of the technology is essential. In 1943, Willem Kolff had the somewhat vague but inspiring goal to construct a working dialyzer prototype and provide life support during recovery from acute renal failure. Despite the many obstacles and shortages of health care during the second world war, he was able to construct the first dialyzer and save a patient's life.

In life sciences, there are many scientists whose work is singularly focused on the design of a new technology for better patient care. Besides the already mentioned research of Robert Edwards on IVF, there are many other examples of technology development research projects that earned the Nobel Prize in Physiology or Medicine. On the list of such winning examples is also the magnetic resonance imaging by Paul Lauterbur and Peter Mansfield (2003); computer-assisted tomography by Allan Cormack and Godfrey Hounsfield (1979); radioimmunoassay of peptide hormones by Rosalyn Yalow (1977); heart catheterization by André Frédéric Cournand, Werner Forssmann, and Dickinson Richards (1956); and the electrocardiogram by Willem Einthoven (1924).

The history of science and technology has many more inspiring examples of highly targeted and very successful technology developments as well. Improvement in transportation safety is one of the 10 greatest public health achievements of the twentieth century according to the Centers for Disease Control and Prevention (CDC 1999). It would be difficult to find a more influential and more productive innovator of automotive safety than Béla Barényi. He became uniquely successful in improving car design and safety with design inventions that are still in use and protect public health today.

Most appropriately, scholarly study of nature should not be contrasted with the development of new technologies. "A scientist is somebody who asks fundamental questions about how things work and tries to figure it out with reason and sometimes with theory, sometimes with experiments. An engineer is someone who has a problem that has to be solved with technology" (Schekman 2013). The many examples mentioned in this discussion highlight blending of the role of scientists and engineers and great benefits from such integrative approach. The deeper understanding of nature and better technology can go hand in hand.

Skillful use of latest technologies and purposeful development of new research methodologies can provide the much-needed tools to gain new insights into the structure and function of the living organisms in nature.

References

Altschul, S.F., Madden, T.L., Schäffer, A.A. et al. (1997). Gapped BLAST and PSI-BLAST: a new generation of protein database search programs. *Nucleic Acids Research* 25 (17): 3389–3402.

Becke, A.D. (1993). Density-functional thermochemistry. III. The role of exact exchange. *The Journal of Chemical Physics* 98 (7): 5648–5652.

Blobel, G. (1999). Transcript from an interview with Günter Blobel, Nobel laureate in Physiology or Medicine 1999. Nobelprize.org. Nobel Media AB 2014. http://www.nobelprize.org/nobel_prizes/medicine/laureates/1999/blobel-interview-transcript.html (accessed 11 February 2017).

Bradford, M.M. (1976). A rapid and sensitive method for the quantitation of microgram quantities of protein utilizing the principle of protein-dye binding. *Analytical Biochemistry* 72 (1–2): 248–254.

Capecchi, M.R. (1994). Targeted gene replacement. *Scientific American* 270 (3): 34–41.

Centers for Disease Control and Prevention (CDC) (1999). Ten great public health achievements – United States, 1900–1999. *MMWR. Morbidity and Mortality Weekly Report* 48 (12): 241.

Chomczynski, P. and Sacchi, N. (1987). Single-step method of RNA isolation by acid guanidinium thiocyanate-phenol-chloroform extraction. *Analytical Biochemistry* 162 (1): 156–159.

Evans, M.J. and Kaufman, M.H. (1981). Establishment in culture of pluripotential cells from mouse embryos. *Nature* 292 (5819): 154–156.

Folch, J., Lees, M., and Sloane-Stanley, G.H. (1957). A simple method for the isolation and purification of total lipids from animal tissues. *The Journal of Biological Chemistry* 226 (1): 497–509.

Gerow, A., Hu, Y., Boyd-Graber, J. et al. (2018). Measuring discursive influence across scholarship. *Proceedings of the National Academy of Sciences* 115 (13): 3308–3313.

Hood, L. (2002). My life and adventures integrating biology and technology. A commemorative lecture for the 2002 Kyoto Prize in Advanced Technologies.

Hood, L. (2011). Future of medicine. *TEDxRainer* (25 December).

Krebs, H. (1981). *Otto Warburg: Cell Physiologist, Biochemist and Eccentric*. Oxford: Clarendon.

Laemmli, U.K. (1970). Cleavage of structural proteins during the assembly of the head of bacteriophage T4. *Nature* 227: 680–685.

Lee, C., Yang, W., and Parr, R.G. (1988). Development of the Colle-Salvetti correlation-energy formula into a functional of the electron density. *Physical Review B* 37 (2): 785.

Lipman, D.J. and Pearson, W.R. (1985). Rapid and sensitive protein similarity searches. *Science* 227 (4693): 1435–1441.

Lowry, O.H., Rosebrough, N.J., Farr, A.L., and Randall, R.J. (1951). Protein measurement with the Folin phenol reagent. *The Journal of Biological Chemistry* 193 (1): 265–275.

Niemann, H. (1994). Béla Barényi: The father of passive safety. Mercedes-Benz.

Office of NIH History (2017). Scientific instruments. In: *Deciphering the Genetic Code, Marshall Nirenberg*. https://history.nih.gov/exhibits/nirenberg/instruments.htm (accessed 11 February 2017).

Pollack, A. (2001). Approaching biology from a different angle. *The New York Times* (17 April).

Riggs, A.D. (1981). Bacterial production of human insulin. *Diabetes Care* 4 (1): 64–68.

Roguin, A. (2006). Scipione Riva-Rocci and the men behind the mercury sphygmomanometer. *International Journal of Clinical Practice* 60 (1): 73–79.

Sanger, F., Nicklen, S., and Coulson, A.R. (1977). DNA sequencing with chain-terminating inhibitors. *Proceedings of the National Academy of Sciences* 74 (12): 5463–5467.

Schekman, R.W. (2013). Nobel Prize talks: Randy W. Schekman (12 December). https://www.nobelprize.org/nobel_prizes/medicine/laureates/2013/schekman-interview.html (accessed 23 June 2018).

Thompson, J.D., Higgins, D.G., and Gibson, T.J. (1994). CLUSTAL W: improving the sensitivity of progressive multiple sequence alignment through sequence weighting, position-specific gap penalties and weight matrix choice. *Nucleic Acids Research* 22 (22): 4673–4680.

Towbin, H., Staehelin, T., and Gordon, J. (1979). Electrophoretic transfer of proteins from polyacrylamide gels to nitrocellulose sheets: procedure and some applications. *Proceedings of the National Academy of Sciences* 76 (9): 4350–4354.

Van Noorden, R., Maher, B., and Nuzzo, R. (2014). The top 100 papers. *Nature* 514 (7524): 550–553.

Walsh, G. (2005). Therapeutic insulins and their large-scale manufacture. *Applied Microbiology and Biotechnology* 67 (2): 151–159.

Wigler, M., Silverstein, S., Lee, L.S. et al. (1977). Transfer of purified herpes virus thymidine kinase gene to cultured mouse cells. *Cell* 11 (1): 223–232.

16

Emulating Nature

To be honest, microorganisms did all the work, and all I did was to organize what they did.

Satoshi Omura (2015)[1]

In natural sciences, the obvious essence of conducting research is investigating and understanding nature. Asking right questions, using precise methodologies, unbiased observations, and rigorous interpretations lead to understanding fundamental structures and functions.

At the same time, it would be simplistic to assume that science is only about studying natural phenomena and discovering its governing laws. Fortunately, there are many other active and also creative approaches to benefiting from research studies.

Emulating nature is emerging as a particularly valuable skill that can give the aspiring researcher a decisive edge. In trying to solve a major practical or scientific problem, the researcher not only studies nature but also tries to learn how nature solves identical or similar problems. Such understanding can provide unparalleled, valuable insight. Another version of this skill is designing an interplay where the experimental intervention and natural responses have carefully coordinated roles in delivering desired outcomes.

Experience and common observation show that every action most likely has generated a reaction somewhere already in nature. For example, when you need a new compound to battle a harmful agent, there is a great likelihood that nature already produced something to counter and uses it effectively somewhere. As the medieval scholar once expressed, "God did not send down any disease without also sending down a cure" (Bukhari 1376). For

---

1 Omura, S. (2015). Agence France-Presse (AFP) Nobel Medicine Prize winner Satoshi Omura speaks in Tokyo (5 October). https://www.youtube.com/watch?v=12dwnBMOHl4 (accessed 12 July 2016).

Innovative Research in Life Sciences: Pathways to Scientific Impact, Public Health Improvement, and Economic Progress, First Edition. E. Andrew Balas.
© 2019 John Wiley & Sons, Inc. Published 2019 by John Wiley & Sons, Inc.

example, harmful bacteria probably already triggered adequate protection somewhere in nature. Similarly, several animal species appear to be well protected against cancer.

Scanning Nature for Valuable Compounds

A good way of learning from nature is to look for potentially useful compounds by screening, selecting, purifying, and, after careful testing, using them for human interventions. In the history of Nobel Prizes in Physiology or Medicine, 2015 was a unique year when two scientists from two different countries working on two separate problems with very similar scanning methodologies were recognized for their landmark discoveries. Both scientists looked for and identified naturally occurring compounds for the development of highly effective pharmaceuticals.

Scanning nature represents a powerful expansion of the methodology developed by Paul Ehrlich nearly a century ago. The Nobel Prize-winning scientist developed the theory that by screening many compounds, a drug could be discovered with desirable therapeutic activities and without causing significant harm to patients. He called such drugs magic bullets.

Ehrlich not only developed the concept but also effectively used it by looking for a chemotherapeutic intervention to treat syphilis. In his laboratory, an enormous variety of chemical compounds were tested for efficacy in killing the bacteria. Ultimately, his famous experiment 606 yielded results and led to the discovery of Salvarsan. It remained the most effective treatment for syphilis until the discovery of penicillin in the 1940s. Modern scanning for natural compounds follows the same logic.

In Japan, Satoshi Ōmura, a professor at Kitasato University, was looking for effective treatment of roundworm parasite infection. The diseases caused by roundworm infections are devastating; among them are river blindness, lymphatic filariasis, and many others. According to international data, parasitic diseases are a threat to an estimated one-third of the world's population, causing much suffering and many deaths. Over many decades of his scientific career, Ōmura developed numerous innovative methods for isolating, culturing, and screening microorganisms and also discovered hundreds of new and important compounds.

Ōmura's starting hypothesis was that microorganisms in the soil are likely to produce effective antiparasitic compounds. To test this hypothesis and identify one or more "magic bullets," he collected many soil samples for screening and evaluation. Ultimately, his landmark discovery came from microorganisms of the soil sample from his favorite golf course. The antiparasitic compound ivermectin was identified and turned out to be highly effective. A modest man with ambitious goals, Professor Ōmura later summarized the fundamental principles of his research, organizing the work of bacteria (Omura 2015).

In 2015, another researcher, Youyou Tu of the Academy of Traditional Chinese Medicine, also received the Nobel Prize for successfully scanning natural compounds. To identify herbs with potential effectiveness against malaria, Youyou Tu chose an unprecedented route by studying ancient medical scripts and interviewing practitioners of traditional Chinese medicine. Over 2000 recommendations and possible compounds were tested for effectiveness. In the Manual of Clinical Practice and Emergency Remedies by Ge Hong of the East Jin Dynasty, Youyou Tu found mentioning of the sweet wormwood (*Artemisia annua*) as recommended to treat malaria. Further analysis identified not only circumstances of efficacy but also its effective compound, artemisinin. Ultimately, this new malaria treatment proved to be uniquely effective, including cases of resistance to previously available drugs.

Scanning nature also worked well in the treatment of noninfectious diseases, like hyperlipidemia. Akira Endo, a research fellow at chemical company Sankyo Co., hypothesized that fungi use chemicals to fight parasitic organisms by inhibiting their cholesterol synthesis. Indeed, fungi have ergosterol in place of cholesterol in the cell membranes. Therefore, inhibiting cholesterol synthesis has great potential in defense of fungi. Endo studied over 6000 naturally occurring compounds and identified mevastatin as an effective agent. Further analysis showed that it primarily inhibits HMG-CoA reductase, a key enzyme in the production of cholesterol.

Shortly afterward, lovastatin was found in the *Aspergillus* mold and became the first commercial statin. Ultimately, the research of Akira Endo led to the development of statin drugs that are among the best-selling and most effective pharmaceuticals in the prevention of cardiovascular diseases and mortality. His notable awards include the Heinrich Wieland Prize (1987), Japan Prize (2006), Massry Prize (2006), and the Lasker-DeBakey Clinical Medical Research Award.

Jennifer Doudna is a biochemist professor and Howard Hughes Medical Institute investigator at the University of California, Berkeley. She did postdoctoral research in the laboratory of Thomas Cech at the University of Colorado, Boulder.

Doudna's research was focused on RNA in its use by plant and animal cells against viral infection. Studies of bacterial DNA revealed that many of them have repetitive sequences that include viral fragments in the genome (i.e. clustered regularly interspaced short palindromic repeats (CRISPR)). The assumption was that this might be the foundation of a prokaryotic immune system that helps bacteria to find, recognize, and eliminate viral attacks.

Doudna started working on further exploration of the role of the CRISPR sequences and also the role of potential RNA intermediates in directing Cas proteins that detect and cut exogenous genetic material. Among the proteins of the CRISPR pathway was the particularly significant Cas9 that can be programmed. The CRISPR/Cas9 technology allows scientists to make precise changes in the genome with the help of this programmable protein. The radically new technology was learned from bacteria and from the way they protect themselves from a viral infection.

Discovery of the CRISPR/Cas9 technology was one of those rare scientific occasions when the breakthrough was relatively quickly embraced. After the discovery of gene editing methodology, patients with major risk factors started approaching Doudna with questions about treatment options, and the news media became interested in controversies of editing human stem cells and potentially embryos. The challenges associated with the quickly embraced discovery also included intellectual property debates (Grant 2015) and heated ethics discussions (Doudna 2015).

In recognition of her work, Jennifer Doudna with Emmanuelle Charpentier and Feng Zhang received the Canada Gairdner International Award in 2016. A year later, she won the Japan Prize with Emmanuelle Charpentier.

In 2016, researchers from the University of Tübingen reported that they identified a potential new antibiotic that is produced by bacteria living inside the human nose. The new antibacterial substance named lugdunin was effectively used to treat skin infections in mice and also against methicillin-resistant *S. aureus* (MRSA) and vancomycin-resistant *Enterococcus* bacteria, some of the deadliest forms of infections (Zipperer et al. 2016). In recent years, there has been significant progress in identifying antibiotics produced by bacteria in the soil. Identification of antibiotics produced by the microbiota adds to the exploration of opportunities.

Studying How Nature Solves the Problem

Since the discovery of DNA structure and subsequently growing interest in the code of life, molecular biologists have learned a lot from bacteria, including methods of cutting DNA, cloning DNA, and copying DNA sequences.

Evolution of the *CRISPR/Cas9* technology also illustrates how **research becomes development** by recognizing the practical potential for discovery in nature. In other words, a process of the living system becomes the starting point of developing breakthrough biotechnology. Together with Charpentier, Jennifer Doudna not only described the pioneering CRISPR gene editing technology (Jinek et al. 2012) but also created a combined "single-guide RNA" molecule that, when mixed with the Cas9 enzyme, can find and cut DNA targets accurately.

The way bacteria fight the flu can be harnessed and changed for a different purpose. The resulting technology is essentially a modified use of tools learned from the study of adaptive immunity in bacteria. It allows scientists to make programmable changes in the genome. "There is always hope that one day I may find something that could be useful either for a therapeutic purpose or technology purposes" – said Emmanuelle Charpentier (L'Oréal Foundation 2016). The CRISPR/Cas9 technology also illustrates how modern science is built on the achievements of a wide range of laboratories, research project, and international efforts.

This genome editing system has a broad range of potential applications from therapeutics to the functional marking of genes. However, Cas9 can cleave off-target sites representing a deviation from the guide and therefore pose significant challenges for genome editing. Using structure-guided protein engineering, MIT scientist Feng Zhang and colleagues designed Cas9 variants with improved specificity to make the system more broadly useful (Slaymaker et al. 2016).

In the development of drugs effective in the central nervous system, the ability to penetrate the blood–brain barrier and delivery of the desired effect represent enormous challenges. Simultaneously, it is an essential safety requirement that most peripherally acting drugs should have physical–chemical properties that prevent them from crossing the blood–brain barrier. Having a clear understanding of the nature of the blood–brain barrier and its functioning is necessary to design new, effective drugs.

In a systematic review for model development, Pajouhesh and Lenz (2005) developed a comprehensive list of the biochemical properties of successful central nervous system drugs. The list was derived from a large number of research studies on thousands of drugs and their observed effectiveness in the treatment of diseases of the central nervous system. Among other requirements, the list includes the following: potent activity at low to subnanomolar, molecular weight <450, minimal hydrophobicity (clogp <5), number of H-bond donors <3, number of H-bond acceptors <7, number of rotatable bonds <8, and many others.

A particular branch of research and development is *biomimicry* – the study and emulation of insects, plants, and animals solving everyday problems. Similarly, human tissues also exhibit physical and mechanical properties that are unique and go far beyond the capabilities of currently manufactured materials. In other words, biomimicry is the science of learning from nature to discover novel structural materials and engineering systems.

In wide-ranging studies, many new technologies have been learned from nature. Studying the burdock plants' hooks led to the development of Velcro. The skin of sharks inspired swimming costumes that cut drag and helped to smash records. Bird bones inspired hollow construction materials that reduce waste without losing any strengths. The Boston Dynamics BigDog robot has four legs comparable with animals, and it can absorb shock and also recycle energy from one step to the next (Raibert et al. 2008). This robot has a remarkable ability to navigate various challenging terrains maneuvering similarly to large four-legged animals.

Among the natural structural materials, collagen stands out as a triple helical structural protein that represents a quarter of all proteins in the body. There are over 20 different types of collagen: in bones and ligaments, the load-bearing connective tissue of rope-like collagen fibers provides resistance to stretching and tearing. In cartilage, collagen provides stiffness and shock-absorbing capability comparable with rubber, while cartilage-on-cartilage sliding interface has less friction than ice sliding on ice. In the lens of the eye, collagen is organized for optical transparency. In the extracellular matrix (ECM) of the amniotic sac, the collagen matrix ruptures when signaling birth (the "breaking of waters").

Mimicking nature has the potential to create new, dynamic, and mechanically functional tissues (Lei et al. 2017). Based on studies of collagen, for example, a biocompatible bonelike material can be developed to repair large bone defects and seeded with cells to form a "tissue-engineered" construct materials for the room-temperature deposition of collagen.

Many naturally occurring tissues and organs are discovered to be engineering marvels. The rapidly growing understanding of engineering properties of tissues is leading to the development of biohybrid artificial organs that are devices substituting for an organ or tissue function and incorporate both synthetic materials and living cells (Nandi et al. 2010).

The promise of this research includes the development of the wearable or implantable bioartificial organs. Notably, the kidney is considered the most remarkable wastewater treatment facility produced by nature. There is already ongoing development of cell therapy devices to replace functions of the kidney lost in both acute and chronic renal failure (Humes et al. 2014).

Learning from the Unexpected Outlier

From time to time, scientists observe natural phenomena contrary to our understanding of nature's laws and established models. Especially in cases of unusual disease patterns, mysterious resistance, or faster than expected recoveries, the outliers can be precious observations. As mentioned before, such out-of-the-ordinary reactions should not be easily discounted as inaccurate or erroneous observations.

One of the most famous studies of peculiar protection was the smallpox immunity of dairy maids as observed by Edward Jenner and consequently the development of smallpox vaccination in 1796. The landmark dairy maid observation led to the accelerated study of smallpox prevention and eventually to the development of the widely used prevention that turned out to be very successful.

Historians also note that the telltale smallpox immunity had been observed and various methods of vaccination had been discovered and rediscovered much earlier throughout the centuries (Riedel 2005). However, Jenner's discovery had far the greatest impact in changing prevention practices.

In **1980,** the World Health Organization declared "The world and all its people have won freedom from smallpox, which was the most devastating disease sweeping in epidemic form through many countries since earliest times, leaving death, blindness, and disfigurement in its wake" (World Health Organization 1980).

Gertrude Elion, one of the most innovative pharmacologists in history, was born and raised in New York. Together with George Hitchings, she is credited with the development of a systematic method for producing drugs based on cell biochemistry and disease processes. The essence of this rational method is identifying an aberrant process in nature that is essential for the development of the disease and subsequently designing a treatment that defeats the aberrant process. The accomplishments of Elion and Hitchings were recognized by the Nobel Prize in Physiology or Medicine in 1988.

At a time when the exact role and structure of DNA were not well understood, DNA was called the stupid molecule because it was everywhere in the body but did not seem to do anything. However, it emerged that synthesis of DNA is an essential part of cell proliferation in the development of cancers and major infections. Correspondingly, preferential antagonists of nucleic acid bases might be able to retard the growth of rapidly dividing cells in various diseases. In 1951, Elion discovered that the substitution of oxygen by sulfur at the 6 position of guanine and hypoxanthine produced effective inhibitors of DNA synthesis (Rubin 2007). Notably, 6-mercaptopurine (6-MP) could produce complete remission in children with acute leukemia.

Ultimately, the list of her pioneering inventions grew very long and included 6-mercaptopurine for the treatment of leukemia and transplant organ rejection; the first immunosuppressive agent azathioprine; allopurinol for gout; pyrimethamine for malaria; trimethoprim for meningitis, septicemia, and bacterial infections of the urinary and respiratory tracts; acyclovir for viral herpes; and many others. The new drugs she invented helped to cure many cases of childhood leukemia, among others.

In the mid-twentieth century, Gertrude Elion, as a woman scientist, had to overcome many obstacles of bias, and she became one of the few Nobel Prize winners without a doctorate. Today, the science of pharmacology and day-to-day patient care continue to rely on the landmark discoveries of Gertrude Elion.

In 1901, the brown teeth mottling phenomenon, a strong colorizing effect that also appeared to protect from caries, was discovered in Colorado. Children developed only stains, but they were "inexplicably resistant to disease." In 1923, it was also recognized that something in the water was causing this effect (Peterson 1997). Researchers playing a key role in the discovery included Frederick McKay, a dentist in Colorado Springs, and Greene Vardiman Black, one of the founders of modern dentistry. In 1931, advanced analysis by ALCOA's chief chemist, H. V. Churchill identified high levels of fluoride in the water samples. Subsequent research identified the optimal level of water fluoridation that prevents caries but does not cause mottling phenomenon.

As result of the promising research, Grand Rapids, MI, volunteered to be the first city to add fluoride to their water (1945). According to follow-up studies, fluoride prevention reduced the incidence of cavities by 60% in 200 000 school-children (1956). Today, fluoridation of drinking water is listed among the landmark public health discoveries of the twentieth century. According to CDC, for every $1 spent on fluoride, it saves $38 in dental treatment and 26% fewer cavities in children <12 years (CDC 2018). Annual per person cost savings resulting from fluoridation range from $15.95 to $18.62 in various communities (Griffin et al. 2001).

In stunningly unexpected ways, nature time and again presents prevention or treatment of otherwise incurable diseases. According to a study by UK researchers, some people appear to have healthy lungs despite a lifetime of smoking (Wain et al. 2015). Analyzing data from the UK Biobank project, the researchers identified genes that appear to be behind remarkable resistance to a well-known harm. Their study improved understanding of the genetic and molecular factors of smoking and lung function and identified potential targets for the development of the therapeutic intervention. While the results are thought provoking, it would be misleading to conclude that cigarette smoking will ever become an acceptable risk factor.

Another example is the *Peto's paradox*, named after Richard Peto, an observation that cancer prevalence is not correlated with body size (Peto et al. 1975). If cancer is an abnormality of cell division, one would assume that any living organism that has more cells should be more vulnerable to cancer. Apparently, cancer risk not only depends on the number of stem cell divisions but also shows tremendous variation depending on anatomical site (approximately 10 000 times) (Noble et al. 2015). Only about 5% of elephants die from cancer as opposed to the roughly 22% human chance of dying from cancer (Callaway

2015). In other words, there must be hugely influential genetic and environmental factors far beyond cell number, cell lifespan, or the number of cell divisions. One of the speculated reasons for the difference is that the elephant genome contains 20 copies of p53, a tumor suppressor gene, while humans have only one (Sulak et al. 2016).

The experimental results from the Cadotte lab at the University of Toronto–Scarborough confirm a statement made by Charles Darwin in his book titled *On the Origin of Species* in 1859: land growing distantly related grasses would be more productive than with a single species of grass. Marc William Cadotte grew 17 different plants in various combinations of one progenitor (Cardinale et al. 2007).

Subsequently, many experiments have shown that multi-species plots are more productive. Cadotte's experiment showed for the first time that species with the greatest evolutionary distance from one another have the largest productivity gain. "If you have two species that can access different resources or do things in different ways, then having those two species together can enhance species function" (University of Toronto 2013).

In his Nobel laureate interview, John Gurdon recalled one of the most valuable outliers for his research (Smith 2012). During his training in Michael Fischberg's lab, a PhD student got some unusual results in a frog experiment that could not be replicated in the next experiment. Fischberg did not assume error by the student. Instead, he told him to bring back the same frog and rerun the experiment. Fortunately, they could find the same frog and got the same result. It led to the discovery of an important, useful genetic marker.

Later, this genetic marker turned out to be exceptionally useful in John Gurdon's experiments leading to his Nobel Prize "for the discovery that mature cells can be reprogrammed to become pluripotent." John Gurdon was happy to highlight the reasoning: "If you get something unusual don't say that's a bad project. Try to work on it, why it is unusual, and follow through because many of the great discoveries come from unexpected results."

Intertwining Ring of Evidence

Nature speaks to us through a variety of channels, and each one has its irreplaceable value. Scientific publications tend to focus on data and observations. Other scientists conceptualize models and principles, i.e. laws of nature. Reconciling differences between observed facts and hypothesized functional models often proves to be a challenge. Typically, stories of people move through the channels of journalism. Not infrequently such stories are inconsistent with well-established models or directly contrary to data from observational research. In other cases, marginal evidence from data analyses becomes made-up or hyped-up story that generates false expectations.

The intertwining *ring of evidence* represents the full circle of essential information that can make life sciences discoveries well grounded in nature. When it can be created, the evidence ring proves to be exceptionally useful for scientific progress, the realization of societal benefits, and communication to people. In the intertwining ring of evidence, the coherence of three essential components is achieved (Figure 16.1):

1) *Model or principle* is describing a particular natural phenomenon. Conceptualizing the law of nature or how it applies is essential to predict the future. The model or principle is needed to make sense of data and explain what people experience.

2) *Observational data* are coming from preferably multiple independent original research studies that are consistent with the model or principle. New model or stated principle has very limited if any value without supporting research data that come from original research showing that it works.

3) *Stories of people* can illustrate the practical significance of a model or principle and also demonstrate how averages of scientific data impact individuals. Case studies and stories are important to show how everything works in reality and communicate otherwise abstract principles to a general audience.

Each of these components is important in its own right and also in its interactions. Having a model without data is often questionable, in particular for the long run. Having data without a model can be confusing and only marginally useful. Having data without actual stories is hard to communicate. Having stories without data can be very manipulative. Having a model without stories is also difficult to understand. Stories without an understanding of the underlying model have limited value.

Fortunately, the consequential effort of a large number of scientists in diverse locations has produced many inspiring, successful rings of evidence in recent history (Figure 16.1). Many of them illustrate how the complicated interplay between nature and intervention can produce amazing practical results. Among the shining examples in the treatment of melanoma is pembrolizumab, otherwise called the famous Jimmy Carter drug.

Connecting Science with People's Life

Principle/model The gradual discovery of PD-1 (programmed cell death protein 1) as an immune checkpoint gave a new opportunity to counter the immune response evasion of tumor cells (Dong et al. 2002). The PD-1 receptor located on lymphocytes is so-called immune checkpoint preventing an immune attack on the body's own tissues. Many cancers hide from the immune system by producing proteins that bind to PD-1 and shut down the ability of the immune system to kill cancer.

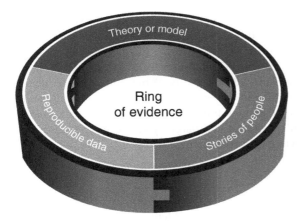

Figure 16.1 Connection with nature: the ring of evidence in life sciences.

In 2014, four researchers shared the William B. Coley Award for distinguished research in tumor immunology for the discovery of PD-1: Tasuku Honjo, Kyoto University (Ishida et al. 1992); Lieping Chen, Mayo Clinic and Yale University (Dong et al. 2002); Arlene Sharpe, Harvard University; and Gordon Freeman, Dana-Farber Cancer Institute. In his seminal publication, Lieping Chen already noted: "These findings have implications for the design of T cell–based cancer immunotherapy."

Observational Data. Pembrolizumab (trade name Keytruda) is a humanized antibody for cancer immunotherapy that targets the PD-1 receptor. Pembrolizumab is an immunoglobulin that binds to and blocks the PD-1 located on lymphocytes. Inhibiting PD-1 action on the lymphocytes prevents the cancer-related shutdown and allows the immune system targeting and destruction of cancer cells.

The drug pembrolizumab was designed by scientists Gregory Carven, Hans van Eenennaam, and John Dulos at Organon BioSciences in the Netherlands. It was initially used to treat metastatic melanoma. Dr. Antoni Ribas, professor of medicine in the Division of Hematology/Oncology at the UCLA School of Medicine, led the successful multicenter clinical trial that generated data substantiating the effective action on PD-1 (Hamid et al. 2013). As a result, the FDA approved the drug Keytruda in 2014.

Stories of People. Following the initial use of the new drug, many astounding stories of unprecedented healing emerged. On the web page describing the accomplishments of Dr. Ribas, the UCLA news website already lists several cases of almost miraculous healings. However, the most compelling story emerged when it became known that former President Jimmy Carter had stage 4 melanoma with brain metastasis, an advanced stage of spreading tumor with historically poor prognosis. After receiving Keytruda treatment, his tumor disappeared, and no further treatment was needed in 2016. Not surprisingly, the case got enormous national and international publicity. It also earned the nickname to the treatment, the Jimmy Carter drug.

The ultimate conclusion is that beyond learning from nature, adopting its solutions can be very effective and advantageous. When life sciences have a vexing problem, most likely nature already solved it somewhere. Research can learn from it, and development can build the lasting solution. Nature has never patented its accomplishments, but you can patent your derivative idea. The creator of the original idea will not file for infringement.

References

Bukhari (1376: tibb, bab 1). Cited by Charles Leslie Medical Anthropology 1976.

Callaway, E. (2015). How elephants avoid cancer. *Nature News* (8 October).

Cardinale, B.J., Wright, J.P., Cadotte, M.W. et al. (2007). Impacts of plant diversity on biomass production increase through time because of species complementarity. *Proceedings of the National Academy of Sciences* 104 (46): 18123–18128.

CDC (2018). *Cost Savings of Community Water Fluoridation.* https://www.cdc. gov/fluoridation/statistics/cost.htm (accessed 23 June 2018).

Dong, H., Strome, S.E., Salomao, D.R. et al. (2002). Tumor-associated B7-H1 promotes T-cell apoptosis: a potential mechanism of immune evasion. *Nature Medicine* 8 (8): 793–800.

Doudna, J. (2015). Genome-editing revolution: my whirlwind year with CRISPR. *Nature* 528 (7583): 469.

Grant, B. (2015). Credit for CRISPR: a conversation with George Church. *The Scientist* (29 December).

Griffin, S.O., Jones, K., and Tomar, S.L. (2001). An economic evaluation of community water fluoridation. *Journal of Public Health Dentistry* 61 (2): 78–86.

Hamid, O., Robert, C., Daud, A. et al. (2013). Safety and tumor responses with lambrolizumab (anti–PD-1) in melanoma. *New England Journal of Medicine* 369 (2): 134–144.

Humes, H.D., Buffington, D., Westover, A.J. et al. (2014). The bioartificial kidney: current status and future promise. *Pediatric Nephrology* 29 (3): 343–351.

Ishida, Y., Agata, Y., Shibahara, K., and Honjo, T. (1992). Induced expression of PD-1, a novel member of the immunoglobulin gene superfamily, upon programmed cell death. *The EMBO Journal* 11: 3887–3895.

Jinek, M., Chylinski, K., Fonfara, I. et al. (2012). A programmable dual-RNA–guided DNA endonuclease in adaptive bacterial immunity. *Science* 337 (6096): 816–821.

Lei, J., Priddy, L.B., Lim, J.J. et al. (2017). Identification of extracellular matrix components and biological factors in micronized dehydrated human amnion/ chorion membrane. *Advances in Wound Care* 6 (2): 43–53.

L'Oréal Foundation (2016). Pr. Emmanuelle Charpentier – L'Oréal-UNESCO Laureate 2016 – Germany (14 March) [Video file]. https://www.youtube.com/ watch?v=xldariJBojY (accessed 23 June 2018).

Nandi, S.K., Roy, S., Mukherjee, P. et al. (2010). Orthopedic applications of bone graft & graft substitutes: a review. *The Indian Journal of Medical Research* 132: 15–30.

Noble, R., Kaltz, O., and Hochberg, M.E. (2015). Peto's paradox and human cancers. *Philosophical Transactions of the Royal Society B* 370 (1673): 20150104.

Omura, S. (2015). Agence France-Presse (AFP) Nobel Medicine Prize winner Satoshi Omura speaks in Tokyo (5 October). https://www.youtube.com/watch?v=12dwnBMOHl4 (accessed 12 July 2016).

Pajouhesh, H. and Lenz, G.R. (2005). Medicinal chemical properties of successful central nervous system drugs. *NeuroRx* 2 (4): 541–553.

Peterson, J. (1997). Solving the mystery of the Colorado Brown stain. *Journal of the History of Dentistry* 45 (2): 57–61.

Peto, R., Roe, F.J., Lee, P.N. et al. (1975). Cancer and ageing in mice and men. *British Journal of Cancer* 32 (4): 411.

Raibert, M., Blankespoor, K., Nelson, G., and Playter, R. (2008). Bigdog, the rough-terrain quadruped robot. *IFAC Proceedings Volumes* 41 (2): 10822–10825.

Riedel, S. (2005). Edward Jenner and the history of smallpox and vaccination. *Proceedings/Baylor University Medical Center* 18 (1): 21.

Rubin, R.P. (2007). A brief history of great discoveries in pharmacology: in celebration of the centennial anniversary of the founding of the American Society of Pharmacology and Experimental Therapeutics. *Pharmacological Reviews* 59 (4): 289–359.

Slaymaker, I.M., Gao, L., Zetsche, B. et al. (2016). Rationally engineered Cas9 nucleases with improved specificity. *Science* 351 (6268): 84–88.

Smith, A. (2012). Interview with the 2012 Nobel Laureates in Physiology or Medicine Sir John B. Gurdon and Shinya Yamanaka (6 December). Nobelprize.org.

Sulak, M., Fong, L., Mika, K. et al. (2016). TP53 copy number expansion is associated with the evolution of increased body size and an enhanced DNA damage response in elephants. *eLife* 5: e11994.

University of Toronto (2013). Productivity increases with species diversity, just as Darwin predicted (13 May). http://www.sciencedaily.com/releases/2013/05/130513152830.htm (17 June 2017).

Wain, L.V., Shrine, N., Miller, S. et al. (2015). Novel insights into the genetics of smoking behaviour, lung function, and chronic obstructive pulmonary disease (UK BiLEVE): a genetic association study in UK biobank. *The Lancet Respiratory Medicine* 3 (10): 769–781.

World Health Organization (1980). *Global Commission for Certification of Smallpox Eradication. The Global Eradication of Smallpox: Final Report of the Global Commission for the Certification of Smallpox Eradication.* Geneva: World Health Organization.

Zipperer, A., Konnerth, M.C., Laux, C. et al. (2016). Human commensals producing a novel antibiotic impair pathogen colonization. *Nature* 535 (7613): 511–516.

17

Scientific Modeling

Understanding the mechanisms of how something works is absolutely fundamental... You can see all sorts of possibilities deriving from it.

John Gurdon (2012)[1]

Invariably, scientific models are created and evaluated when a living mechanism is explored and understood. Scientific papers and scholarly books are only the media of communication, but the new substance, the scientific discovery, needs more specific identification. In this context, scientific models represent the new way of understanding nature. They are the most valued end products of research projects. In other words, modeling is the most ambitious goal of research on how something works in nature.

For the most part and with few exceptions, the Nobel Prize in Physiology or Medicine has been awarded for discovering profoundly important mechanisms of health and disease. Among them are key regulators of the cell cycle (2001), RNA interference (2006), role of dendritic cell (2011), positioning system in the brain (2014), G proteins in signal transduction (1994), regulation of cholesterol metabolism (1984), peptide hormone production of the brain (1977), and many others. Awards for identifying pathogens of major diseases also fall into the category of understanding mechanisms (e.g. *Helicobacter pylori* and its role in gastritis and peptic ulcer disease in 2005).

In the toolbox of research, modeling nature and functions of the living organisms have a special, privileged significance. Research based on such understanding appears to be the ultimate rational approach. Obviously, models of physiology and pathophysiology are also essential for the successful prevention, diagnosis, and treatment of diseases. Developing interventions without understanding the system or modeling sounds like hoping for a lucky strike, counting on jackpot

---

1 Gurdon, J.B. and Yamanaka, S. (2012). Interview with the 2012 Nobel laureates in Physiology or Medicine Sir John B. Gurdon and Shinya Yamanaka (6 December). https://www.nobelprize.org/mediaplayer/index.php?id=1858 (accessed 27 December).

Innovative Research in Life Sciences: Pathways to Scientific Impact, Public Health Improvement, and Economic Progress, First Edition. E. Andrew Balas.
© 2019 John Wiley & Sons, Inc. Published 2019 by John Wiley & Sons, Inc.

in the lottery, crossing fingers, or randomly shooting into the sky. In a wide-ranging and sometimes automated screening, testing without modeling might work, but these achievements are probably the exceptions and not the rule.

In recent years, it is somewhat limiting that the term model became synonymous with selecting the most appropriate animal for a particular experiment. Choosing an animal model for an experiment has little to do alone with a better understanding of nature and the powerful tools of scientific model development. Defaulting conversations about modeling to choosing *animal models* is not a good service to the progress of research and the science of understanding.

This chapter should conceptualize *scientific modeling* in promoting a better understanding of nature and gaining recognition by the scientific community. Conversely, many landmark and accidental research discoveries in health sciences have saved millions of lives but never had the slightest chance of getting the Nobel Prize. Linking tobacco to lung cancer, basic technologies of transportation safety, and fluoridation of drinking water are among the 10 Great Public Health Achievements in the twentieth century according to CDC, but none of them has reached the highest scientific recognition usually reserved for the discovery of new models (CDC 1999).

In the race between theorists and experimentalists, the theorist usually has a tremendous advantage by building influential and widely used models. In other words, research projects of experimentalists would always have a very hard time to match the monumental impact of Albert Einstein's theory of relativity or Charles Darwin's theory of evolution and natural selection. This chapter describes the birth and upbringing of many successful models that have changed the understanding of life and also our way of life.

Hallmarks of the Scientific Modeling

The three-dimensional cardboard cutout and wireframe model of the DNA double helix developed by Watson and Crick is far from being the same as reality, but it is an elegant, compelling, and useful portrayal of the actual molecular structure.

In his classic textbook published in 1869, the Russian scientist Dmitri Mendeleev tabulated elements according to their chemical properties, particularly atomic weight. He noticed patterns of chemical similarity that led him to postulate the periodic table where all elements fell into place. By using this model, he was successful in predicting the discovery of many new elements unknown in his time (e.g. germanium, gallium, and scandium).

There is enormous variability in the scientific problems that can be addressed by models and also in the actual technologies of modeling (Figure 17.1). Not all scientific models are created equal. Obviously, there is room for quality judgment when it comes to consideration of models. However, some basic principles of modeling apply to nearly all of them (Table 17.1). These principles can be very helpful in developing models for maximum benefits.

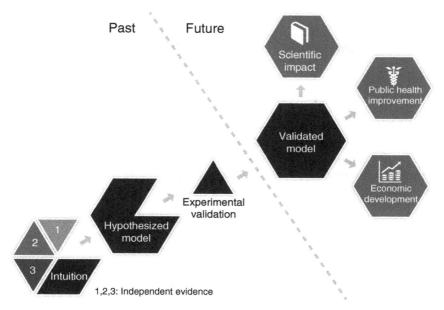

Figure 17.1 Development of a scientific model: building on past evidence and application to predict the future.

Table 17.1 Principles and benefits of scientific modeling.

1) Models are built on already available compelling and published evidence, from multiple sources
2) The model is constructed before starting the measurements or experimentation
3) The model is a clear statement how an important part of nature or society works
4) The model can be a brief paragraph, diagram, mathematical formula, or physical object
5) Complexity is controlled by the researcher. The sport is leaving out unnecessary details
6) Initial models tend to be wrong as most research is weeding. The next model will be better
7) Effective research has specific, meaningful testable hypotheses based on a compelling model
8) Accurate models can be tested from multiple directions, and the results should be confirmatory
9) Testing disconnected hypotheses without modeling is shooting in the sky
10) Effective modeling is one of the strongest defenders against false positives

Models are **greatly simplified but useful representations of reality**, actual objects, phenomena, and processes. Successful models help us in understanding how something important works in nature or society. The *simplified representation* is intended to clarify complex concepts and bring them to an easily understandable context or space.

Models always simplify. In other words, models are not the same as reality as they are reduced reflections of some specific aspects of reality. Therefore, the sport in modeling is to leave factors out instead of including everything, which would make the model unmanageable and not useful.

Preference for simplicity in establishing cause–effect relationships and in scientific modeling has been a well-recognized principle over many centuries since Ptolemy and Aristotle to Thomas Aquinas and Bertrand Russell. One of the most famous versions is the so-called Occam's razor, stating that among competing assumptions, the hypothesis with the fewest assumptions should be selected. Simpler explanations should be preferred over more complex ones whenever possible. In other words, sophistication of methods and analyses should not overwhelm the pragmatics of causal inferences.

To be functional, the models absolutely must ignore many less relevant factors. The sport in modeling is what to leave out and not futile attempts to include everything that would make the model unmanageable and useless. "With four parameters I can fit an elephant, and with five I can make him wiggle his trunk" – stated John von Neumann, the founder of modern computing (Dyson 2004). Of course, the challenge is finding numerical solutions when the complexity is climbing high. As Mike Levitt, a computational chemist from Stanford, summed it up, "look for representation that is accurate enough but still computable to make predictions" (Levitt 2013).

Models are invaluable in guiding measurements and making testing replicable. There are endless opportunities of measurement in nature, but not all of them make sense in attempts to understand essential structures and functions. For obvious reasons, the "lets-measure-something" research builds on a weak and unstable project planning without the benefit of having a compelling model solidly based on already available scientific evidence. Additionally, such research results are only protected by levels of statistical significance, again a frequent source of errors and non-repeatable results.

Models must be *consistent with known facts* and knowledge derived from past data and observations. Already known reliable facts about certain structure or phenomenon must be in conformity with the model. As described above, the model is greatly simplified representation of processes in nature but of course never knowingly false representation of reality.

High-impact models tend to have a recognizable region and period of validity where consistency with known facts applies. The basic laws of Isaac Newton apply in the movement of objects in gravitational space but do not apply in a larger context. With the advancement of science, most but not all models tend to become obsolete and replaced by other better models. One may say, all models are wrong, but some are much sooner than others.

One of the most illuminating examples of an exceptionally useful model with a narrowly limited region of validity is the Koch postulates (Koch 1880). This model serves determination of the cause of most infectious diseases. In order to be identified as causative organism, the infectious agent must (i) always be present, in every case of the disease but never in healthy people; (ii) be isolated from a host containing the disease in pure culture; (iii) cause the same disease when inoculated into a healthy, susceptible animal in the laboratory; and (iv) be isolated from the infected animal and identified as identical to the agent from the originally diseased patient.

Limitations of the Koch postulate model are numerous. Asymptomatic carriers of cholera, typhoid fever, hepatitis C, or polio show the limitations of the first postulate. Currently, prions causing mad cow (Creutzfeldt–Jakob) disease cannot be grown in culture, effectively limiting the second postulate. It is well known that not all infections with *Mycobacterium tuberculosis* or *Vibrio cholerae* will cause disease, limiting the third postulate. As a result of limitations recognized today, meeting the requirements of Koch postulates is considered sufficient but not necessary for determining the cause of infectious diseases.

Data are about the past, while **models *lead to the future*.** Data, observations, and artifacts are always about the past, something that has already happened. On the other hand, the purpose of models is to predict important effects and outcomes. Triple value models can help in prevention and treatment of diseases, better target future scientific experiments, and also generate economic activity.

When widely published, greater understanding and new models can not only change usual practice but also serve as launch pads for further successful research. Modeling unimportant or marginal issues does not make much sense but still happens, not infrequently.

To serve a useful purpose, scientific models should be prepared for future use and should help in predicting consequences of changes in nature or society. Implementation of the model is called simulation. "Once something works it makes a big difference. It is worth doing" – with the words of John Gurdon (2012).

Avedis Donabedian, a professor at the University of Michigan, created his landmark model of quality of health care by highlighting the structure, process, and outcome as essential dimensions of quality evaluation (Donabedian 1966). Before the development of his model, characterizing the quality of health care was one of the most chaotic cavalcade of individual perceptions, differences between patient and clinician expectations, and scholarly debates. His landmark paper became one of the most brilliant and influential examples of successful narrative modeling. Developed half a century ago, the Donabedian model is still the foundation of health-care quality improvement today. The model helps to identify the advantages and limitations of health-care quality measures and also encourages the gradual shift toward outcome-based measurement and reimbursement.

Shinya Yamanaka is a former orthopedic surgeon who became an exceptionally successful stem cell researcher in Japan. After completing his studies in Osaka and practicing medicine for a short time, he went for a postdoctoral training at the Gladstone Institute in San Francisco. The switch from clinical work to research was a refreshing change as he later remembered: "Being a surgeon is totally different from being a scientist. You have to follow textbooks. You have to follow instructions from your mentors... As soon as I started science, I found it very interesting. It is like a whiteboard, I can write whatever I want to write, I can draw whatever I want to draw" (Gurdon and Yamanaka 2012).

His scholarly intuition came from the recognition that "we cannot keep destroying embryos for our research. There must be another way." In his evolving stem cell research, he was further encouraged by the tadpole nuclei transplantation studies (Gurdon 1962), mouse fibroblast transformation (Davis et al. 1987), and Dolly the sheep studies (Campbell et al. 1996). Eventually, he was able to induce pluripotent stem cells from embryonic and adult fibroblast cultures by defined factors (Takahashi and Yamanaka 2006). In his adult cell transformation model building, there were 24 most promising genes based on the published research of other scientists. Eventually, he identified four genes that worked well in reprogramming human skin cells. In 2012, John B. Gurdon and Shinya Yamanaka received the Nobel Prize "for the discovery that mature cells can be reprogrammed to become pluripotent."

Typically, **models need *validation and refinement*.** Not infrequently, theorists or model developers themselves make recommendations for experimental validation. In 1916, Albert Einstein proposed three classical tests to confirm general relativity experimentally: the perihelion precession of Mercury's orbit, the deflection of light by the sun, and the gravitational redshift of light (Einstein 1916).

Due to the overwhelming and consistent experimental evidence pointing to the configuration, the *DNA double helix* model developed by Watson and Crick was already largely validated at the time of publication. In 1958, Matthew Meselson and Franklin Stahl conducted a test confirming the hypothesis that DNA replication was semiconservative (i.e. when the DNA is replicated, each of the two new double-stranded DNA helixes consisted of one strand from the original helix and one newly synthesized) (Meselson and Stahl 1958). Essentially, it also provided additional validation of the DNA double helix model.

Naylor and Finger (1967) suggested a widely accepted three-step process for model validation: (i) build a model that has solid face validity, (ii) subsequently validate model assumptions, and (iii) finally compare the model input–output transformations to corresponding transformations for the real system. Breakthrough models tend to have implications in multiple directions and therefore diverse attempts to use them become multiple validations.

Historically, the concept of validity was divided into content, face, and construct validity: (i) Content validity refers to the extent of representation of all the key aspects of a given construct. (ii) Face validity means that the model is a reasonable imitation of a real-world system to people who are familiar with the real-world system. (iii) Construct validity is the extent to which a model accurately portrays what it supposed to represent (Messick 1995). Construct validity is often viewed as the overall validity of a model.

All models are wrong beyond their limits, but the location where good models break down is a source of new, powerful insight. Conducting research studies based on models, as opposed to isolated hypotheses, creates opportunities to either extend the validity of a model in important directions or, on the contrary, lead to its rejection and the discovery of more accurate models.

More rigidly defined models appear to be the most promising candidates for breakdown and discovery of new insight. For example, the central dogma of molecular biology was defined by Nobel laureate Francis Crick, one of the codiscoverers of the structure of DNA. According to the central dogma of molecular biology, transcription flows only one direction on DNA → mRNA → protein pathway. This was a clear but somewhat rigid model of the DNA role in dictating the protein synthesis.

Discovered by Peyton Rous in 1911, the Rous sarcoma virus has proved to be a uniquely effective tool for cancer research. This virus caused tumor in chicken and was noted for its extreme malignancy, a tendency of widespread metastasis and ability to be transmitted by a filtrate free of tumor cells. Howard Temin observed that mutations of the virus changed the structural characteristics of the infected cell: long fusiform cells produced long fusiform cells, and rounded cells produced more rounded cells. In his later studies, Temin identified that certain tumor viruses carried the enzymatic ability to direct the transfer of information from RNA back to DNA with the help of reverse transcriptase (Temin and Mizutani 1970). Today, this enzyme is recognized for influential role in some tumors and retroviral diseases like HIV, hepatitis B, and many other diseases (Blattner 1999).

Scholarly Intuition and Triple Evidence

Every scientific model development has its unique challenges. However, a closer examination of the evolution of several scientific models shows some interesting common characteristics. One may speculate that the common pattern is not necessarily reflective of the modeling process itself but the processing capacity of the human brain in imagining nature or society.

Scholarly intuition and *triple evidence* (SITE) appears to be the pattern behind many landmark discoveries (Figure 17.1). A recurring pattern of human thinking is the 1 + 3 launch of scientific modeling: a spark of scholarly intuition, sometimes called Eureka, amplified by three independent measurement results. Subsequently, the SITE launch propels the research toward the scientific modeling achievement. Finally, the model is tested, and, if successful, a validated scientific model is ready for general practice.

In the case of Harald zur Hausen's discovery, papillomavirus can cause cancer, the simple but powerful model received the Nobel Prize in Physiology or Medicine in 2008 (zur Hausen 2008). He summed up the scholarly intuition by saying, "I was very convinced that I was on the right track from the beginning." The three independent amplifying sources of evidence included the following: (i) animal studies in 1930 showed that papillomavirus can cause cancer; (ii) based on patterns of epidemiology, cervical cancer has long been suspected of being caused by an infectious agent; and (iii) there were anecdotal reports of malignant conversion of genital warts that contain typical papillomavirus particles.

The work of Nobel laureate Ralph Steinman on dendritic cells also illustrates the SITE launch (Nussenzweig 2011). His scholarly intuition originated from the animal studies in 1967: lymphocytes alone were not sufficient to produce immune responses; accessory cells were required to initiate immunity and to present antigen to lymphocytes. The three independent measurement techniques that played a key role in the identification of dendritic cells included the following: (i) Palade's fixation and electron microscopy of cells, (ii) de Duve's centrifugation methods to separate subcellular components, and (iii) Steinman and Cohn found through phase contrast microscope a different cell with dendritic processes but no prominent phagocytic vacuoles. The ultimate confirmation came with Steinman's method for removing macrophages by adding antibody-coated sheep red blood cells that formed rosettes and went to the bottom of gradient centrifugation and dendritic cells were collected from the light interphase.

The development of the double helix DNA model by Watson and Crick also shows somewhat similar pattern of thinking (Watson 1968). The scholarly intuition and shared passion that the structure of DNA is one of the most important keys to understanding genes and evolution brought together James Watson and Francis Crick in Cambridge. Among other important experimental observations, the following particularly important measurement findings were influential: (i) The Hershey–Chase experiments suggested that the DNA is the genetic material. (ii) The amount of guanine is equal to cytosine, and the amount of adenine is equal to thymine (by Erwin Chargaff). (iii) The double helix structure was likely based on X-ray diffraction studies (by Rosalind Franklin).

Rosalind Franklin was a brilliant chemist, experimentalist, and X-ray crystallographer in England. She came from a family that highly valued education. During the war years, she received PhD in Cambridge and later learned crystallography in Paris. After returning to Cambridge, her experiments played a key role in understanding the molecular structures of DNA, RNA, tobacco mosaic virus, poliovirus, coal, and graphite.

Unknown to Franklin, Watson and Crick saw some of her unpublished data, including the famous "Photograph 51" that provided particularly valuable insight into the structure of the DNA. Ultimately, Franklin's article as supporting evidence appeared in the same issue of *Nature* where Watson and Crick's landmark DNA model was published. Remarkably, one of her postdoctoral colleagues, Aaron Klug, who expanded on the studies she started, got the Nobel Prize in Chemistry.

Franklin was an excellent, passionate, skillful researcher with a vast capacity for taking frustration frequent in scientific investigations. When Franklin saw the Watson–Crick DNA model, she commented: "It is very pretty, but how are they going to prove it?" She was highly critical of speculation, which allowed others to get ahead (Olby 1972).

On the other hand, the somewhat dismissive tone, use of her unpublished results, and unwelcome references to "Rosy" by male scientists of the neighboring laboratory have been a source of controversies ever since. Due to her tragic and untimely death as a result of ovarian cancer at the age of 37 in 1958, the Nobel Prize for DNA could not have gone to Franklin. Over time, Rosalind Franklin emerged as a powerful symbol of talented but unfairly marginalized women in life sciences.

The case of DNA modeling also illustrates substantial advantages of modelist over experimentalist. Watson and Crick were massively focused on creating the DNA model. Meanwhile in the neighboring laboratory, Rosalind Franklin probably also suspected the correct DNA structure but was primarily looking for experimental or confirmatory measurements. As a result of the broader, fact-driven but theorist approach, Watson and Crick got ahead and first published the DNA structure. In the competition among various research methods, the modelist approach has advantages over observational research.

Validation and Refinement

In the age of non-repeatable research, developing scientific models also appears to be a highly valuable defense options against accidental artifacts. First, the well-developed model is already partially validated by integrating all previous results and by being consistent with everything we already know with occasional limitations of course. Second, the development of hypotheses based on a comprehensive model makes the measurements fewer and more selective. Third, every measurement serves the testing of only one underlying hypothesis, the investigated model and therefore less susceptible to statistical error. Fourth, the model can serve as the ideal springboard for the development of a reliably functioning prototype, the queen of repeatability.

The reliance on independent evidence can give a remarkably strong foundation for scientific modeling. In the age of vast amounts of non-repeatable research, having three or more independent, compelling sources of evidence is a huge safety net, especially when those have already been validated in successful replications. Future mathematical and computational generalizations are likely to break down current limitations. The search for multiple evidence pointing toward a common underlying model is at the heart of forthcoming big data and computational research in life sciences.

Well-constructed models can have a tremendous and long-lasting impact on further research. Certainly, the DNA double helix model is a classic example. In making a choice for future research, Randy Schekman describes in his *Nobel Lecture* (2014) the tremendous influence of the work of Jonathan Singer at UC San Diego, particularly his widely cited fluid mosaic model for the organization and structure of the proteins and lipids of biological membranes (Singer and Nicholson 1972).

Choosing Models to Understand How Nature Works

The foundation of constructive modeling was eloquently explained by Albert Einstein in his 1919 article. "When we say that we understand a group of natural phenomena, we mean that we have found a constructive theory which embraces them." Constructive theories attempt to "build a picture of complex phenomena out of some relatively simple proposition" (Einstein 1919). Most models described in this chapter are in the category of constructive theories.

By far, the most frequently used type is the *narrative model*. Obvious advantages include convenient development, well-established routes of dissemination, and ease of access. Due to the ubiquitous use of peer-reviewed publications in the communication of research results, the narrative model is considered a convenient choice in life sciences. Frequently, reports on narrative models also include graphs and tabulated research results.

Generally speaking, narrative models can serve as a straightforward description of cause-and-effect relationships accurately and easily. The model itself tends to be relatively simple. Explanation of *H. pylori* causing peptic ulcer does not require sophisticated modeling techniques, just a simple description.

However, this profoundly important discovery, now easy to say but previously difficult to prove, led to the Nobel Prize of Barry Marshall and Robin Warren in 2005. It also saved millions of lives from suffering by identifying the appropriate antibiotic treatment.

In widespread research practices, most models are intended to remain narrative descriptions with a few tables in a peer-reviewed scientific publication. Unfortunately, this level of modeling is not particularly effective to facilitate utilization of the results afterward. Currently, presenting an underlying model is not part of scientific publication requirements, and many studies do not have a model at all. All too often, many scientific publications reflect a high level of uncertainty and confusion and the model is unclear even at the basic narrative level.

In recent decades, the model of *evidence-based medicine* as one of the simplest and most attractive concepts of inference has gained broad acceptance in health care. This model is very simple, in fact, the simplest possible: past performance guarantees future results. Particularly, the randomized controlled clinical trial, the most rigorous clinical experimental design, is highlighted as the ultimate source of reliable evidence on past performance. Only clinical interventions that successfully pass this test are to be considered evidence based.

Of course, there are major limitations. If the clinical trial intends to predict the future, the randomized sample of experimental subjects must be appropriately representative (i.e. age, gender, and race). It is also well known that the simplest of all models – past performance guarantees future results – is not universally trustworthy (e.g. it is to be avoided in stock market investment). Most importantly, the model of evidence-based medicine is applicable only to the final assessment of clinical interventions but not for the scientific understanding and modeling of complex functions in the living system and nature.

Graphs and charts can serve as *two-dimensional models*, a popular and widely used approach to modeling. Chemical reactions, transport processes, and many functional anatomic models can be easily and conveniently modeled in graphs or flowcharts. Such graphic models are also well liked for educational purposes.

An excellent example of the power of two-dimensional modeling, the Krebs cycle or otherwise called tricarboxylic acid (TCA) cycle, is the principal metabolic pathway in all aerobic organisms. It is a cycle of eight chemical reactions that produces ATP by breaking down glucose in the mitochondria. As an initial step, pyruvate is converted into acetyl-CoA that enters the cycle of oxidative changes. The result is carbon dioxide and chemical energy in the form of guanosine triphosphate (GTP). Also, the cycle provides precursors of several amino acids and the reducing agent NADH for other important biochemical reactions. The NADH generated by the TCA cycle enters into the oxidative phosphorylation pathway, resulting in usable chemical energy in the form of ATP.

Several critical components and reactions of the citric acid cycle were identified by the Nobel laureate Albert Szent-Györgyi in the 1930s. The complete TCA cycle itself was finally identified in 1937 by Hans Krebs for which he received the Nobel Prize in Physiology or Medicine in 1953. Beyond scientific significance, the Krebs cycle has major practical consequences as well. For example, polylactic-co-glycolic acid (PLGA) is one of the most successful biodegradable polymers because its hydrolysis leads to lactic acid and glycolic acid that are easily metabolized by the body via the Krebs cycle (Danhier et al. 2012). Consequently, PLGA has minimal systemic toxicity when it is used for drug delivery or biomaterial applications.

In the history of life sciences, one of the most celebrated *three-dimensional physical models* is undoubtedly the double helix of the DNA, the genetic material. It is elegant and central to understanding life and has many profound implications for further research, clinical diagnosis, and therapies. It was developed by Nobel laureates James Watson and Francis Crick. This model went through countless changes and improvements until it reached the final and well-recognized expression. The resulting the DNA model was so elegant

---

Hans Krebs was born in Germany. After studying medicine at the Universities of Göttingen, Freiburg-im-Breisgau, and Berlin, first, he worked at the Kaiser Wilhelm Institute for Biology at Berlin-Dahlem and later at the Medical Clinic of the University of Freiburg-im-Breisgau. There, he discovered the urea cycle that removes ammonia, a poisonous by-product of other internal biochemical reactions, in the liver. In 1933, the German government terminated his appointment, and, subsequently, he was invited and went to the School of Biochemistry in Cambridge. Over the years, he became uniquely experienced in modeling sequential reactions of chemical cycles.

In 1935, he became lecturer in pharmacology at the University of Sheffield, a place where his Nobel Prize-winning discovery was made. In Sheffield, he completed the experimental work that resulted in the formulation of the tricarboxylic acid (TCA) cycle, today known as Krebs cycle, in 1937 (Stubbs and Gibbons 2000).

Interestingly, his landmark manuscript on the TCA cycle was rejected by editors because they already had "sufficient letters to fill correspondence columns of Nature for seven or eight weeks." As a result, Krebs published his paper in the Dutch journal *Enzymologia* (Krebs and Johnson 1937). After 19 happy years in Sheffield, he went to Oxford and identified the glyoxylate cycle, the third of the landmark metabolic cycles discovered in his laboratory. At a meeting of the American Philosophical Society in 1970, he suggested elimination of wasteful and unproductive research, much of which seemed to be supported as "occupational therapy for the university staff."

and also scientifically compelling that it was published in a single-page article without peer review in the journal *Nature* on 25 April 1953. Based on reconstruction largely from its original pieces, today the three-dimensional Watson and Crick DNA model is exhibited in the National Science Museum in London.

Mathematical modeling is widely considered the queen of all modeling. In its most developed and elegant applications, accurate representation, the simplicity of description, and readiness for replication are invaluable assets. According to Einstein (1919), the starting point and foundation of theories of principle are "empirically observed general properties of phenomena, principles from which mathematical formula are deduced of such a kind that they apply to every case which presents itself." Their particular merit is "their logical perfection and security of their foundation."

In the exploration of vesicle traffic, a major transport system in cells, Randy Schekman choose another type of model, *in vitro replication*. He recognized that understanding of the full potential of the sec mutant collection needed the development of a cell-free reaction to replicate at least a portion of the pathway *in vitro* (Schekman 2013). His talented graduate student suggested that membranes prepared by a gentler lysis procedure, like quick freeze–thaw of yeast spheroplasts, can preserve membrane organization well. Applying this methodology, they observed the production of a heterogeneous spread of low electrophoretic mobility forms of a radioactive precursor and reproduced the transport reaction in separate incubations.

Today, *computational models* are increasingly available especially in cases of large and complex datasets. The method of computer modeling or simulation is sometimes referred to as *in silico* experimentation. Generally, two types of assumptions can be made about a computational model. Assumptions made about the system operation and arrangement of the components are called structural assumptions. Data assumptions are about system processes and availability of data to build and validate a conceptual model.

For validating input–output transformations, the test compares system outputs to model outputs for the same input conditions. Observing the system must generate sufficient data to perform this test. For development and validation of computational models that are developed by analyzing large datasets, there is a need to have two separate datasets for avoiding overfitting to one particular dataset: the first one is for estimating the models (development dataset), and the second is for validating the models (validation dataset) in a process called cross-validation. The validation dataset should be different but coming from the same population as the estimation dataset.

The validation process examines residuals that are differences between the predicted output from the model and the measured output from the validation dataset. The residual analysis includes the whiteness and independence tests. The whiteness test criteria require that the residual autocorrelation function be inside the confidence interval of the corresponding estimates (i.e. the residuals are uncorrelated). The independence criteria require that residuals be uncorrelated with past inputs. Otherwise, the model does not adequately describe how the output relates to the input. Good computational models pass both tests with few or no exceptions. Statistical models can be tested using the goodness-of-fit tests and other techniques (e.g. Kolmogorov–Smirnov test or chi-square test). Outliers must be carefully checked.

The ultimate scientific and practical model is the *working prototype*. Genuine three-dimensional scientific innovations give rise to this type of scientific model, the working prototype. When a scientific discovery is non-obvious, novel, peer reviewed, replicable, generalizable, and beneficial, the result has immediate practical value. Occasionally, the research process will reach this very advanced stage under the leadership and with the limited resources of the original research investigator. More frequently, transfer of research turns to those who are responsible for practical application and delivery.

When someone sits on an airplane, and it takes off, the passenger is personally assured that the fundamental Kutta–Zhukovsky mathematical model of aerodynamic lift, the foundation of modern airplane design, is working perfectly. When someone takes penicillin, and it heals the *Streptococcus* infection, the repeatability of Alexander Fleming's research is demonstrated yet again.

The examples are countless, but the conclusion is the same; the ultimate scientific model that proves the point and expresses the practical side of scientific observations is the working prototype or product. Not all valuable scientific results can or should be translated into practice through prototypes and products. However, when prototyping is possible, the repeatability can be proven with a very reasonable gold standard.

Understanding Processes Leads to Practical Solutions

In summary, the scientific model and its continuous refinement are at the heart of proper research. It is the first and strongest indication of someone's talent in choosing an area of interest and reflects the level of understanding nature. When it is ready and validated, the scientific model becomes the substance, the centerpiece, and also symbol of someone's research accomplishment.

Is modeling nature an effective way to three-dimensional innovation? In his Nobel Banquet speech, Hans Krebs summed it up: "The research I have been doing - studying how foodstuffs yield energy in living cells - does not lead to the kind of knowledge that can be expected to give immediate practical benefits to mankind...I am convinced that an understanding of the process of energy production will eventually help us in solving some of the practical problems of medicine" (*Nobel Lectures* 1964).

Despite its tremendous promise and repeatedly proven value, description of a model or reference to a model is not required part of publication in the current practice of reporting research results. Even when a model description is provided, quite often the model itself remains indistinguishable or buried under the massive amount of technical details in the description of the research project.

Modeling of nature and the living organism should be better recognized as one of the greatest tools and skills of ambitious research. Particularly in basic research, the era of isolated hypotheses is probably over. Using an underlying model surpasses the junior investigators' desire to just "see what happens." In research planning, there is a pressing need for expository models that are built on already available data and previous research. Subsequent study or experiment should be validation of the hypothesized model.

The most eloquent guidance on modeling came from Harold Varmus, Nobel laureate and former director of the NIH National Cancer Institute (Varmus 2011): "All basic scientists who look to the NCI for funding should know that I will tolerate no retreat on the study of model systems and the pursuit of fundamental biological principles."

References

Blattner, W.A. (1999). Human retroviruses: their role in cancer. *Proceedings of the Association of American Physicians* 111 (6): 563–572.

Campbell, K.H., McWhir, J., Ritchie, W.A., and Wilmut, I. (1996). Sheep cloned by nuclear transfer from a cultured cell line. *Nature* 380 (6569): 64–66.

Centers for Disease Control and Prevention (CDC) (1999). Ten great public health achievements – United States, 1900–1999. *MMWR. Morbidity and Mortality Weekly Report* 48 (12): 241.

Danhier, F., Ansorena, E., Silva, J.M. et al. (2012). PLGA-based nanoparticles: an overview of biomedical applications. *Journal of Controlled Release* 161 (2): 505–522.

Davis, R.L., Weintraub, H., and Lassar, A.B. (1987). Expression of a single transfected cDNA converts fibroblasts to myoblasts. *Cell* 51 (6): 987–1000.

Donabedian, A. (1966). Evaluating the quality of medical care. *The Milbank Memorial Fund Quarterly* 44 (3): 166–206.

Dyson, F. (2004). A meeting with Enrico Fermi. *Nature* 427 (22 January).

Einstein, A. (1916). The foundation of the general theory of relativity. *Annalen der Physik* 49: 769–822. See, also, original works in (1952) *The Principle of Relativity: A Collection of Original Papers on the Special and General Theory of Relativity*.

Einstein, A. (1919). Time, space, and gravitation. *London Times* (28 November).

Gurdon, J.B. (1962). The developmental capacity of nuclei taken from intestinal epithelium cells of feeding tadpoles. *Journal of Embryology and Experimental Morphology* 10 (4): 622–640.

Gurdon, J.B. and Yamanaka, S. (2012). Interview with the 2012 Nobel laureates in Physiology or Medicine Sir John B. Gurdon and Shinya Yamanaka (6 December). https://www.nobelprize.org/mediaplayer/index.php?id=1858 (accessed 27 December 2016).

Koch, R. (1880). *Investigations into the Etiology of Traumatic Infective Diseases*, vol. 88. London: New Sydenham Society.

Krebs, H.A. and Johnson, W.A. (1937). The role of citric acid in intermediate metabolism in animal tissues. *Enzymologia* 4: 148–156.

Levitt, M. (2013). Nobel Prize in chemistry press conference. https://www.youtube.com/watch?v=bW8RkFBmYeI (accessed 27 December 2016).

Meselson, M. and Stahl, F.W. (1958). The replication of DNA in *Escherichia coli*. *Proceedings of the National Academy of Sciences* 44 (7): 671–682.

Messick, S. (1995). Validity of psychological assessment: validation of inferences from persons' responses and performances as scientific inquiry into score meaning. *American Psychologist* 50: 741–749.

Naylor, T.H. and Finger, J.M. (1967). Verification of computer simulation models. *Management Science* 14 (2): B-9.

Nobel Lectures, Physiology or Medicine 1942–1962 (1964). Amsterdam: Elsevier Publishing Company.

Nussenzweig, M.C. (2011). Ralph Steinman and the discovery of dendritic cells. *Nobel Lecture* (7 December).

Olby, R. (1972). Rosalind Elsie Franklin's biography. In: *Dictionary of Scientific Biography* (ed. C.C. Gillespie). New York: Charles Scribner's Sons.

Schekman, R. (2013). Genes and proteins that control the secretory pathway. *Nobel Lecture* (7 December).

Schekman, R.W. (2014). Nobel lecture: genetic and biochemical dissection of the secretory pathway. *Nobelprize.org*. Nobel Media AB 2014. http://www.nobelprize.org/nobel_prizes/medicine/laureates/2013/schekman-lecture.html (accessed 23 June 2018).

Singer, S.J. and Nicholson, G. (1972). The fluid mosaic model of membrane structure. *Science* 175 (4023): 720–731.

Stubbs, M. and Gibbons, G. (2000). Hans Adolf Krebs (1900–1981). His life and times. *IUBMB life* 50 (3): 163–166.

Takahashi, K. and Yamanaka, S. (2006). Induction of pluripotent stem cells from mouse embryonic and adult fibroblast cultures by defined factors. *Cell* 126 (4): 663–676.

Temin, H.M. and Mizutani, S. (1970). RNA-dependent DNA polymerase in virions of Rous sarcoma virus. *Nature* 226 (5252): 1211–1213.

Varmus, Q.A.H. (2011). NIH cancer chief wants more with less. *Nature* 475: 18.

Watson, J.D. (1968). *The Double Helix: Being a Personal Account of the Discovery of the Structure of DNA*. New York: Atheneum.

zur Hausen, H. (2008). Interview with the 2008 Nobel laureates in Physiology or Medicine Harald zur Hausen, Françoise Barré-Sinoussi and Luc Montagnier (6 December). https://www.nobelprize.org/mediaplayer/index.php?id=1046 (accessed 27 December 2016).

18

Mastering Bioentrepreneurship

We file a patent about the same time we write the paper.
Robert Langer (2012)[1]

When a young researcher has compelling research ideas, and grant reviewers endorse the proposal, the project may receive significant external funding for several years. In such cases, the researcher will need be able to manage resources into the millions of dollars, which may surpass even the annual budget of his or her own department chair. The receipt of such grant funding is a pleasant situation, but not without risks and daunting challenges. Now what?

Many young researchers suddenly find themselves being the project director or manager responsible for large budgets, hiring and firing associates, battling political attempts to take over part of the resources while trying to focus, and achieve the desired research results. In the background, there is the unmistakable expectation that the successful researcher will generate significant scientific discoveries, remain well funded through further major grants, and provide long-term employment for associates.

Prosperous biomedical researchers also need to be successful leaders, managers, and occasionally risk-taking entrepreneurs. Selling ideas and efficient administration of people and resources are skills that are not only very different from scholarly understanding of human biology but also not taught by many doctoral and research training programs. Financially competent, politically savvy, and effective project management needs a high level of leadership professionalism and strong operative skills.

No wonder, many scientists become hard-charging operators of project budgets and forceful implementers of research ideas. In many projects, ambitious and challenging goals must be achieved with limited resources.

---

1 Langer, R. (2012). Interview with Robert Langer, Biotechnologist and Entrepreneur. Academy of Achievement, Washington, DC. http://www.achievement.org/achiever/robert-s-langer-ph-d/#interview (accessed 21 January 2017).

Innovative Research in Life Sciences: Pathways to Scientific Impact, Public Health Improvement, and Economic Progress, First Edition. E. Andrew Balas.

A good set of operative skills can effectively supplement research skills, smooth the rough edges, and lead to milestone achievements sooner. This chapter highlights some of the skills, from the basics to the higher level.

Catching the Semi-truck: Researcher Gets the First Large Grant

Managing a million-dollar (or more) project is not a trivial task and requires a great deal of professionalism in management and leadership. Believing that spending such large budgets is nothing more than implementation and enforcement of the original research proposal is probably one of the gravest mistakes that a researcher can ever make.

The history books of academia are loaded with stories of mismanagement and non-repeatable results. One-time wonder researchers get one big grant and nothing major ever again. Others pile up major extramural support and fabulous recruitment packages but unable to spend, slow to hire, and constantly blame everybody else. In some cases, domineering academic supervisors try to divert precious resources of the researcher to meet unrelated institutional needs. If the researcher does not understand the basics of finances and human resource management, then the search for excuses will not help. In the work of these researchers, the amount of waste, ineffectiveness, and frustration can become unbearable.

"Everything that was easy and inexpensive to discover has already been discovered. Therefore, our research projects are always very expensive" – said an anonymous research director. Of course, this statement is not necessarily accurate. Today and always, the best research projects achieve discoveries that have a public health and monetary value far exceeding the amounts invested in the research project itself. Among the *high-value discoveries*, there are also the results of surprisingly inexpensive projects (e.g. landmark meta-analyses, modeling projects, major clinical trials, and many others).

The *managerially competent* researcher knows how to turn the wheels and tighten the screws. As the research growths, so does the budget. Most frequently, it is a complicated meshwork of research grants, free resources made available by the institution, startup funds, and various in-kind contributions. Management skills of the funded researcher are tested every day. Here are four simple rules to guide initial steps in learning biomedical research project management.

The first rule is to **know the project budget and where the money goes**. Being able to read and write budgets are necessary first steps. Especially, vet lab biomedical research and clinical studies can be very expensive and complex from a budgetary perspective as well. Comprehensive one-page overviews of revenues and expenses are important when multiple sources of funds are used. Familiarity with budgeting is also important in preventing unauthorized use for unrelated purposes. Especially in the case of government funds, diverting to other purposes is highly irregular but occasionally happens and may even be illegal.

Certainly, researchers should learn how to use spreadsheets for budgets and market calculations. Believing that anybody can prepare a budget and the researcher does not need to pay attention to the lines can quickly become a costly mistake. In initial financial calculations, professionalism of an accountant is not needed, but an ability to do back of the napkin calculations is a must. The engaging book by Weinstein and Adam (2009) provides an excellent introduction to the all-important role of quick guesstimations and back of the envelope calculations. Guesstimation is more than a wild guess but less than a precise calculation.

The second simple rule is that the researcher should **use** *free resources* **as much as possible**. In academia, the availability of free resources is almost endless. The university has funds for graduate research assistants that the researcher can access. The neighboring laboratory or a central instrument pool may offer the use of their equipment periodically. PhD students want to come and work with you on an exciting project. Visiting fellows from other universities may come to the laboratory at their expense. The company is willing to donate some supply or equipment because they are interested in the progress of the research.

The third rule is to **learn basic** *management skills* or get access to someone who is highly competent and trustworthy. The following subject areas can be particularly highlighted for study and professional development: leadership and strategic planning, continuous quality improvement, motivating people and teamwork, fundamentals of budgeting and financial management, basics of human resource management, and marketing and commercialization. Most institutions provide short courses; there are many good books, and online courses are also good options. Some researchers go above and beyond by enrolling into formal graduate education, Master of Business Administration, Master of Public Health, Master of Health Administration, or comparable educational programs.

In moving beyond basics, it is important to turn our attention toward those research specific management and entrepreneurship competencies that prove to be characteristic traits of successful serial research innovators. In reading many stories of success, it is almost impossible not to recognize that several highly successful biomedical research innovators are not just brilliant scientific minds but also skilled entrepreneurs. In fact, some of their business and entrepreneurial skills surpass the outstanding skills of shrewd business professionals.

The rule of *1-4-7 and 7-4-1* in seeking grant support summarizes one focus of interest for seven grant applications and seven grant applications for one funded project. In many agencies and foundations, the chance of success is 15% or less. Therefore, the average researcher has to produce seven grant applications to have a chance for one award. Obviously, shooting seven applications into seven directions will go nowhere. Therefore, all grant application efforts should be invested toward the one primary interest of the researcher. Simultaneously, the seven applications should explore seven sides of the primary interest.

Every research project is the antechamber of the next. Many years ago, Mark Skolnick, discoverer of the BRCA breast cancer genes and the founder of Myriad Genetics, made the comment that NIH grant applications are becoming highly competitive. You cannot describe the plans convincingly enough without already having a large part of your studies completed. Therefore, every grant-funded project becomes the pilot study of the next grant application.

Getting industry funding heavily relies on the researcher's corporate competence. Just like in obtaining support from someone's employing institution, it is advisable to remember the simple *bicycle principle* of presenting major requests: the first cycle is confidential conversation about the request, and, in case of success, the second cycle is the written or public presentation. Most high-ranking leaders do not like to be surprised.

When a project proposal gets funding, the researcher has three technical objectives: (i) completion of the project to the satisfaction of the funding agency, (ii) achieving ambitious scientific goals that can be published in well-ranked journals, and (iii) generating sufficient progress to make the next grant application successful. Typically, the pilot studies should be much deeper and more extensive than what is just described in the preliminary study section of a grant application.

Effective management of research projects and laboratories is the practical foundation of experimental science. When research becomes productive, the next challenge is the effective transfer for use and benefit of society. There are several ways to achieve such transfer, and commercialization is one of them.

Entrepreneurial life scientist **Raymond Schinazi** is Frances Winship Walters professor of pediatrics and director of the Laboratory of Biochemical Pharmacology at Emory University. His area is the development of antiviral therapy, particularly the treatment of diseases caused by the human immunodeficiency viruses, herpesviruses, HBV, HCV, and dengue virus. He published over 500 research articles and has over 90 patented innovations. With coinventor, Dennis Liotta, professor of chemistry at Emory, several landmark drugs came out of his laboratory, including Emtriva and others. According to worldwide statistics, over 90% of AIDS patients take at least one drug developed by Dr. Schinazi.

Dr. Schinazi is also spectacularly successful research leader and entrepreneur. He launched the HIV DART: Frontiers in Drug Development for Antiretroviral Therapies, an international biennial conference series. The license for Emtriva was sold by Emory University for $540 million in 2005. Dr. Schinazi founded several very successful antiviral drug companies, among them Pharmasset, Inc. This company was later purchased by Gilead for over $11 billion. Press reports indicated that Dr. Schinazi's share was $400 million from the sale. Dr. Schinazi is also an active philanthropist and serves on the board of International Centre for Missing & Exploited Children.

Commercialization of Research Products and By-products

While effective management of project resources is a universal demand for the skills of successful researchers, *commercialization* of research products goes beyond and mostly involves applied scientists. The problem-driven research appears to have three drivers: focused interdisciplinary approach, engagement with industry, and entrepreneurial faculty (Joly et al. 2012). Table 2.2 provides a side-by-side comparison of the challenges in basic research and applied research.

The *erudite entrepreneurs* have a special gift for extracting transformational innovation from the often-complex results of the scientific literature and research laboratories. Aspiring researchers and people interested in commercializing scientific achievements need to have sharp reading skills to recognize the value and separate from waste. When it happens, many remarkable scientific and business successes come out of astute reading of the scientific literature.

Good ideas deserve to be protected. The best discoveries of university researchers warrant support when they can be socially beneficial as products and services. In commercialization, one of the most important rules is the first protection of intellectual property and publication of results only afterward. Considering the billions of dollars invested in research, there is a significant societal interest in seeing a good return on investment, not just regarding better health care but also economic progress.

There is an often-mentioned synergism between scholarly productivity and commercialization activities. According to a recent study of 15 institutions, faculty entrepreneurs' productivity is not only greater than their colleagues but also does not decrease following the formation of a startup (Lowe and Gonzalez-Brambila 2007).

The occasional interpretation that the only potential value coming out of research is the resulting peer-reviewed publication is not only wasteful but also very limiting. It is important to avoid being ignorant of the commercial value of research and all its products and by-products along the way. One may say that if you do not value your own discovery, then why would you expect appreciation by others?

Some researchers view technology transfer, the ultimate step of applied research, as a limitation of academic research freedom, impediment of the well-rounded education, marginalization of curiosity-driven basic research, and significant threat to research integrity (Joly et al. 2012). The desire to understand the world and the desire to make money are different objectives and sometimes collide. Focusing only on points of divergence fails to recognize the many valuable products of the research and their tremendous societal value.

Recognizing the potential market for a particular innovation may require vision and courage beyond common imagination. An example from biomedical research is *in vitro* fertilization and the work of Robert Edwards. At the beginning of his efforts, not only did the public appeared to be disinterested, but also the profession was directly hostile to his research. This all changed after the discovery of functioning human *in vitro* fertilization. Analogously, Apple CEO Steve Jobs was also renowned for recognizing market needs long before people themselves understood what they want.

Development of a *working prototype* is recommended as an initial step toward commercialization. It may take significant amounts of resources, time, and energy, far beyond the scope of the original research project. This can be a major endurance test for faculty researchers. On the other hand, a working prototype is one of the best ways to demonstrate the replicability of research results. One may say, customers are the ultimate peer review (Fletcher and Bourne 2012).

Monetization is generating revenue from available assets like data sources, methodologies, and research discoveries. One of the great challenges of research is prompt and full recognition of the many great values produced in the scientific laboratory.

The first potential asset of every scientific project is the new *research databases* generated in the process. Researchers spend an enormous amount of time on measuring, collecting, cleaning, and organizing new data. The resulting databases often have multiple significance. First, they substantiate the findings and discoveries of the research project. Therefore, they should be carefully preserved if someone would like to cross-check the results in the future. Second, several studies documented that research articles publishing datasets receive a higher level of trust and citations (Piwowar et al. 2007). Third, a clean database has the potential of becoming a research resource for other projects as well. Material transfer agreements can regulate and monetize the transfer of such databases among academic institutions. Fourth, some datasets may also have commercial value and can be considered for sale to corporations, if pertinent human subject and other data protection requirements permit.

One of the most celebrated serial innovators is **Robert Langer** who is the David H. Koch Institute professor at the Massachusetts Institute of Technology (MIT). He has written over 1200 articles and has over 1000 issued and pending patents worldwide. His patents have been licensed to over 250 pharmaceutical, chemical, and medical device companies. He is the founder of many biotechnology companies and one of the very few people ever elected to all three United States National Academies. Bob Langer received the Queen Elizabeth Prize for Engineering on 2015.

In large part, Dr. Langer's research has focused on the biological "shipping business," controlled drug delivery systems. When a pharmaceutical company has new, complex drug, they need to have the delivery system in the body, and the Langer lab has many breakthrough discoveries: polymers to deliver drugs, particularly genetically engineered proteins; long-term delivery systems for insulin, interferon, growth hormones, and vaccines; systems that can be enzymatically, magnetically, or ultrasonically triggered to increase release rates; biodegradable polymeric delivery systems; and delivering drugs across complex barriers such as the blood–brain barrier.

Some of his other projects are focused on different problems, like developing drugs that specifically inhibit the process of neovascularization in cancer, retinopathy, rheumatoid arthritis, and psoriasis and biodegradable polymer systems to be used in cell transplants for engineering new organs. The Langer lab at MIT has about 100 associates and $10 million budget annually.

Ingenious research often requires the development of *new methodologies* to measure previously untested aspects of human physiology and pathology. The greatest discoveries in the history of life sciences illustrate the skillful use of suspected biological principles and latest results of new measurements. If the natural phenomenon is important and the procedure is safely replicable, the methodology may have great practical significance and needs to be disclosed to others for use. Of course, the disclosure can be through peer-reviewed publications or protected intellectual property. Either way, the methodology can attract beneficial attention and may contribute to the progress of science in addition to the pioneering research discovery.

For obvious reasons, the discoveries of biomedical research can represent tremendous value, including targeted and also accidental discoveries. Making discoveries and recognizing practical value are two distinct challenges. Talented investigators can meet both of them. Researchers should always remember the words of Thomas Edison "Just because something doesn't do what you planned it to do doesn't mean it's useless" (cited by Finn 2002). Discoveries are typically protected as intellectual properties through one of the well-established protection channels (patent, copyright, trade secret, material transfer agreement).

Licensing intellectual property means that the company takes over the further exploitation of the idea and its commercialization. Licensing can be exclusive, limited to one company, or nonexclusive. For obvious reasons, exclusive licenses tend to be more expensive. The licensing payment can be one lump sum up front, or it can be prolonged and based on sales revenues.

When purchasing licenses, companies need to apply due diligence to make sure the stated value matches the actual value and all conditions are given for successful commercialization of the licensed intellectual property. In the age of prevalent non-repeatable research, such due diligence makes a lot of sense. In considering commercial value and protection, considerations typically include uniqueness, societal demand, and the investment necessary for completion of the project. Purchasing license from a university is essentially getting commitment from the university that it will not sue the company for infringement. Typically, the university licenses the patent "as is" meaning that all consequential liabilities are taken by the purchaser of the technology license.

This list of values would not be complete without adding that successful commercialization often starts with more inconspicuous sharing of faculty skills and unlicensed *process know-how*. Based on a sample of 11 000 professors, a study found that a surprisingly large share of academic entrepreneurship occurs outside the university intellectual property system. About two-thirds of those businesses that were launched by academics are not based on disclosed and patented inventions (Fini et al. 2010). Most of these are based on unique know-how, trade secret, or exclusive production processes.

Outreach, Elevator Pitch, and Marketing

"Business has only two basic functions: marketing and innovation," wrote Peter F. Drucker, the father of business development consulting (Drucker 1954). It has happened many times in the recent history of innovation that a scientific result or new technology was discarded or sold undervalue by the developers only to be picked up by someone else who turns them into huge success.

Especially when dedicated and passionate researchers enter into the marketplace with innovative ideas, there is a tendency to focus and spend more on research and production. Research commercialization initiatives have to carefully watch and fine-tune the ratio of marketing and development. Venture capital is not a research grant, and the startup company is not simply continuation of someone's research project. Under-marketing is one of the grave dangers of small companies led by researchers with big ideas.

Many years ago, two faculty members, a junior and a more senior, visited together a small but very successful and innovative medical software company in St. Louis. Both faculty members were first-time visitors at the company that had about 120 employees at that time. As they were standing in front of the building, the more senior professor made the comment: watch out, there will be about 100 people marketing and supporting the products, and, somewhere in the backrooms, there will only be about a dozen software developers. As soon as they entered the building, they were directed to the backroom for the meeting with the dozen developers, precisely as predicted.

In the excellent book of Steven G. Blank, titled *The Four Steps to the Epiphany Successful Strategies for Products That Win*, readers can find informative description of the problem of overdevelopment and under-marketing (Blank 2013). One of the great dangers of startup companies launched by researchers is the excessive focus on research and development and the lack of attention to the acquisition of customers.

Outreach is extending interactions, collaborations, and services beyond current limits. Successful researchers show many examples of successful outreach and its many benefits. Pasteur was known to visit wineries with his students, and these visits led to an understanding of the problem of fermentation and later developed the technique of pasteurization (Hook 2011).

A good start to successful outreach is contacting a new person, laboratory, or company every week. The convenient starting point can be a phone call, email, personal visit, or handshake in the hallways of a conference. Many of these attempts will get nowhere, but some will generate substantial benefits.

Innovative researchers can also put to effective use the resources and networks of their institutions. Particularly, academia offers many resources, and the successful researcher knows how to get them and benefit from them. This is the first circle of network development within the entrepreneurial researchers who successfully mobilize synergistic interests, collaborating departments, occasional student projects, travel support, and many other freebies of academia.

Especially in applied research, the largest part of the outreach effort should connect the researcher with ultimate users of the targeted results. Understanding their needs and priorities can be very stimulating in the research process. Increasingly, reviewers of grant applications are also asking for representation of patients, patient organizations, and practitioners from the targeted disease area, often as members of an advisory board. If the ultimate customers are involved, results are more likely to be connected to reality and responsive to societal needs.

In efforts of commercialization, the researcher will need to develop an **elevator pitch,** a brief compelling summary explanation of the research results and their practical benefits. It has to be concise and compelling to the extent that the researcher can successfully communicate to someone else while riding the elevator between the second and the sixth floors.

In preparing an elevator pitch, it is good idea to think about the wording carefully, use language that is clear to the educated layperson, and test it repeatedly in conversations with friends not familiar with the peculiarities of the particular research area. No one should assume that the terms and acronyms researchers use every day are meaningful to the layperson. Good communication skills are essential for successful scientific entrepreneurship.

Sometimes, even the elevator speech of the smartest people does not work, and the inventors have to roll up sleeves and launch a startup company. When Larry Page and Sergey Brin invented the Google search engine, none of the leading Internet search companies were interested in buying the license. Ultimately, the inventors decided to create a startup company that later became Google, and now the holding company Alphabet, with a multitude of interest areas that range from Internet services to electric car, mobile technologies, venture capital management, and others.

The development of targeted **marketing** for a particular research discovery requires substantial business skills. In some cases, the starting point can be relatively straightforward by analyzing the epidemiology of certain diseases or conditions and estimating the need for particular tests or treatments. In other cases, it is harder to specify the anticipated market. In any case, the visionary leader foresees the societal need. No matter how simple or complex is the market estimation, it is always a good idea to consult people with industry and marketing experience.

Classic functions of the marketing include identification of the customer needs, product design that meets those needs, promotion and communication of the benefits to attract purchases, pricing that considers consumer inability and unwillingness to buy, and making the product available at the right place and the right time. All these functions must be accomplished at a profit to reward investors and support future product development. Obviously, many of the necessary skills are far beyond the competence and interests of most research scholars.

Commercialization of university innovation requires special expertise and focused effort. The Technology Transfer Office (TTO) of the university can protect and market intellectual property developed by university employees and bring together industry partners, entrepreneurs, and investors. Working with startup and established companies, TTOs move new technologies from the university laboratory to the marketplace and to ultimately benefiting society.

In a world where typical institutions protect intellectual property as their own, faculty researchers need to have a good understanding of their rights and responsibilities (Fletcher and Bourne 2012). In exploring potential partners and buyers, the Technology Transfer Office of the university can play a particularly valuable role. When the TTO office gets an intramural innovation disclosure, they may send letters to potentially interested companies telling them the predictable benefits like enhancing the value of products or services, saving expenses, making services faster, and other benefits.

Most universities require that faculty inventors assign all intellectual property rights to their employer. One notable exception is the University of Waterloo that assigns the intellectual property rights of all scholarly inventions to the inventor, with the requirement to disclose it to the university (Hoye 2006). In other words, responsibility for commercialization is the task of the inventor while conflicts of interest can be monitored.

Several universities, when they decline intellectual property protection for a faculty idea, return non-commercialized ideas to the faculty and allow them to protect and commercialize on their own. Such policies directly challenge the prevailing practice that faculty ideas are automatically assigned intellectual properties of the employing university. In 2015, Virginia Tech adopted the general policy stating that faculty members retain their intellectual property rights if the university does not wish to pursue the idea.

Sergey Korolev was a pioneering Soviet aerospace engineer who gained recognition as the chief designer of complex technical projects like transporting humans into space and back. This was the age of famous space race between the Soviet Union and the United States. Among others, Korolev is credited for leading the development of Sputnik, the first artificial satellite around the globe, first spaceflight by a dog, Yuri Gagarin's first human spaceflight in 1961, and the first spacewalk. During his leadership of the space program, his name was a closely guarded state secret, and he was only referred to as the "chief designer."

Far ahead of his time, Korolev skillfully applied advanced systems engineering methods to develop large, complex systems with less advanced Soviet technologies and interactions with a complicated party bureaucracy. He was thoroughly familiar with the entire planning and production process frequently personally visiting the machine shops, astronaut training events, construction projects, and design bureaus.

"I was always amazed by Korolev's working method. Even before the goal of the current activity had been attained, preparations were already under way for the next step, and the steps beyond that," remembered Abram M. Ghenin, one of his medical research associates (Rhea 1995). At the age of major uncertainties regarding the survivability of weightlessness and cosmic radiation, his pioneering work opened new frontiers in human physiology and space exploration with a remarkable safety record. Fifty years later, Korolev's rocket was still the most accessible and most reliable human transportation to space. His scientific management credo was that "I have the responsibility not only for the technology but the people and the whole project."

Planning for Customer Acquisition

Development of a sound business plan that, among others, identifies customers is essential for success. As it has often been stated, if it does not work on paper, it will not work in reality. Like in every major human endeavor, planning for commercialization is essential. Especially when startup companies are being formed, there is a need to develop a business plan. The business plan is not just a roadmap for the startup developers, but equally importantly it is a requirement for presentation to venture fund managers.

The plan does not have to be lengthy. Some authors advocate one-page business plans (Horan 2004). Successful business plans for startup companies tend to include 10 essential components: (i) executive summary giving a very brief but compelling description (~elevator pitch); (ii) company description including its mission and vision, what it wants to be in a few years; (iii) products and services, with a description of readiness for commercialization; (iv) market analysis in terms of size, structure, growth, trends, and sales potential; (v) competitors in the market and the market share the startup would like to achieve; (vi) organization and management of the startup company, including skills of key leaders; (vii) sales strategies to generate revenues from the product or service; (viii) strengths, weaknesses, opportunities, and threats characterizing the startup company; (ix) projected budget for several years including revenue and expense projections; and (x) funding requirements for growth and expansion. Addressing these items does not have to be an excessively lengthy business plan.

In the planning process, *customer acquisition* **to product development** ratios are especially useful (sales–marketing people to researchers–developers; S&M-R&D). Often startups are launched by a researcher–developer and a business person, reflecting one-on-one ratio. However, this should change quickly. The lack of go-to-market strategy and being overloaded with research technicalities are absolute killers of university startups.

For first-time technology entrepreneurs, a 3 : 1 S&M–R&D ratio is recommended, meaning three times more money should be spent on customer acquisition than on product development (Cummings 2011). Others advocate a 2 : 1 S&M–R&D ratio in the software as a service sector (Tunguz 2013). The right ratio is probably sector dependent, and biotechnology is special in many ways. However, the pressing need to focus more money and attention on marketing and sales is undeniable.

From the beginning of the commercialization effort, researchers are advised to work with the Technology Transfer Office (TTO) of their employing university. Business negotiations and minutia of paperwork needed for patenting, licensing, material transfer agreements, startup businesses, and many other commercialization activities are not for the stomach of most researchers. It is also helpful to coordinate such efforts with the employer to be consistent with pertinent policies, minimize chances of disagreements, and maximize benefits from institutional support. Owning a patent is one thing but being able to defend it is another. Most researchers do not wish to be out in the wilderness alone with a potentially expensive and protracted legal defense of a patent when someone challenges the rights.

Technology transfer organizations can be part of shared services at the university, a separate foundation dedicated to managing research funds and intellectual property, and finally can be a for-profit company sustaining itself through commercialization revenues. The following examples illustrate the various arrangements, and also the multitude of ways success can be achieved.

Established in 1970, the Stanford University Office of Technology Licensing is a successful example of the shared services model within a nonprofit institution. Its mission is to transfer Stanford technology to society's use and benefit. It takes pride on socially responsible licensing that encourages technology development and use.

At the other end of the spectrum is Yissum Research Development Company of the Hebrew University of Jerusalem Ltd. In 1964, it was founded as a for-profit company owned by the university to protect and commercialize the university's intellectual property. At the time of writing this book, Yissum generates over $2 billion revenues in annual sales.

Founded in 1925, the Wisconsin Alumni Research Foundation (WARF) is the private, nonprofit patent and licensing organization for the University of Wisconsin–Madison but separately incorporated. Built from decades of licensing and investment revenues, it manages $2.6 billion endowment to support research and scholarship at the university. Being a separate nonprofit organization, WARF can more easily partner with entrepreneurs, startup companies, and equity investments and then shared services TTOs.

When a faculty innovation disclosure is received, the university TTO will go through a systematic process. Typical steps of evaluation prior to attempts of monetization include the following: checking that the potential patent meets the legal qualification criteria; for patentable innovation, listing probable competitors including the status quo; identification of potentially interested buyers; exploring interest by marketing to potential buyers; and negotiation for licensing and closing the deal (Bloch 2013). Patent protection is not an obstacle to publication, but patenting must precede publication.

In launching startups, universities may contribute their know-how in exchange for equity shares in partnerships and companies. There are various monetization arrangements for the commercialization of university intellectual property. The university may get sponsorship of research in exchange for transferring intellectual property rights arising from the sponsored project.

University guidance can be especially valuable in preparing for the so-called "valley of death," an often-used term to highlight the initial, highly vulnerable phase when startup companies tend to perish. After receiving the first round of

> **Patrick Soon-Shiong,** son of Chinese immigrants and former University of California at Los Angeles surgeon, is a billionaire biomedical innovator and entrepreneur. He published over 100 research articles and has over 170 issued patents. In 1987 he performed the world's first full pancreas transplant.
>
> He developed the cancer drug Abraxane; the first FDA-approved protein nanoparticle albumin-bound delivery technology. Throughout his career, he launched several companies, among them most notably American Pharmaceutical Partners. At one point, it was the only safe source of heparin in the United States. He developed and sold two multibillion-dollar pharmaceutical companies. He is an influential promoter of individualized cancer vaccines based on genome sequencing.
>
> Forbes estimated his fortune above $12 billion. He is an active philanthropist and part owner of the Los Angeles Lakers. Patrick Soon-Shiong is an unusually effective communicator of scientific ideas and promises. Once Steven Salzberg, a professor of biomedical engineering at Johns Hopkins University, said about him that "I've heard him make claims that are clearly wild exaggerations" (Baker 2015). In 2011, he launched NantWorks to create transformative global health information and advance diagnostics to better identify and target specific disease characteristics. Today, Patrick Soon-Shiong is a trend-setting advocate for broader and much faster use of cancer genome sequencing to guide individualized treatment.

support, the startup company tends to incur very significant expenses – spending on offices, staff, and operations – while sales revenues are likely still low or nonexistent. When the initial money is already spent, the valley of death comes in sight. The startup firm needs to manage its investor relations very effectively to avoid falling victim to negative cash flows in the valley of death.

Pipeline of Life Sciences Innovation: Assets to Returns

In recent decades, research on university entrepreneurship has been steadily growing, and we have been able to get a much deeper understanding of the processes of scholarly creativity and research innovation (Rothaermel et al. 2007). These studies have been largely focused on the entrepreneurial research university, new firm creation, environmental context including networks of innovation, and productivity of technology transfer organizations. In comparison to various scientific fields, the largest number of startups comes from the field of computer sciences and electrical engineering and only about one-third from the biomedical research disciplines (Kenney and Patton 2011). Statistics of research on university entrepreneurship not only indicate the particular challenges facing commercialization of life sciences achievements but also highlight opportunities to overcome the challenges.

Particularly, the extensive experience of pharmaceutical companies is valuable in assessing and positioning research for success. More specifically, leading pharmaceutical companies have to continuously run on an innovation treadmill. Patents create the intellectual property right for 20 years after the patent application date. After 20 years, the company loses the product often completely but sometimes partially. It is also estimated that the time needed from scientific discovery to product launch is approximately 15 years. Correspondingly the intake of new products must match or surpass the expiration of established products.

Consequently, pharmaceutical companies are eager to convert intellectual property assets – new scientific discoveries and promising new technologies – into successful products that generate financial returns, the *pipeline of innovation*. In this process, companies have to carefully evaluate, forecast, and manage scientific uncertainties, regulatory changes, the variability of health insurance coverage, and policy changes (Remnant and Lesser 2014).

In testing new pharmaceutical interventions, clinical trials are conducted in sequential phases, and each phase tests different assumptions (Figure 18.1). In addition to the phases described earlier, the *phase IV trial* is conducted after approval and marketing. It evaluates the effect in various populations and gains further information about side effects with longer-term use.

New pharmaceutical products go through the following pipeline after scientific discovery: preclinical testing, phase I clinical trial, phase II clinical trial, and phase III clinical trial for product launch. Development from discovery to the completion of phase II clinical trial is called *early-stage pipeline*. Products that pass phase II enter into the *late-stage pipeline* that should launch as a product within the next three to four years.

Pipeline momentum, the ratio of late-stage products to early-stage assets, is a useful concept in understanding the transformation of intellectual property assets into commercially viable products (Figure 18.1). At every stage of testing and development, assets can progress to the next phase successfully, stall/fail entirely, or be terminated due to underperformance. Since 2010, out of 236 early-stage assets, approximately 143 products were launched by 12 life sciences companies, giving an average of 0.61 pipeline momentum (Remnant and Lesser 2014).

Correspondingly, for every five dollars generated by newly launched products, there is an estimated two-dollar loss due to product failures (Remnant and Lesser 2014). Extending the calculations into the early phases of the pipeline makes much larger losses likely (i.e. for every product launched there are many more failures). Ultimately, the process of new drug introduction is enormously costly, often exceeding a billion dollars. The upshot of this cost projection is that there is less financial motivation to pursue commercialization of pharmaceutical discoveries that have less than several

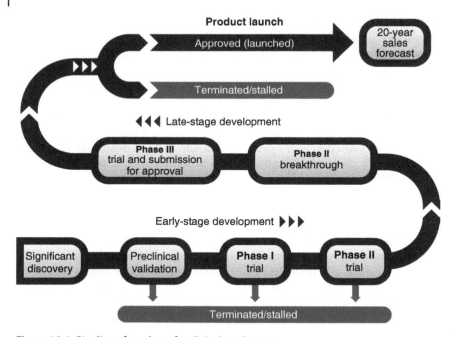

Figure 18.1 Pipeline of products for clinical application.

hundred million-dollar market projection. For controversies surrounding this threshold and its implications for diseases in developing countries, please refer to the humanism chapter of this book.

In managing the R&D process, companies are eager to keep pipeline momentum within a reasonable range. If the momentum is very weak, the company is busy in innovating but struggling to commercialize a very costly position. If the momentum is very high, the company is successfully commercializing, but the low rate of innovation undermines future position when the currently owned patents expire. In making the R&D process successful, effective management of the assets to returns transition is essential (Remnant and Lesser 2014). The asset to return performance statistics suggest that companies focusing on fewer therapeutic areas can achieve higher ratios. Interestingly, externally received assets tend to generate better returns than internally developed ones.

In summary, bioentrepreneurship adds tremendous value to life sciences. Many practically meaningful research results get a chance to enter into sustainable practice and the economy. Mastering bioentrepreneurship involves not only a firm understanding of the systems of support available to the researcher but also managing potential pitfalls in the process. A growing number of successful researchers shows skills of commercialization to target research, position discoveries, create enterprise, and benefit society.

References

Baker, P. (2015). If cancer becomes Biden's cause, a bold but polarizing doctor is on call. New York Times (1 November).

Blank, S. (2013). *The Four Steps to the Epiphany: Successful Strategies for Products That Win*. Cork: BookBaby.

Bloch, D.S. (2013). Monetization of patent and other IP rights: an introduction. *Fordham IP Conference*. http://fordhamipconference.com/wp-content/uploads/2010/08/David_Bloch_Monetization_of_Patent_and_Other_IP_Rights.pdf (accessed 21 December 2010).

Cummings, D. (2011). The 3:1 customer acquisition to engineering spend ratio. Blog. https://davidcummings.org/2011/04/25/the-31-customer-acquisition-to-engineering-spend-ratio (accessed 29 January 2017).

Drucker, P.F. (1954). *The Practice of Management*. New York: Harper and Row Publishers.

Fini, R., Lacetera, N., and Shane, S. (2010). Inside or outside the IP system? Business creation in academia. *Research Policy* 39: 1060–1069.

Finn, C. (2002). *Artifacts: An archaeologist's Year in Silicon Valley*. Cambridge, MA: MIT Press.

Fletcher, A.C. and Bourne, P.E. (2012). Ten simple rules to commercialize scientific research. *PLoS Computational Biology* 8 (9): e1002712.

Hook, S.V. (2011). *Louis Pasteur: Groundbreaking Chemist & Biologist*, 8–112. Minnesota: ABDO Publishing Company.

Horan, J. (2004). *The One-page Business Plan*. Berkeley, CA: The One Page Business Plan Company.

Hoye, K.A. (2006). University intellectual property policies and university-industry technology transfer in Canada. PhD dissertation. Systems Design Engineering, University of Waterloo Ontario, Canada.

Joly, Y., Livingstone, A., and Dove, E.S. (2012). *Moving Beyond Commercialization: Strategies to Maximize the Economic and Social Impact of Genomics Research*. Ottawa, ON: Genome Canada.

Kenney, M. and Patton, D. (2011). Does inventor ownership encourage university research-derived entrepreneurship? A six-university comparison. *Research Policy* 40 (8): 1100–1112.

Langer, R. (2012). Interview with Robert Langer, Biotechnologist and Entrepreneur. Academy of Achievement, Washington, DC. http://www.achievement.org/achiever/robert-s-langer-ph-d/#interview (accessed 21 January 2017).

Lowe, R.A. and Gonzalez-Brambila, C. (2007). Faculty entrepreneurs and research productivity. *The Journal of Technology Transfer* 32 (3): 173–194.

Piwowar, H.A., Day, R.S., and Fridsma, D.B. (2007). Sharing detailed research data is associated with increased citation rate. *PLoS One* 2 (3): e308.

Remnant, J. and Lesser, N. (2014). *Measuring the Return from Pharmaceutical Innovation*. London: DeLoitte Center for Health Solutions.

Rhea, J. (1995). *Roads to Space: An Oral History of the Soviet Space Program*, 197–199. New York: Aviation Week Group.

Rothaermel, F.T., Agung, S.D., and Jiang, L. (2007). University entrepreneurship: a taxonomy of the literature. *Industrial and Corporate Change* 16 (4): 691–791.

Tunguz, T. (2013). The ratio of engineers to sales people in billion dollar SaaS startups. Blog. http://tomtunguz.com/saas-spend-allocation-benchmarks (accessed 29 January 2017).

Weinstein, L. and Adam, J.A. (2009). *Guesstimation: Solving the World's Problems on the Back of a Cocktail Napkin*. Princeton, NJ: Princeton University Press.

19

Art of Scientific Communication

In science credit goes to the man who convinces the world, not the man to whom the idea first occurs.

Francis Darwin (1914)[1]

Research without effective communication has little scientific or societal value and no chance for funding. Today, written reporting and oral presentation of results are not just minor closing steps of a lengthy research project. Development of manuscripts is often viewed as the tangible objective, continuous guideline, and one of the most important products of research projects.

Scientists write articles, books, and proposals and give talks routinely. Researchers have to be good communicators to get grant and contract funding, to convince people who know nothing or perhaps very little about the particular area of research, and to interact with colleagues to bounce ideas and get advice.

Obsession with a desire to understand and accuracy have always been the hallmarks of the best scientific communicators. Albert Einstein was famous for his passion for short and exact wording. After more than 30 years, Einstein was asked to recreate a handwritten copy of his landmark paper originally published in June 1905. While copying, he suddenly stopped and critically noted "Das hätte ich einfacher sagen können" [I could have said it simpler] (Pais 1982).

Research is not talking business. The work of the scientist should first and foremost focus on the collection and processing of data. Plenty of communication is part of active research from the beginning to completion. However, the overwhelming majority of effort should always be concentrated on the development of hypotheses, experimental design, collection of evidence, data producing measurements, setting up methodologies, analyses of facts,

---

1 Darwin, F. (1914). Francis Galton. *The Eugenics Review* 6 (1): 1.

Innovative Research in Life Sciences: Pathways to Scientific Impact, Public Health Improvement, and Economic Progress, First Edition. E. Andrew Balas.
© 2019 John Wiley & Sons, Inc. Published 2019 by John Wiley & Sons, Inc.

and development of new models. The priority should be achieving a new level of understanding. Beautiful words can never replace substance or reliance on evidence.

Researchers not only routinely review each other's grants and papers as peer reviewers but also often receive critical feedback from colleagues, either anonymously or publicly at various forums like conferences and seminars. Intelligent debating is not only about defending someone's position, but it is also learning from the comments and criticism received.

"My brain is open!" – this was the standard greeting by Paul Erdős, one of the most prolific mathematicians of the twentieth century and the author of 1500 scientific papers (Schechter 2000). A quick start into developing communication skills can be having a dozen or so personal discussions with people about the ongoing or planned research. These people can be researchers working in different areas, laypeople, or family members. The conversations should include explanations of the overall goals of the study, primary points, methodologies, and anticipated results. Watching reactions and getting advice can be very educational. It can also show which messages resonate well and what needs to be more clearly articulated.

When scientists with their findings and discoveries enter into the social and political discourse, conflicts emerge quite frequently between truth and politics for reasons described in Hannah Arendt's famous essay (Arendt 1967). In this context, the *concept of truth* is defined as something "beyond agreement, dispute, opinion, or consent" that does not change based on the number of people accepting or rejecting it. Due to the rules of their profession, scientists observe regularities in nature as noted by Schrodinger and Barth (1932).

Such regularities or laws of nature, if confirmed, can be identified as truth according to the above definition. Not surprisingly, scientists can be stressed and challenged when their replicable "truthful" observations become just one of the many opinions in a societal discussion. The controversies surrounding Nobel laureate Robert Edwards' work on *in vitro* fertilization illustrate the challenges (Johnson 2015). It requires stamina and excellent communication skills to win the desired societal support.

Willingness to challenge dogma or accepted "truth" has tremendous scientific and practical significance. This task can be particularly daunting for young investigators who are especially dependent on funding and the peer-reviewed process. Their success in communicating rationale for pursuing alternative mechanisms can be critical to the success of novel studies and scientific progress in new ways.

Ultimately, great communication skills represent tremendous value in selling almost anything what research and science can produce. Among the many skills, scientific writing represents one of the most valuable tools in the hands of an experienced researcher for specifying research plans and communicating results of original studies effectively.

Yardstick of Communication: Original Research Report

Original research reports are not only the ultimate deliverables of scientific projects but also represent a highly significant point of reference in technical writing. Reliance on data, instead of speculations, is increasingly demanded, and it has become a distinguishing feature of original research. Statements of the trusted researcher should always have roots in actual observations and scientific evidence. The reliance on quantitative evidence has significant implications for the development of research proposals and reports.

Being *data driven* and having a preference for quantitative evidence has to be the preferred approach in research (e.g. detailed list of observed data, results of statistical point estimates, and confidence intervals). As people often say, in God we trust, all others bring data. In the area of life sciences, conclusions of any research study are expected to be strictly based on the collected data and subsequent careful, objective, and, ideally, blinded analysis. In discussing the observations of other investigators, the quantitative evidence is again the most valuable, and therefore the data with practical significance should be abstracted primarily (e.g. in the introduction and discussion section of research reports).

Beginning researchers are sometimes surprised by the major differences between the literary/journalistic style and *scientific writing style.* Many early career PhD students and postdocs use a "flowery" style or flowing prose but, ultimately, learn the art of scientific writing needed for original research reports.

The four most glaring, distinguishing qualities of scientific writing include (i) consistently and repetitively using the same word for the same concept, (ii) insisting on evidence behind every major statement, (iii) avoidance of emotional and colorizing words, and (iv) placing every message into rigorously defined sections.

When speed is desired, advance writing can help to get valuable comments on the reasons and methodology of the study before it becomes too late. By definition, the scientific investigation starts well before data collection, with identification of the original question, hypotheses by which the question can be pursued, and experimental design to test the hypotheses. Typically, all of this groundwork will be laid before the study is conducted and data are collected.

In other words, the development of the final report, or at least its first part, can start before the actual data collection and analysis. Writing the introduction and methods sections and defining the planned tables can streamline the entire research process and help focus on the data that you need.

To preserve *objectivity*, scientific writing is expected to keep emotions out and away. Passionate stories, colorful qualifiers, and inspiring quotations are incompatible with required objectivity in original research reports. Expressions like brilliantly stated, heroic effort, or astonishing achievement might work well in personal conversation but have no place in scientific writing. Emotional words are not only irrelevant to the scientific evidence but often viewed as deliberate attempts to manipulate the reader and distract attention. Abstinence from emotional words is a fundamental requirement of technical writing.

In original research reports, statements by other investigators are often mentioned, but such statements are supposed to be based on evidence and data and not just speculation or authority. Unsubstantiated reliance on the views of renowned authorities is perceived as another expression of strong emotional commitment in the interpretation of observations. The demand for novelty and data-driven approach is becoming the norm of original research reports.

If something can be misunderstood, it will be misunderstood. In many cases, you can predict what reviewers will criticize in the manuscript or grant application. Numerous types of skeptical comments can be forecasted and, consequently, prevented by making the manuscript airtight. Answering the criticism before you get it can further streamline the manuscript acceptance. As mentioned above, studies that challenge exiting dogma can be very meaningful, but communicating these ideas or acquiring funding based on a review by peers who may have a vested interest in existing dogma can be challenging.

The general structure of scientific articles reporting original data is a series of well-defined sections (title, abstract, keywords, introduction, methods, results, discussion, references, tables, and graphs). To communicate all critical aspects of the study, a slot is defined for the description of each principal components of the research project and for all pertinent information that might be needed to generalize or repeat the results.

Learning from Examples and Feedback

To develop a bespoke research manuscript, you need to select a suitable *target journal*. This choice comes from the definition of the target audience. Data may be of interest to both clinicians and basic scientists, so deciding which way to go is a key first step. Bottom line is who will find your data and conclusions most relevant to their progress.

In the life sciences field, it is always advisable to go for publication in an ambitious, high-impact, MEDLINE-indexed, peer-reviewed journal. One way to identify such a journal is to look for articles similar to yours and determine their journal. Obviously, the careful reading of author guidelines and adaptation to the requirements of the target journal is the first, most fundamental step. It may be helpful to contact the editor or a member of the editorial board to gain insight on whether your article is appropriate for their journal. This will guard against an article being returned without even being submitted for scientific review.

Today authors also need to consider the relative merits of paper versus open-access online journals. More and more journals are going to open access with an associated article processing charge (or page charge) instead of relying on journal subscription fees. Open-access journals actually reach a broader audience, and you still have PDF versions of the article for distribution.

When you have the journal selected, identify about 3–5 *comparator articles* that resemble the type of article you want to write. In preparing an article submission, it is best to develop a manuscript that looks like, tastes like, and smells like other articles recently published in the targeted scientific journal.

After selecting the comparator articles, make your manuscript very similar to them in terms of (i) overall word length, (ii) subheadings and lengths of various section, (iii) number and style of figures and tables, (iv) number of references in each section, and (v) style and length of paragraphs throughout the manuscript. As a result, you should have a manuscript that becomes a good-looking, perfect style fit for the initially targeted journal.

At the time of the first submission, it is also advisable to develop a *journal priority list*, typically about 5–10 sequential options. Peer-reviewed and journal editorial decisions have a relatively high degree of entropy, and rejections happen quickly and easily. Just as most experiments fail many times before success, researchers need to deal with rejections frequently. As mentioned, the Nobel Prize-winning paper of Hans Krebs was rejected by a leading journal because they already had too many articles waiting for publication.

Most researchers are deeply frustrated when reading rejections. On the other hand, a critical review, even if it calls for additional experiments, can be very valuable and lead to a much stronger paper. Sometimes a cursory review is more concerning than the one in which the reviewers critically assessed the paper and made valid suggestions. Obviously, scientists/authors sometimes get too close to their work and miss something obvious that a trained third party sees right away.

If comments come back, there may be an opportunity to make some improvements in the manuscript. However, the rejection is often editorial decision without any explanation. Having the list of target journals will help to manage the stress and expedite the resubmission process to the next journal as soon as the rejection is received from the previous journal. In any case, each resubmission should tailor to the formatting and style requirements of the particular journal (e.g. type of the manuscript, length, referencing style).

In developing project proposals and grant applications, *pointers for reviewers*, comparisons with stated review criteria, and quality expectations might be very helpful. Reviewers are busy people and typically evaluate many applications in a very limited amount of time, sometimes minutes. When their eyes are rushing, the evaluation may miss key points very easily. Therefore, an application that makes the response to the review criteria clear can have advantages. When submitting a grant application, the review criteria are often available, and you should do your best to support the benevolent reviewer.

When your grant application is in a relatively advanced stage but not yet finalized, it is a good idea to take the review criteria and develop a clear and well-substantiated response to each of them. Subsequently, the responses should be strategically placed in the grant proposal to make them readily visible, without making them intimidating. As stated previously, glowing marketing words and commercial sounding self-praise can do great harm to credibility and should always be avoided. Important evidence and factual presentation should be the scientific writing style.

Getting your grant proposal reviewed internally before submission also has major benefits. This is particularly important for young investigators. Many academic departments have peer and grant review committees for this purpose. It is advisable to approach the entire process with humility, integrity, and purpose.

Rejection is a normal, routine part of scientific work. In scholarly debates, being questioned or even attacked is customary. A PhD student confronted with his/her first scientific controversy often gets disoriented and sometimes takes it personally. In reality, critical and skeptical feedback should be seen as the opportunity for the finest hour in the pursuit of understanding.

Debating skills are essential components of scientific communications. Quite frequently, the feedback is highly critical and indeed hurtful and appears to collapse fundamental assumptions. It is important to learn how to deal with criticism, how to listen to it, how to respond with facts, and, absolutely most importantly, how to learn from it. The productive scientist is ready to discuss and argue for ideas in a courteous but vigorous way. Scientific debates test someone's resolve, ability to stay positive, and intelligence in mastering the facts. Only the evidence of reality matters.

In the current world of science, studies almost always include multiple collaborators/authors, with each having a particular role. These individuals are coinvestigators in grants and co-authors on paper. Their many discussions, vigorous debates, and collaborative work create the greatest value. For a paper, the sequence of authorship is important with convention being that the first author, who is often a trainee, has done much of the work and the last (senior) author is his/her mentor and has provided direction for the studies.

Paul Lauterbur, a chemist by education, received the Nobel Prize for his discoveries concerning magnetic resonance imaging in 2003. Most of his work leading to the development of MRI diagnostic testing was conducted at Stony Brook University. His Nobel lecture describes the journey of discovery of this revolutionary imaging technology for the diagnosis of a wide range of human diseases (Lauterbur 2003b).

Some of the remarks also shed light on the many rejections he faced during his scholarly endeavor. After discovering one of the important principles, he wrote an article for the journal *Nature*, which was "summarily rejected. I felt this was a mistake, not because I foresaw all of the medical applications that would follow, but because of the physical uniqueness of the concept... My appeal to Nature was followed by submission of a revised version of my manuscript containing references to cancer and other more obviously relevant topics, and this time it was accepted." Somewhat ironically, he also added that "almost thirty years later, Nature publicly celebrated its appearance there."

Not only were Lauterbur's manuscripts subjected to rejection but also his ideas of commercialization. "Before I began describing it in detail everywhere, however, the University's agent rejected the patent application because they felt that it could not generate enough funds to pay for the application process."

Evidence-based Writing: Sentences and Paragraphs

Researchers are also in the business of influencing people's minds and way of thinking. Scientific writing is a heavily engineered process with great attention to the details.

The fundamental building blocks of scientific communication are the paragraphs that also serve as bones in the skeleton of the manuscript. Each paragraph is supposed to focus on one particular message. The sequential messages of paragraphs have to be carefully selected and ordered. Overstating the importance of clear paragraph structuring is impossible. The best interest of authors is to take advantage of all these tools.

The first sentence of the paragraph, or *paragraph sentence*, summarizes the message of the entire paragraph. Subsequent sentences should give further details, justification, and evidence substantiating the paragraph sentence. For obvious reasons, each paragraph should have one overarching message and one paragraph sentence that is brief and compelling. It is widely recognized that concise, practical statements have a better chance to gain and keep the attention of readers.

The sequence of paragraphs should follow a clear and understandable logic. The progression of messages or paragraph sentences has to persuade the reader that the ultimate conclusions are valid. By reading only the paragraph sentences, the reader should get a good understanding of the main messages of the entire manuscript. It is useful to start writing a new section with the definition of paragraph sentences. Often subheadings can be valuable, especially in the results or methods section of a paper. In this way, the reader understands when one set of results begins and ends.

The length of a paragraph has to be proportional to the relative significance of the message. Not only wording but also length, placement, and structure of paragraphs send messages. By providing more details, longer paragraphs give the impression of a more important message and vice versa. Rarely, a paragraph may consist of a single sentence (e.g. statement of the objective in an abstract). On the other hand, paragraphs longer than two-thirds of a type-written page are often considered lengthy or excessive.

Recently, the use of a series of very short paragraphs (bullets) is gaining popularity. The advantages of bullets include the use of very short statements, specific points, and well-separated lists of critical items. Approximately equal length of bulleted descriptions can help the readers to follow a list of equally important items. In documents prepared for busy practitioners or persons who have little familiarity with research, it is highly recommended to use very short paragraphs and bullets.

It is optional but often helpful when the last sentence of a paragraph provides the transition to the next paragraph. When a paragraph is complicated, or a switch to the message of the next paragraph would be difficult to follow, the use of transition sentence can be particularly useful. Unlike paragraph sentences, transition sentences are not considered necessary components of paragraphs. However, such sentences can help readers in following the logic of the paper.

In scientific writing, a knockout sentence is a group of words that makes an unusually strong and well-received introductory statement (Table 19.1). The development of scientifically sound, powerful messages can be greatly facilitated by conversations with other professionals and even laypeople. When genuine interest lights up their face, or they say "that is fascinating," then you might have something. When you fully acquire the skills of developing such powerful sentences, the writings will make a tremendous impact.

Table 19.1 Seven criteria for effective paragraph sentences.

1)	New, advanced, and stays away from banalities
2)	Typically, the first sentence of the paragraph
3)	Captures a major message of the manuscript
4)	Evidence based, either from literature or data presented
5)	Insightful and eye opening, immediately appreciated
6)	Succinct, carefully worded, and levelheaded
7)	Logical guide or skeleton for the entire manuscript

Günter Blobel, a professor at Rockefeller University, comes from a Silesian-German family. After studies at the University of Tübingen (MD) and University of Wisconsin–Madison (PhD), his research focused on the synthesis and transportation of proteins. He studied the signal peptide sequences attached to each protein that serve not only as address tags but also as a cellular positioning system that guides to a particular location and also makes the transfer possible over the otherwise impermeable membrane. The large protein unfolds itself and passes through the membrane channels as a long chain of peptides.

In addition to major scientific significance, his discoveries had tremendous clinical and economic implications. Today, the pharmaceutical industry uses cultured cells to produce proteins for drugs (e.g. insulin, growth hormone, coagulation factors). Formerly, insulin came from animal tissue extracted in slaughterhouses. In addition to the Lasker Award, Dr. Blobel also received the Nobel Prize in Physiology or Medicine in 1999, the King Faisal International Prize in 1996, and the Albert Lasker Award for Basic Medical Research in 1993.

His work also illustrates the tremendous value of skills in communicating complex biological phenomena to often unfamiliar audiences. For example, Blobel got funding from fellowships and grants from the National Institutes of Health; the American Cancer Society; the Damon Runyon-Walter Winchell Cancer Fund; the Helen Hay Whitney Foundation; the Jane Coffins Child Foundation; the Leukemia Society of America; the Life Sciences Research Foundation; Boehringer Fond, Deutsche Forschungsgemeinschaft; EMBO Fund; Humboldt Stiftung; Thyssen Stiftung; Human Frontier Science Program; NATO; Canadian, French, Japanese, Swedish and Swiss research organizations; and the Howard Hughes Medical Institute. It is impossible that the reviewers of these foundations and sponsors were all experts of protein signaling or even physiology.

Being evidence based also implies staying away from unsubstantiated speculations, a grave risk of being an award-winning scientist. As he summed it up, "one thing I want to be very careful about, I don't want to give opinions that are not based on facts, I don't want to discuss atomic disarmament and all sorts of other problems now" (Blobel 1999).

Time-honored Structure of Scientific Manuscripts

In attempts to understand nature, the well-established steps of the scientific method drive actions of the researcher. These steps have been defined and refined over many centuries, and historians trace the most fundamental concepts back to Aristotle. In the following list, the essential steps of the scientific method are described with an added indication of the corresponding section in original research manuscripts:

1) Identifying interesting and important questions to advance our understanding of nature or society; synthesizing what has already been known about the particular subject; and consequently establishing a research goal to understand what is missing (Introduction).
2) Development of a detailed, replicable methodology for the collection of new data based on an appropriate study sample; design of testable hypotheses and analyses of collected research data; and identification of appropriate quality measures to assure accuracy (Methods).
3) Description of all findings of the study including the sample used; presentation of collected data and analytical results accurately without interpretation or other influencing bias; and attachment of tables and graphs depicting results (Results).
4) Narrative summary and explanation of the principal findings; exploration of practical implications and alternative interpretations; potential threats to validity; and recommendations for future actions, including research (Discussion).

Introduction. The purpose of the introduction is to describe the problem addressed by the research and highlight significance, i.e. the gaps documented but unfilled by previous research:

- The introduction is usually a brief review of what is already known based on the published literature that includes 5–20 recently published and peer-reviewed articles as references but does not include tables or graphs.
- The first paragraph tends to give the overall picture: general problem and practical significance of the area selected for the research study (e.g. escalating cost of laboratory tests, difficulties of moving molecules through the blood–brain barrier).
- Subsequent paragraphs describe the gaps that have not been filled by any previous research studies (= the great opportunity of your study...), controversial views, various conflicting statements in the literature, and previous, failed attempts to solve the problem.
- The last paragraph of the introduction is a clear and explicit statement of what needs to be done and the stated objectives of the particular study. It has to be very short and also practical. Tools of research should never be stated as research objectives (e.g. the purpose of the study is to understand wellness priorities of elderly nursing home residents).

Methods. The role of the methods section is to specify research design and all data processing activities to help readers in understanding and, more importantly, replicating the study. Especially when clinical interventions are tested, the practical applicability depends entirely on the replicable description of the tested interventions (for example, in the landmark Diabetes Control and Complications Trial Research Group (1993)):

- Typically, it is helpful to have an opening paragraph with a summary of the overall research design, especially in the case of studies with a complicated methodology. Such an opening can provide an overview of the entire methodology and can help readers to understand the big picture before going into the technical details.
- The sample is the source of data, and representative sample selection is essential for generalizability of results. Subjects selected for the sample can be patients (in clinical studies), providers (in health services research), animals (in basic sciences), or reports (in meta-analyses). Sample description has to specify the eligibility criteria for inclusion/exclusion. Reviewers and readers are reassured when power analysis justifies the sample size.
- Description of the study site has particular significance in clinical and health services research. It includes the type of facility (inpatient or outpatient), size, clinical specialty (department), patient population served, for-profit or not-for-profit environment, and reimbursement mechanisms (e.g. fee-for-service, capitation, or prospective payment).
- Many research studies evaluate an innovative intervention (e.g. a new drug, surgical technique, financial incentive, or digital information). Based on the report of such studies, readers have to be able to apply the intervention. Therefore, the specification of all technical details is needed to support adequate replication elsewhere.
- Variables define what is measured in the study. All variables and measurement techniques have to be fully specified in the methods section (e.g. tests, reagents, questionnaires). It is useful to check that all variables mentioned in the text, tables, or graphs are defined here. In case of complicated but established methods, it may also be practical to refer to published papers versus using valuable space to go into great detail on technicalities of the reported study (e.g. these methods are published elsewhere with a reference).
- Observation includes all data collection and quality control activities, which make sure that complete and valid data are collected (e.g. timing and administration of tests, documentation of results, manual and computerized error checks, blinding procedures, databases). The process of observation is distinguished from the mere specification of variables.
- Description of data transformations and statistical analyses usually includes techniques of estimating parameters and hypothesis testing. It is advisable to work with competent colleagues throughout the process of research and also in the description of analyses.

The discovery of penicillin by **Alexander Fleming** is one of the most recognized scientific tales. In 1928, Fleming went on holiday with his family and upon return his attention turned to a stack of staphylococci cultures at a corner of his laboratory in St. Mary's Hospital, Paddington. Apparently, one culture was contaminated with a fungus, and the colonies of staphylococci immediately surrounding the fungus had been destroyed. Subsequently, Fleming grew the mold in a pure culture and found that it produced a substance that killed the bacteria. He identified the source as the *Penicillium* genus and published the results (Fleming 1929).

In the popular culture today, Fleming gets the almost full credit for the discovery of penicillin. This interpretation is not even consistent with the Nobel Prize as credit for the discovery was shared with Ernst Chain and Howard Florey. After his discovery, Fleming did not do anything with the finding for a decade. When Chain and Florey read Fleming's paper, they became interested in potential practical applications. First, they treated a patient with a severe facial infection, achieved real improvement, but ran out of penicillin. As a result, they became interested in large-scale industrial production. Further clinical tests and efforts to produce larger quantities became very successful (Goldsworthy and McFarlane 2002).

It should also be noted that the bactericidal effect of the mold was known for centuries. In other words, Fleming's achievement was not so much the discovery of a new bactericidal effect but documenting this fact in the scientific literature. The case was also another good example of quick reading of scientific publications by someone who understood the practical value of their discovery and had the ability to organize industrial production. The discovery and efficient production of penicillin undoubtedly has saved millions of lives worldwide. In scrutinizing the tale of penicillin discovery, historians also note that Florey eschewed publicity, while Fleming actively sought it (Goldsworthy and McFarlane 2002).

Results. The section is usually stuffed with facts, numbers, tables, and graphs testing the attention span of the readers. The role of the results section is to give an unbiased description of all new observations without comments, interpretation, or any other form of speculation. Essentially, this is a section for numeric results, statistical estimates, and hypothesis tests without entertaining speculations. Such a dry approach is needed to help differentiate between observed factual results and subjective interpretation in the next section. To aid the reader, subsets of results can be highlighted by a bolded heading, such as "imaging" or "histology":

- Often, results go beyond what can be presented in a narrative description and need to be communicated in tables and graphs. Usually, editors prefer tables when the amount of data is too much to include in the text but not too overwhelming to be displayed in a table. The use of graphs is often discouraged because of the inaccuracies of presentation and at least on paper the higher costs of reproduction. Therefore, figures should only be used when the amount of data is overwhelming and prohibits the use of tables (e.g. time series of multiple measurements).

By definition, figures typically are used to illustrate something that can only be appreciated visually, as with histologic findings, versus numerical data. Figures usually are described by accompanying written legends that detail the salient findings that are being illustrated. Typically, arrows or asterisks are added to clearly identify the key features. Many journals limit the number of figures and/or tables that can be included in the published paper but allow additional material to be presented as "supplemental material." Here, it is critical to decide which illustrative material should be included in the published paper versus the abridged supplemental material.

- Usually, the first paragraph of the results section is a brief description of the study sample. It should offer an overview of baseline characteristics of the selected subjects (e.g. age, sex, and diagnosis distribution of patients, the number of years in practice, the number of cases, and specialty of the participating providers). In randomized controlled clinical trials, an additional baseline comparison is needed to document that the various groups of subjects, usually intervention and control, are comparable.
- Subsequent paragraphs describe the observed data and findings of the statistical analyses. Frequently, a separate paragraph refers to each table or graph of the report. It is a basic requirement that the narrative description of tables and graphs should add additional information coming from the study without duplicating data from the table. Technical notes are often needed to understand the results and should appear in the text rather than as a complicated footnote at the end of tables (e.g. a particular table includes data only from a subgroup of patients).
- The last component of the results section often focuses on the reliability of findings. It is increasingly usual that the face value of reported data is scrutinized and the potential effect of various biasing factors needs to be evaluated (e.g. agreement between observers can be assessed by Cohen's kappa, tolerance calculation can be used to get the number of negative studies needed to overturn the conclusions of meta-analysis).

Discussion. The role of this section is to interpret the results presented in the previous section. After a probable headache caused by the dry methodology and number-driven results, it can be refreshing to learn about the practical meaning of the study results. The discussion is closer to the way people talk about the results of research and its practical implications. Many researchers follow a non-sequential reading of scientific articles: first abstract, if it looks promising, the introduction, then discussion, and, if everything goes well, methods and results. In the discussion section, mainly the following issues need to be addressed:

- The first paragraph is a concise but very practical summary of the most significant findings. As life sciences research studies tend to generate a vast number of numeric results, it becomes very difficult to figure out the practical meaning of data. Therefore, this paragraph may be the most important paragraph in the entire paper. No numbers, no limitations of the study, and only the major practical messages need to be summarized here. Authors should never expect that the readers will figure out the significance of the study. This is not a place to be modest; only accuracy is required.

- Frequently, there are alternative explanations and limitations regarding the generalizability of results. These aspects of interpretation need to be discussed in a subsequent paragraph (e.g. different underlying models, technical limitations of the measurement of effect, or possible confounding effects). It reflects the investigators' depth of understanding of the involved issues if they can specify more than one of those explanations for the results. To avoid possible misinterpretation, readers can be assisted by a discussion of the various assumptions behind the interpretation and various limitations of the study. While limitations should never be overemphasized, a brief discussion of potential concerns with the results can strengthen the paper.

- Usually, a justification for the methodology is necessary to convince readers and reviewers that the selected technology was right. Most frequently, reviewers attack your methodology by calling it unusual, unjustified, incomplete, or inconsistent. It is true that the specifics are described in the methods section, but this part of the discussion is the place to argue with the anticipated comments of the reviewers. Such explanations can be very helpful for readers not thoroughly familiar with the methodology of the particular subject area.

- Finally, a summary of the practical recommendations and return to the big picture can wrap up the report on the research study. The results of research studies can often substantiate specific practical recommendations, but these are not always obvious after reading the many numeric results or the interpretation at the beginning of the discussion. Therefore, an exact message to the practitioner can be a valuable addition. This closing part can also refer to the big picture mentioned in the introduction and place the reported study with its results.

Authors of original research reports often like to mention some of the promises and expectations generated by their study. A particular consideration is the potential for future studies. Just as no dissertation addresses all issues or questions, even a well-designed study and paper may end with more questions than were evident at the beginning of the study. Therefore, future studies can be important. The format of some journals calls for a separate section at the end of the discussion for conclusions that are a succinct, sometimes bulleted list of key findings.

References. The uniform editorial guidelines represent a concise and clear approach to the specification of referencing style and recommend a chronological numbering for referencing. In original research reports, the list of references usually includes between 20 and 60 articles. Most frequently, the introduction needs about 5–15 references; some articles include 2–3 references in the methodology section, and the rest is required for the discussion section. It is also recommended to have at least 50% of references from the last three or four years. Otherwise, the literature base might be labeled as outdated and irrelevant.

Authors should double check that all references are cited in the manuscript and refer to the particular point of interest. The order of references and points at which they are cited in the manuscript typically changes during the many revisions that a paper undergoes. Automated reference managers that can be set for a particular journal's format for the order and number of authors to be included, journal and volume, year of publication, and pagination facilitate this process and help ensure accuracy.

Quality Control: Testing the Manuscript Before Release

The final step of manuscript development has to be quality control, a search for common deficiencies to reduce the chances of rejection or major criticisms of the manuscript. Obviously, the most important tests are those that check the scientific soundness of the presented observations. There are six *basic tests before releasing* the research results and the manuscript:

1) *Non-obvious*: Ensure that there are no common, inconsequential, superficial, meaningless, marginal, and trivial statements that an uneducated person from anywhere would probably know or guess about the subject. Equally, there should be no emotional, inflated, pompous, grandiloquent, marketing sounding, exaggerated, and larger-than-life statements that would go against scientific objectivity.

2) *Novel*: All statements of the manuscript should be new and compelling. Researchers cannot afford to give the impression of intellectual weakness in broader communications. If readers get bored or irritated, the author is lost. Repeating what has already been stated many times before does not make much sense and fails to add value to the literature.

3) *Reproducible*: There are no significant statements and speculative paragraphs without substantiating evidence referenced from the literature or

rooted in the data presented in the results section. This test also includes the description of quality control measures and the efforts made by the researchers to reproduce the results.

4) *Self-reviewed*: This test is an opportunity to check against the review criteria of the particular journal or granting agency. One may also call it the sleep-on-it test. When you allow yourself at least one but preferably more days, you will be able to read the manuscript more like an outsider and get a better impression of how others will interpret it. While papers are now written almost exclusively with word processors, there can be value in reviewing a printed version at this stage.

5) *Generalizable*: This is the time to cross-check clarity of identifying the population at large where your research findings and conclusions are applicable by extension from the study sample. If an intervention was evaluated in the research, it is paramount to have a well-identified description sufficient for subsequent replication.

6) *Useful*: Every research project has beneficiaries, most frequently future researchers and also some product, like a scientific model, relevant factual observation, or intervention evaluation. This test makes sure that all information is adequately presented in the manuscript or offered in simultaneously disclosed datasets or multimedia to help future researchers and other users.

Additionally, structural deficiencies commonly occur in many original research reports submitted for publication. Most of these deficiencies not only confuse readers but also trigger highly critical comments of reviewers and may lead to rejection or require substantial rewriting.

Completeness of reporting is certainly a crucial issue of quality. There are several common deficiencies that indicate a lack of information in the manuscript (e.g. the purpose of the study is unspecified, description of the intervention is unclear and prevents replication, the variables of evaluation are not defined clearly, and exact numbers are not reported in the result section).

The completeness of the study design and methodology description is often scrutinized by reviewers. If significant aspects are unjustified or reflect a lack of planning, the study is headed for rejection. For example, a sample not representing the targeted population is a fatal defect of an experimental design that cannot be remedied by rewriting the manuscript.

In some cases, confusion about the role of various sections leads to the impression of incomplete reporting (e.g. the numeric results are mixed with interpretive comments; textual description repeats the content of the table or graph; or paragraphs are structured poorly and sized randomly).

Consistency is another major issue of quality. In the process of writing the manuscript, author attention is usually focused on the section under development. This creates opportunities for inconsistencies among sections. Statements need to be in their rightful sections instead of being spread out in various or mismatched sections. In other words, no methods should be presented in the results section, no interpretive discussion in the results section, no new study results in the discussion section, etc.). Something as simple as inconsistently identifying sections of a paper with bolding, underlining, or italics can be distracting for the reader.

Unlike the authors of manuscripts, reviewers of original research papers and grant applications tend to focus on the structure and consistency of the entire study. Reviewers often highlight discrepancies or controversies between various parts of the manuscript. One of the most frequent problems is the lack of relationship between the numeric results of the study and the enthusiastic interpretation in the discussion section. It is true that the discussion section is the section for reasoning, ideas, and various consequential assumptions of the investigators. However, all parts of the interpretation should be based on evidence and reported data.

Publishing a Research Paper Is Not "Mission Accomplished"

Studies of innovation dissemination show that publishing innovative discoveries and interventions in scientific peer-reviewed journals is rarely enough information for practitioners (Balas and Chapman 2018). Adoption health sciences of innovation happens in sequential waves and certain practices are leaders while others are lagging adopters (Figure 19.1). The sequence of waves also coincides with repeated validation and successive knowledge transformation. Ultimately, adoption requires many steps of transformation, and researchers striving to achieve practical impact need to consider what can or should happen after the publication of their scientific results.

Scientists need to be prepared for a two-way interaction with diverse audiences. Most research takes place in a so-called *triple helix environment*, intertwining academia, industry, and government (Shinn 2002). In such an environment, reverse communication also happens when society communicates expectations and preferences for the new knowledge. A particular challenge in communicating with others is recognizing the culture and level of comprehension by the other party.

In certain, politically more appealing areas, scientists are challenged to communicate effectively with a lay audience and politicians (e.g. *in vitro* fertilization, cloning, or health effect of global climate). At Stony Brook University, the Alan Alda Center for Communicating Science works to train scientists and health professionals to communicate more effectively with the public, public officials, media, and others outside their discipline.

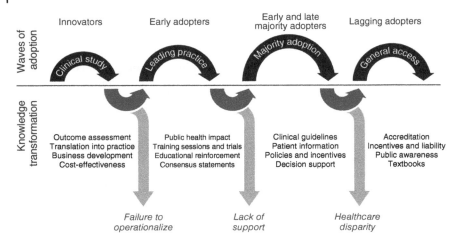

Figure 19.1 Waves of knowledge transformation and adoption of innovation in health care.

With the emergence of digital and social media, scientists get new channels of communication. Creating access to online databases, recorded video publication of methodologies, and open-access digital publishing are just some of the emerging opportunities to strengthen original research reporting after appropriate peer review. Being driven by evidence, original data, and replicable science remains just as relevant in the digital age.

Access to broader audiences can be greatly facilitated by the use of blogs, video lectures, electronic newsletters, Twitter messages, and many other emerging channels. There are significant differences among various scientific disciplines in the preferential use of multiple digital channels (Holmberg and Thelwall 2014). The practical use of these non-peer-reviewed channels of communication requires excellent presentation skills, understanding of online reputations, and solid evidence base. With the explosion of digital information and the proliferation of fake news, the scientific community needs to be prepared to gain visibility and maintain credibility.

In communicating with broader and lay audiences, exaggeration of the potential implications of research findings is a particular danger. A systematic review of animal studies of acute ischemic stroke found that out of 500 recommended "neuroprotective" treatment strategies, only two turned out to have confirmed benefits in human studies (Van der Worp et al. 2010). As a result of the many translational failures, the Laura and John Arnold Foundation calls journalists to stop publishing studies conducted with mice.

In the digital age, many risks of original scientific communications get elevated to unprecedented significance. The move toward inexpensive digital and open-access journal publishing created opportunities for abuses by bogus peer reviews, unscrupulous posting, and exorbitant fees for publishing. In response, the University of Colorado Denver librarian and researcher Jeffrey Beall created a list of potential or probable predatory scholarly open-access publishers (Butler 2013). Plagiarism is a concept that crystallized with the emergence of copyrights laws (Angélil-Carter 2014). Digital technologies also elevated the risks of easy, unattributed copying. Fortunately, it is also becoming easier to detect and deter unauthorized use of copied material with the help of new software tools.

Not infrequently, journalists ask renowned scientists about global, total, and intercontinental problems that are far beyond and outside of their competence and just unanswerable at present. What is the origin of life? What makes humans really humans? Of course, the answers tend to be like the king's sword: long and flat. Once, Nobel laureate Paul Lauterbur was asked: "What is going to happen in the future that you cannot imagine today? Obviously, this was a quite impossible, self-contradictory question" (Lauterbur 2003a).

In communications, the scientists are always expected to present substantial and novel observations. Unfortunately, we are all familiar with the cautiously sounding but typical useless statements in many research publications: "it may improve," "it has the potential to," and "there is a need for further research in this area." If you cannot measure it and if you cannot replicate it, you fail to improve it, and you must not promise it – that is the expectation by practitioners and informed readers in these days.

The bottom line is that communication skills are essential to progress in science. Not infrequently, several researchers make the same or almost identical discovery, but only the communicator gets the full recognition. As people say, without marketing, it does not exist. The best scientists tend to be great communicators as well.

References

Angélil-Carter, S. (2014). *Stolen Language? Plagiarism in Writing*. New York: Routledge.

Arendt, H. (1967). Truth and politics. *The New Yorker* (25 February).

Balas, E.A. and Chapman, W. (2018). Roadmap for diffusion of innovation in health care. *Health Affairs* 37 (2): 198–204.

Blobel, G. (1999). The Nobel Prize in physiology or medicine 1999. Interview Transcript. http://www.nobelprize.org/nobel_prizes/medicine/laureates/1999/blobel-interview-transcript.html (accessed 14 January 2017).

Butler, D. (2013). The dark side of publishing. *Nature* 495 (7442): 433–435.

Darwin, F. (1914). Francis Galton. *The Eugenics Review* 6 (1): 1.

Diabetes Control and Complications Trial Research Group (1993). The effect of intensive treatment of diabetes on the development and progression of long-term complications in insulin-dependent diabetes mellitus. *The New England Journal of Medicine* 1993 (329): 977–986.

Fleming, A. (1929). On the antibacterial action of cultures of a penicillium, with special reference to their use in the isolation of B. influenzae. *British Journal of Experimental Pathology* 10 (3): 226.

Goldsworthy, P. and McFarlane, A. (2002). Howard Florey, Alexander Fleming and the fairy tale of penicillin. *The Medical Journal of Australia* 177 (1): 52.

Holmberg, K. and Thelwall, M. (2014). Disciplinary differences in Twitter scholarly communication. *Scientometrics* 101 (2): 1027–1042.

Johnson, M.H. (2015). Robert Edwards: Nobel laureate in physiology or medicine. In: *The Nobel Prizes 2010* (ed. K. Grandin), 379. Stockholm: Nobel Foundation.

Lauterbur, P.C. (2003a). Interview with Paul C. Lauterbur. https://www.nobelprize.org/mediaplayer/index.php?id=552 (accessed 14 January 2017).

Lauterbur, P.C. (2003b). All science is interdisciplinary – from magnetic moments to molecules to men. Nobel Lecture 2003 (8 December). *Nobelprize. org* – Nobel Lecture – The Nobelprize in Physiology or Medicine.

Pais, A. (1982). *Subtle Is the Lord: The Science and the Life of Albert Einstein: The Science and the Life of Albert Einstein.* New York: Oxford University Press.

Schechter, B. (2000). *My Brain Is Open: The Mathematical Journeys of Paul Erdos.* New York: Simon and Schuster.

Schrodinger, E. (1932). *UberIndeterminismus in der Physik.* Leipzig: J. A. Barth.

Shinn, T. (2002). The triple helix and new production of knowledge prepackaged thinking on science and technology. *Social Studies of Science* 32 (4): 599–614.

Van der Worp, H.B., Howells, D.W., Sena, E.S. et al. (2010). Can animal models of disease reliably inform human studies? *PLoS Medicine* 7 (3): e1000245.

Part Four

Atmosphere of Excellence

20

Quality and Performance Improvement

The step from the laboratory to the patient's bedside...is extraordinarily arduous and fraught with danger.

Paul Ehrlich (1908)[1]

Productive, successful research depends not only on the talent of individual scientists but also on the supportive environment both regarding physical facilities and intellectual resources of innovation. In seeking optimal milieu for innovative research, organizational determinants and range of possible actions need to be examined.

The fundamental ingredient of scientific success is creative thinking. Nothing can replace the power and value of brilliant minds, elegant methodologies, and breakthrough discoveries whose time has come. The unique talents of Albert Einstein, Louis Pasteur, Marie Curie, Howard Temin, George Palade, Rosalyn Yalow, Robert Edwards, or César Milstein cannot be replaced by any organizational hocus-pocus. Creativity, unpredictability, and impulsiveness of great scientific adventures need respectful support.

Few universities can afford to attract and support Nobel Prize winners, but many research universities and institutions have the responsibility to achieve major progress in critical areas of life sciences. In many ways, the productivity of research can make or break the financial future of a university. It is an overriding interest to find roads to success and embrace research not just with a sense of the academic mission and societal purpose but also with a good understanding of business realities.

Well-funded research can bring millions of dollars to universities, making them prosperous, world renowned, and prominent. Many universities have grown tremendously as a result of research funding. At the same time, many other universities are struggling with the research mission. According to a widely

---

1 Ehrlich, P cited by Reitz, A.B. (2012) Future horizons in drug discovery research. *ACS Medicinal Chemistry Letters* 3: 80–82.

Innovative Research in Life Sciences: Pathways to Scientific Impact, Public Health Improvement, and Economic Progress, First Edition. E. Andrew Balas.

held perception, research costs money, and the revenues never compensate for the many infrastructural expenses.

It is important to look at the issue of research productivity not just with a focus on creative minds and independent laboratories but also at the level of the employing organization. Research universities and independent research institutes represent the essential organizational background of good science. In this discussion, the term university will be used, but most of the points are applicable to large independent research institutes as well.

In their efforts to improve performance, many university leaders observe surprisingly similar shortcomings and defects in a variety of laboratories. At large numbers, obvious patterns of quality can be recognized, and several clear needs for correction emerge. Research universities and institutes are organizations that have the responsibility to support hundreds of research laboratories.

Among the particular concerns are the shortage of rigor, high frequency of non-repeatable research, and prevalent quality defects. In recent years, retractions have increased 15-fold according to Thomson Reuters Web of Science. As mentioned earlier, about 50–89% of biomedical research is considered irreproducible, which is equivalent to a $28 billion waste in the research enterprise each year (Freedman et al. 2015). In academia, a place where people should understand, value, and seek evidence-based and measurable outcomes, there exists a need to periodically re-examine and reiterate the mission.

Other chapters already provided abundant information about the various quality deficiencies and performance improvement needs of research in life sciences. The ultimate answer to these questions must be a change in the research laboratories. Before such change can happen, the improvement initiatives need to be systematic, respectful, and, at least to some extent, organizational.

Quality Control Looks at Populations and Organizations

In recent years, there have been many signs of coming significant changes in quality and performance expectations of life sciences. The World Health Organization released quality guidelines to instruct basic biomedical research about the need for improvement (World Bank and WHO 2006). Even more telling that NIH is starting to hold institutions accountable for the quality of clinical research. "NIH will withhold clinical trial funding to grantee institutions if the agency is unable to verify adequate registration and results reporting from all trials funded at that institution" (Hudson et al. 2016).

To explore how quality and performance improvement can be promoted, the structure and function of life sciences research need to be fully appreciated. Essentially, there are two major levels of organization that should be appropriately distinguished (Figure 20.1).

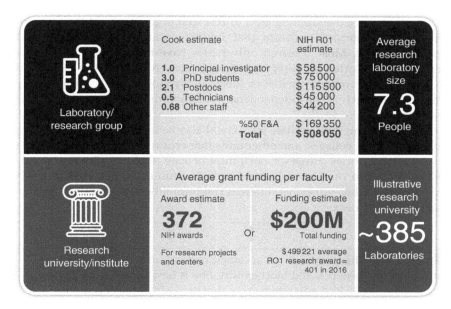

Figure 20.1 Illustrative size of research laboratories and research universities in the United States and United Kingdom. *Source:* Adapted from Cook et al. (2015) and NIH RePORT (2017).

The basic unit of research activities is the *laboratory or research group* under the leadership of a lead researcher or principal investigator. The laboratory or research group is funded by extramural grants or intramural support; the first one is typically in the range of an NIH R01 or comparable research grants. In many ways, the funded research laboratory headed by the independent investigator is the ultimate source of creativity and productivity.

Based on UK experience, the average research group size is 7.3 people (Cook et al. 2015): 1.0 principal investigator, 3.0 PhD students, 2.1 postdocs, 0.5 technicians, and 0.68 other staff (mostly research associates). This definition can be compared with the US experience with NIH R01 grant-funded research laboratories and groups.

Using the current NIH recommendations for research personnel compensation, the following ballpark figures can be developed for supporting the average research group as defined by the Cook et al. (2015) estimate: principal investigator, $58K; 3 PhD students, $75K; 2.1 postdocs, $115K; 0.5 technicians, $45K; 0.68 other staff like research associates, $44K; and 50% F&A reimbursement, $169K – totaling $506K. This total is very comparable with the $499 221 average R01 research award in 2016. In other words, the average R01 grant in the Unites States funds a research group very similar to the average in the United Kingdom.

The other more overarching level of research organization is the *research university* or *research institute* that houses numerous laboratories and research groups. Most research universities have hundreds of such independent laboratories to observe, support, and grow. It is the enormous challenge of research universities to substantially grow scholarly productivity while fully respecting the independence and unpredictable creativity of their researchers.

Occasionally, an intermediate level could also be identified regarding large research centers, colleges, and other units that create clusters of collaborating research laboratories. However, such intermediate level units have responsibilities similar to the universities/research institutions. Obviously, all levels have significant but somewhat different responsibilities in quality and performance improvement.

Emmanuelle Charpentier is an accomplished, award-winning researcher in microbiology, genetics, and biochemistry. She is best known for her research related to the CRISPR/Cas9 system in collaboration with Jennifer Doudna's laboratory. Notably, she made major progress in repurposing the CRISPR system for genome editing by combining Cas9 with synthetic "guide RNA" molecules (Jinek et al. 2012).

Charpentier was born in France but defines herself as a mobile researcher. After receiving her research doctorate, she worked as postdoc at the Institut Pasteur from 1995 to 1996; postdoc at the Rockefeller University in New York from 1996 to 1997; assistant research scientist at the NYU Medical Center from 1997 to 1999; research associate at the Skirball Institute of Biomolecular Medicine in New York from 1999 to 2002; guest professor and lab head at the Institute of Microbiology and Genetics, Vienna University, from 2002 to 2004; lab head and assistant professor at the Department of Microbiology and Immunobiology from 2004 to 2006; private docent and lab head at the Max F. Perutz Laboratories from 2006 to 2009; and lab head and associate professor at the Laboratory for Molecular Infection Medicine Sweden (MIMS) of Umeå University from 2009 to 2014. Since 2015, she is the director of the Max Planck Institute for Infection Biology. The long line of countries, universities, and laboratories illustrates not only the broadening horizons of talented researchers but also the enormous challenges of modern universities when it comes to recruitment and retention of scientists.

Beyond promotions to new positions, Emmanuelle Charpentier received numerous awards and recognitions, including the Breakthrough Prize in Life Sciences shared with Jennifer Doudna, the Otto Warburg Medal, the Canada Gairdner International Award with Jennifer Doudna and Feng Zhang, and the Japan Prize jointly with Jennifer Doudna in 2017.

There are two different ways to estimate the number of research laboratories at a university (NIH RePORT 2017): First, a particular university received 372 NIH awards suggesting a similar number of research projects and centers in the selected year, and second, using an alternative estimation method, we can start with the fact that the same university received $200M in total NIH funding in 2016. Dividing this number by the average $499 221 funding amount of an R01 research award, the estimated number of research laboratories or groups comes to 401. In other words, the university has about 372–401 research groups/laboratories headed by principal investigators.

In the terminology of the Small Business Administration, it can be concluded that most research laboratories and groups can be called microbusinesses (firms with 1–9 employees). Regarding their budget, the approximately $500K per principal investigator per year number is also consistent with the fact that NIH spends 10% of its $32B budget on 6000 intramural researchers, an average of the $538K amount per researcher.

The above-described institutional realities of research and also basic facts of quality evaluation highlight the significance of having an organizational and population-based approach to improvement. In other words, addressing the problem-isolated quality defects needs comparisons with the larger community of scientists and population data (Drexler et al. 2003).

For example, research laboratory A uses six leukemia–lymphoma cell lines, and two of them turn out to be cross-contaminated, misidentified, or misclassified. This problem undermines the accuracy and reproducibility of research results. To clarify the situation, the leader of laboratory A talks to the director of research group B who mentions that they had two contaminated among twelve cell lines. However, there is a huge difference in cell line quality: critically substandard in laboratory A and better than the national average in research group B.

Not surprisingly, the problem of research quality improvement is very similar to the ongoing quality improvement initiatives of health-care organizations. In his classic paper, Avedis Donabedian successfully conceptualized the quality of health care by focusing on structure, process, and outcomes (Donabedian, Donabedian 1966). Today, the US health care has one of the most developed and sophisticated measurements and continuous quality improvement systems in the world that is among others shaped by the *Donabedian concepts of quality* (e.g. HEDIS quality measures, Joint Commission, Vizient measures, CMS initiatives, etc.).

Part of the mission of CDC is to control health-care-associated infections. For this purpose, the National Healthcare Safety Network was launched to track and ultimately eliminate health-care-associated infections. In its annual publication, the National Healthcare Safety Network publishes benchmark data for institutions to compare their performance with expectations. The benchmarks are appropriately adjusted to the type of facility and patient population.

Based on the seminal model and evidence-based intervention of Peter Pronovost (Pronovost et al. 2006), the central line-associated bloodstream infection (CLABSI) is a widely recognized indicator that can be mentioned to exemplify the concepts of health-care quality improvement. NHSN (2017) provides an illustrative example of using CLABSI indicator for improvement: a particular hospital may experience five counted, 2.365 predicted CLABSI events over 1850 exposed days. The resulting 2.114 standardized infection ratio (SIR) is very much within the 0.775–4.686 SIR confidence interval showing the quality performance of the particular hospital in line with national expectations.

The bottom line is that individual research laboratory groups would have a hard time to recognize and understand the relative significance of lapses in quality and also would have a hard time to keep up with the ever-increasing quality defect recognitions and improvement interventions. Any organized attempts to improve quality and productivity of research have to consider the need to learn from the experiences of hundreds of research laboratories and set the goal of improving practices, university wide.

Cycle of Quality Improvement in the Research Enterprise

Continuous quality improvement is a concept that has been successfully spreading across a wide range of industries and professions, including health care. It is a concept that recognizes every defect as an opportunity for improvement, and progress requires well-coordinated teamwork. In research and innovation settings, there are many encouraging initiatives, but still, there is a scarcity in the application of quality improvement methods. One may reasonably assume that CQI also has the potential to transform the research enterprise, similarly to many other industries including patient care.

In organizational efforts, the quality improvement cycle is considered a classic tool to achieve positive change. The cycle is a general framework to identify problems, development of interventions to achieve change, and application of measures to assess progress.

In the classic quality improvement cycle, the first step is recognition of a problem or a defective pattern and in response setting an overall goal for improvement (Figure 20.2). Subsequently, the overall goals need to be translated into specific indicators to measure progress. After selection of quality and performance indicators, the baseline data should be collected to evaluate current or initial performance.

The next task is launching the quality improvement team that is assembled for this particular purpose and collaboratively designing the intervention that is expected to improve quality. The intervention should comprehensively

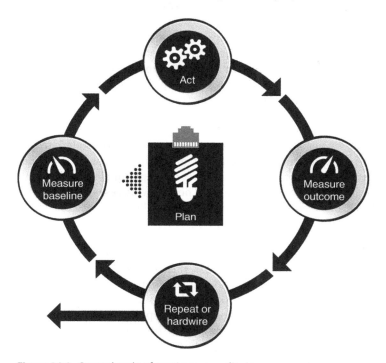

Figure 20.2 General cycle of continuous quality improvement.

address the defect and hold the promise of resolving all the key issues and producing measurable, better outcomes.

The intervention can be a line of action or a variety of activities that are known to change organizational performance (e.g. educational sessions, new reporting requirements, feedback reports, reminder messages, purchasing new supplies, new team structure for certain patient care procedures, and many others). Again, successful implementation of these actions needs teamwork.

After an appropriate time for implementation, the next step is repeated measurement of performance based on the quality indicators selected. When this measurement happens, there will be an opportunity for comparison with baseline performance. As a result, the observed value of the quality improvement initiative can be demonstrated.

When the before and after values show significant, beneficial difference, the quality improvement project can be considered a success. In such cases, it is essential to make sure that the achieved improvement is sustained for the long run. Most likely, it requires a larger range of additional measures to solidify changes and make them lasting. Often, this process is called hardwiring.

When the before and after intervention measurements of quality do not reflect sufficient change, the quality improvement project needs a continuation and the team should go back to the drawing board. In such cases, the measurements and particularly the quality improvement interventions will be revisited to make sure that everything was on target when the performance improvement issue was addressed. Most likely some new and hopefully more effective interventions can be developed and applied.

A welcome sign of progress is that several interventions have been tested in real-life situations, either in before-after or randomized controlled trials. Developing and implementing organizational interventions for quality and performance improvement is not just an abstract concept but already the reality in many areas of research and life sciences. The following research improvement interventions and their published impact should illustrate the concepts:

Many clinical research projects are heavily dependent on biospecimen donation, but relative participation is far less than desirable. The challenge can be much greater in certain cultures where such donation may not be viewed as consistent with traditional values. To address such need, Tong et al. (2014) developed a culturally appropriate educational seminar on biospecimen collection for Chinese Americans. As a result of this intervention, participants showed greater increases in willingness to donate biospecimen.

The usual slow response from peer review to authors of submitted manuscripts has been a growing problem in scientific publishing. Some scientific publishing change was needed to address the average 48 days delay until the final decision. In response to this need, the early editorial manuscript screening was developed and tested. According to Johnston et al. (2007), this intervention effectively reduced time to final decision to an average of 18 days, while impact assessment of manuscripts remained unchanged.

With the tremendous advances in technologies, genetic sequencing is rapidly spreading in scientific investigations and clinical practice. The number of research projects is skyrocketing in the areas of studying human DNA, viral and microbial genomes, and fragments of nucleic acids found in various places. However, it turns out that nearly 72% of previously published microbial and viral metagenomes had contamination (Schmieder and Edwards 2011). Frequently, human DNA contamination is the overriding concern. In response to this need, the DeconSeq technology was developed for the rapid, automated identification and removal of sequence contamination, particularly human DNA contamination.

Complete reporting of clinical trials is necessary to support adequately not only the scientific review but also the practical use of results. On the other hand, reporting the methodology and results of clinical trials is an exceptionally demanding and often confusing task. To address this need and achieve improvement, the *Consolidated Standards of Reporting Trials (CONSORT) guidelines* were developed. The wide-ranging study by Kane et al. (2007) demonstrated significant and consistent improvements in all the key aspects of RCT reporting after adopting the CONSORT guidelines.

Continuous research quality and performance improvement is the opportunity for organizations to support the many researchers they employ. Concepts of research quality and performance improvement include several important action lines: setting targets for improvement, providing education of researchers, monitoring research quality and outcomes, disseminating information about improvement opportunities, and developing and implementing interventions that can improve quality and performance.

The list of organizations that can make a difference in research quality and performance is substantial: research universities, independent research institutes, clinical coordinating centers of the major clinical trials, program project grant cores that serve the joint needs of several independent research laboratories, coordinating and specialized centers, research funding agencies, and science-oriented philanthropic foundations.

Quality Disposition of Research Projects

In many ways, the general research approach of the scientific laboratory or principal investigator also defines what can be done regarding quality improvement. There are two types of research disposition: observer and explorer.

Observer. The descriptive or observational research watches nature and living organisms either passively without intervention or by experimenting, i.e. changing some circumstances. Either way, the essence of research is the collection of data and fact. You may not understand the underlying mechanisms very well, but try to observe functions, and draw some useful conclusions. Due to the complexity of diseases and unknown underlying mechanisms, many clinical studies are in the observational category.

Typically, observational research is the chosen pathway of the applied researcher. In the quest for reproducibility, observational research relies exclusively on rigor, transparency, and self-replication of the observations and statistical analyses. The researcher draws conclusions based on data, analyses, and P values. Obviously, relying solely on the accuracy of methodologies is a slim defense line, and non-reproducible observations are frequent.

In terms of publication numbers in the general scientific literature, descriptive studies hugely outnumber exploratory research. Collecting data and unleashing sophisticated statistical analyses appear to be much more prevalent in the literature.

Explorer. The exploratory approach is entirely different by centering on principles and laws of nature. This disposition can also be called the modelist approach: the ambitious goal is the identification of a principle or development of a model that most accurately describes the particular phenomena.

Being an explorer means that you are trying to understand a complex natural phenomenon with a purpose.

Generally, exploratory research is the chosen pathway of the basic scientist. In addition to the quality control mechanisms of observation, the modeling/exploratory approach can apply a third, very important level of quality protection, the so-called triangulation (Balas and Ellis 2017). As mentioned earlier, the essence of triangulation is that the researcher repeatedly attempts to test the same model but from different angles, using different samples, methodologies, and analyses. Such an approach can help to understand how the modeled principles work from various directions.

Undoubtedly, descriptive research is often useful in accumulating facts for future theoretical or practical use. On the other hand, the failing of many descriptive research reports is one of the reasons why reading science is becoming more difficult than publishing research results.

Measures of Success in Scientific Research

The traditional approaches to measuring success and improvements in innovative research rely on most readily accessible data sources: research expenditures, number of publications, citations, publications in "high-impact" journals, and number of patent awards maintained by institutions (Mongeon et al. 2016). Faculty awards, promotion and tenure policies, and scientific reputations of institutions typically use these metrics.

On the other hand, the wide-ranging societal value of research cannot generally be assessed by these traditional indicators. For example, scientists citing each other have an unclear relationship with the actual societal impact of research results. There is a need for measuring the much broader range of societal value.

Peter Higgs is physicist professor at the University of Edinburgh. His pioneering work focused on subatomic particles. Published in 1964, his seminal work on the Higgs mechanism first indicated the existence of a new elementary particle with mass. The existence of such particle, called Higgs boson, was essentially confirmed by CERN's Large Hadron Collider in 2012. For his fundamental work and significant discoveries, Higgs received the physics Nobel Prize in 2013.

Interestingly, his many major discoveries did not come out of a usual faculty publication stream. Famously, Peter Higgs noted that he "wouldn't be productive enough for today's academic system" (Aitkenhead 2013). After his groundbreaking discovery of subatomic particles in 1964, Higgs published fewer than 10 papers. In his department a message was circulated every year: "Please give a list of your recent publications." Higgs' response to the research assessment exercise was simply "none" in most years.

While indicators of quality and productivity are useful, they are just proxies measuring progress toward meaningful organizational goals. Indicators of quality and productivity can promote the progress of hundreds of laboratories at major research universities. However, proxies of productivity should never be equated with actual productivity. As the case of Prof. Higgs illustrates, some Nobel Prize-winning scientists do not fit very well the widely used metrics of scientific productivity. What may be a useful assessment for a university may not be as good in evaluating individuals.

In a three-dimensional space, the following *performance indicators* not separately but together can illustrate a more comprehensive and balanced assessment of the value to society: (i) scientific dimension – per faculty average of peer-reviewed publications, citations based on publications in the top 10% from WoS Core Collection, and competitive federal research grants received; (ii) public health dimension – completed clinical trials, scientific contributions to FDA-approved products, and scientific contributions to clinical practice guidelines; and (iii) economic dimension – number of joint publications with industry, per faculty average number of startup companies, and per faculty average gross income received from licensing of protected intellectual property. There can be many other meaningful configurations of scientific value indicators, but the point is that the measurement needs to be multidimensional and go beyond traditional indicators.

University ranking systems can also provide some but limited support for societal impact analysis. Research universities often measure their value based on national and international ranking systems. Making references to ranking and improvement in ranking can be an attractive point in advertising brochures and in requesting government funding. On the other hand, it is important to recognize that research quality improvement initiatives need precisely targeted measurements of defects (Vernon, 2018).

For performance improvement initiatives, some university ranking systems can be sources of ideas. There are many university ranking systems that have various levels of credibility and construct validity: Carnegie Classification, Leiden Ranking, Academic Ranking of World Universities (Shanghai), QS World University Ranking, Center for World University Ranking (CWUR), The Times Higher Education Supplement (Times), SCImago Institutions Rankings World Report (SciMago), University Ranking by Academic Performance (URAP), Thomson Reuters Innovative University Ranking (TR), US News and World Report – Global Ranking (USN&W), EU Multirank, Round University Ranking (RUR), and Webometrics (Web).

In general terms, most ranking systems heavily rely on research performance indicators and therefore attach oversized significance to the scholarly activities of educational institutions. The most widely used scientific indicators include number of publications, number of citations (may be normalized), number of articles as corresponding author, number of articles in highly cited journals or top 25% of journals, number of articles with external or international collaboration, number of articles with industry collaboration, number/percent of articles within the topmost cited field, and others. A more concerning aspect of many university ranking systems is the excessive reliance on peer assessment that can be very subjective and misleading.

In university rankings, various indicators are used for the assessment of economic impact: patents filed, patents awarded, patents filed globally, number of publications cited in patents applications, co-patents with industry, startups initiated, R&D expenditures, R&D from industry, papers per research income, research/institutional income per staff and students, science and engineering (S&E) staff, ratio of R&D to institutional revenue, and others.

In continuous quality improvement efforts, the best quality indicators are those that are targeted at the selected defect and convey a clear, compelling message to all researchers about the need for improvement. If the quality improvement is targeting cell line contamination, then the appropriate indicator is the ratio of contaminated samples as established by independent measurement. If the quality improvement initiative is focused on recruitment in clinical trials, then the appropriate indicator is accrual rates in ongoing trials of the university. Similar measures of defects are essential in documenting and achieving lasting improvement.

Selection of the most appropriate quality indicators is not a trivial exercise in the research organization. The indicators should be chosen to reflect the seriousness of the targeted defect. Anybody thinking about simple choices, like impact factor, should be appropriately warned by Leslie Sage, senior editor in *Physical Sciences, Nature*: "Impact factors are created by narrow-minded bureaucrats, bean counters who like to have a single metrics that they can look at… It is complete nonsense, completely ignore it" (Sage 2011).

The above non-exhaustive list of major challenges illustrates that quality improvement opportunities and obligations of research universities are rapidly expanding. Accordingly, the research leadership of the future is likely to assume a much more active and substantial role. It is reasonable to anticipate that the traditional focus on regulatory compliance efforts and resource allocation will not be sufficient level of research leadership at major universities and institutes. In other words, supportive, motivating, and substantial research leadership will be increasingly needed to achieve quality and performance improvement at major research universities and institutes.

References

Aitkenhead, D. (2013). Peter Higgs: I wouldn't be productive enough for today's academic system. *The Guardian* 7.

Balas, E.A. and Ellis, L.M. (2017). Preclinical data: three-point plan for reproducibility. *Nature* 543 (7643): 40–40.

Cook, I., Grange, S., and Eyre-Walker, A. (2015). Research groups: how big should they be? *PeerJ* 3: e989.

Donabedian, A. (1966). Evaluating the quality of medical care. *The Milbank Memorial Fund Quarterly* 44 (3): 166–206.

Drexler, H.G., Dirks, W.G., Matsuo, Y., and MacLeod, R.A.F. (2003). False leukemia–lymphoma cell lines: an update on over 500 cell lines. *Leukemia* 17 (2): 416–426.

Freedman, L.P., Cockburn, I.M., and Simcoe, T.S. (2015). The economics of reproducibility in preclinical research. *PLoS Biology* 13 (6): e1002165.

Hudson, K.L., Lauer, M.S., and Collins, F.S. (2016). Toward a new era of trust and transparency in clinical trials. *JAMA* 316 (13): 1353–1354.

Jinek, M., Chylinski, K., Fonfara, I. et al. (2012). A programmable dual-RNA–guided DNA endonuclease in adaptive bacterial immunity. *Science* 337 (6096): 816–821.

Johnston, S.C., Lowenstein, D.H., Ferriero, D.M. et al. (2007). Early editorial manuscript screening versus obligate peer review: a randomized trial. *Annals of Neurology* 61 (4): A10–A12.

Kane, R.L., Wang, J., and Garrard, J. (2007). Reporting in randomized clinical trials improved after adoption of the CONSORT statement. *Journal of Clinical Epidemiology* 60: 241–249.

Mongeon, P., Brodeur, C., Beaudry, C., and Larivière, V. (2016). Concentration of research funding leads to decreasing marginal returns. *Research Evaluation* 25 (4): 396–404.

NHSN (2017). *The NHSN Standardized Infection Ratio: A Guide to the SIR*. CDC Updated June 2017. https://www.cdc.gov/nhsn/pdfs/ps-analysis-resources/nhsn-sir-guide.pdf (accessed 11 June 2018).

NIH RePORT (2017). *NIH RePORT Data*. https://report.nih.gov (accessed 9 June 2017).

Pronovost, P., Needham, D., Berenholtz, S. et al. (2006). An intervention to decrease catheter-related bloodstream infections in the ICU. *New England Journal of Medicine* 355 (26): 2725–2732.

Sage, L. (2011). How to publish in nature – Leslie Sage (SETI Talks). https://www.youtube.com/watch?v=ys8wHUPd6Vo (accessed 17 June 2017).

Schmieder, R. and Edwards, R. (2011). Fast identification and removal of sequence contamination from genomic and metagenomic datasets. *PloS One* 6 (3): e17288.

Tong, E.K., Fung, L.C., Stewart, S.L. et al. (2014). Impact of a biospecimen collection seminar on willingness to donate biospecimens among Chinese

Americans: results from a randomized, controlled community-based trial. *Cancer Epidemiology and Prevention Biomarkers* 23 (3): 392–401.

Vernon, M.M., Balas, E.A., and Momani, S. (2018). Are university rankings useful to improve research? A systematic review. *PloS one* 13 (3): e0193762.

World Bank and WHO (2006). *Handbook: Quality Practices in Basic Biomedical Research*. Prepared for TDR by the Scientific Working Group on Quality Practices in Basic Biomedical Research. http://www.who.int/tdr/publications/quality_practice/en (accessed 24 June 2018).

21

Institutional and National Strategies

As long as...scientists are free to pursue the truth wherever it may lead, there will be a flow of new scientific knowledge to those who can apply it to practical problems.

Vannevar Bush (1945b)[1]

The progress of science is dependent not just on the talent of individual researchers and their material resources but also on the intellectually nutritious environment and supportive policies. The institutional, local community, and national policymakers have great, intertwining responsibilities. Effective policies supporting research go far beyond the need for funding and include a large variety of regulatory actions, transparency and reporting of research results, intellectual property policies, regulation of entry when launching start-ups, and many others. Ultimately, it will take the convergence of priorities, leadership actions, and resources that can together lead to extraordinary research productivity and discoveries.

Essentially, life sciences research is the curiosity enterprise for production of discoveries. As a result of the ever-increasing number of researchers and many ongoing research projects at any given time, there is a need to recognize the trends of productivity in life sciences research. By and large, research has some characteristics of cottage industry, while others look like mass production.

Setting strategic priorities and developing policies that promote productive research require an intimate understanding of the scientific process, careful deliberations, and decisive actions. When it comes to growing research, much depends not only on the research faculty but also on the institutional and broader environment. This chapter draws attention to some opportunities to leap forward in policy support for life sciences research as well as translation and implementation of scientific discoveries.

---

1 Bush, V. (1945b). *Science, the Endless Frontier: A Report to the President*. Washington, DC: US Government Printing Office.

Innovative Research in Life Sciences: Pathways to Scientific Impact, Public Health Improvement, and Economic Progress, First Edition. E. Andrew Balas.
© 2019 John Wiley & Sons, Inc. Published 2019 by John Wiley & Sons, Inc.

Institutional Opportunities to Support Research

Scholarly productivity of a research university refers to how efficiently it uses people, funding, and other resources to transform ideas into outputs useful to society. While the research productivity in some institutions may have remained at a predictable trajectory as reflected by steady ranking, other institutions experienced an accelerated growth or decline over the past decade. Overall, the landscape of most productive research universities is changing.

In writing about the five fundamental principles of government support for scientific research and education, one of the most influential science administrators in history, Vannevar Bush (1945b), effectively defined the action lines of universities and research institutions. In providing support for basic research in colleges, universities, and research institutes, the government must respect the complete independence of institutional control of policy, personnel, methods, and scope.

The principle of faster, better, and cheaper applies to scientific research as well. It is worthwhile to consider the Vilfredo *Pareto principle*, also known as the 80-20 rule (Pareto 1896). Researchers and also research administrators can benefit from the recognition that in many undertakings, 80% of the effects come from 20% of the causes. Contrary to prevalent assumptions, research and scholarly innovation can be self-sustaining and revenue generating by thinking Pareto.

Effective research leaders and most improved institutions can do the basics but also much more by serving as a source of inspiration, exploring resourceful new opportunities, and promoting a creative environment for research success. The following list provides examples of institutional and local community actions that appear to effectively facilitate research productivity.

Resource Development. While thinking remains the most significant asset of productive research, most scientific projects need appropriate physical resources, instrumentation, offices and laboratories, animal facilities, computing resources, libraries and information services, and many other types of infrastructural support. Small grants and travel support are usual components of resource development. The same applies to intellectual resources of infrastructural support (e.g. biostatistics, preferential access to specific data sources, human subject protection system, and many others). An emerging new opportunity is fundraising for science buildings that promote convergence and interdisciplinary interactions.

Support for Creativity. The government agencies retain the discretion in the allocation of funds among institutions. According to Vannevar Bush, the government should promote research through contracts or grants to organizations instead of running laboratories of its own. Accordingly, support for grant and contract applications, grant writing and entrepreneurship development, and administrative support for the management of awards to complete research projects are among the primary action lines of research institutions. The support for creativity also needs to include actions that generate excitement and appreciate the accomplishments. Seminars, speaker series events, hosting workshops, competitions, newsletters, press releases, recognition events, and many other ways can lead to the creation of an energetic and successful research culture.

Developing Research Talent. Another widely recognized opportunity and responsibility of research institutions is the triple task of promoting competent research: (i) advancing the professional development and research competence of already hired faculty members and other research personnel; (ii) hiring successful researchers with fitting expertise, great scholarly promise, and often transferable external funding; and (iii) particularly in educational institutions, reassuring teaching faculty members about their continued significance and viability, so unnecessary anxiety and pushback can be avoided.

Research Partnerships. Strong and well-functioning networks and ecosystems can greatly enhance the success of the innovative research. Typically, network relations consist of a limited number of contacts that are based on a high degree of trust and informal exchanges. Social networks can be intramural (within the department or university) and extramural (companies, financing, and other research institution) components and can greatly increase opportunities for such research partnership development and also make them successful. Strategic development of a critical mass of researchers with diverse but synergistic interests is often part of the strategic action.

Innovation Networks. Science parks and incubators play a great role in facilitating commercialization of innovation. University entrepreneurship networks, innovation networks, science parks, and incubators appear to greatly facilitate a more stable flow of higher-impact patents. Park-based startups tend to have significantly improved access to equipment, R&D, and qualified personnel, are more involved in cooperation with universities, and have a higher level of network activities (Lindelof and Lofsten 2004). Such resources can give the firm with a distinct competitive advantage. Innovation networks extend the concept of a research partnership to bridge the gap between science and practice.

Technology Transfer. Historically, universities were not very much involved in the commercialization of innovation. The turning point was the Bayh–Dole Act of 1980 that encouraged and streamlined patenting of publicly funded research discoveries. With the growing emphasis on innovation and industry partnerships, the trajectory of successful research career has been effectively redesigned in many institutions. Creating an institutional culture that motivates productive research and practical innovation is a lengthy but essential process.

Colyvas (2007) chronicled the early institutionalization of technology transfer at Stanford University. While the interests in the production of economically useful knowledge steadily increased, the process proved to be lengthy and convoluted. Researchers had to separately interact with the Office of Sponsored Programs and Office of Technology Licensing. The supportive policies, like customary revenue sharing arrangements or patenting process, were developed gradually. Inventor credit, recognition of team effort, and ownership arrangements evolved.

Startup Support. As illustrated by the practices of Columbia University, MIT, Stanford University, and others, a standardized approach to university startup licensing can be particularly helpful. Such consistency can not only dramatically simplify the lives of both university and faculty but also expedite processing and licensing. In such policy setting, most of the terms are fixed (equity, up-front payments, and patent expenses); others are variable based on the technology type (royalty, milestones). Consistency in the policies implies commitment that "no one gets a better deal." If a startup wants a reduction in one term (e.g. equity), they have to offer a corresponding increase in another term (e.g. equity reduction for larger up-front expense share). In other words, the deal cannot become an arrangement that is much more desirable to all other faculty members.

Prototype Development Support. As mentioned before, functioning prototypes based on scientific discoveries represent a particularly prominent milestone of research as they (i) validate reproducibility of a research observation, (ii) represent functioning model of particular phenomenon, and (iii) cross into the sphere of commercialization. With advanced technologies, prototyping can be significantly accelerated, including 3D printing, web service development tools, computer-aided design (CAD) software, or prototype tooling that may not be efficient for production but fast enough for producing pilot products.

According to former MIT professor John Meada, "if a picture is worth a thousand words, a prototype is worth a thousand meetings" (cited by Banfield et al. 2017). In putting science to work, one of the critical tasks is the development of the minimally viable product or prototype (MVP). This is the first reality check. To facilitate reaching the phase of MVP, research institutions can provide targeted support, expertise, and tools.

Policy Actions. Research institutions can also apply many policy actions to enhance, encourage, and expand productive research and effective research agenda setting. These actions or interventions can include research time allocation policies, interdepartmental appointments, revised promotion and tenure expectations, setting salary recovery expectations, IP revenue sharing policies, streamlined administrative services like IRB approval, and many others.

Not infrequently, high-ranking research administrators limit themselves to the very minimum of reactive leadership. At the very basic level, they process grant applications, manage sponsored programs, manage available laboratory space and resources, support the Institutional Review Board process, enforce compliance with research regulations, and distribute small amounts of seed funding to struggling researchers. The presumed justification for such approach is that research is an inherently creative and independent line of activities. No department chair, dean, or vice president can dictate the research interest of faculty members; it is only their treasured choice.

While regulations are needed, and compliance is important, any research administration that reduces its role to the compliance enforcer does a disservice to intellectual creativity and scientific productivity. Few things can be more discouraging than institutional administration constantly talking only about the bureaucratic tasks of research: preparing for monitor visits, what to expect during an audit, indirect (F&A) costs and waivers, regulatory and patient binders, electronic sponsored route, preparing and submitting a new IRB project, recognizing and reporting reportable events, cost transfer, understanding a notice of award, and many others.

There are many challenges to face, but they are all addressable. At the institutional level, occasionally grandiloquent research strategic plans are put forward without substantial follow-up or specific measures of progress. High performance, after all, is not achieved by press releases and a few random institutional decisions but by all-encompassing actions to promote innovation and creativity in research with the uncompromising long-term support of top management. Institutional research leaders need to take an active, strategic, and inspiring role in supporting their independent research laboratories.

Vannevar Bush was an electrical engineering professor at Massachusetts Institute of Technology (MIT) who became one of the most influential science administrators of all times. During World War II, Bush headed the US Office of Scientific Research and Development (OSRD), through which wartime military R&D was supported, including the Manhattan Project. He is also credited with bringing about the National Science Foundation.

As a professor, Bush produced a remarkable string of inventions and innovative ideas. He is considered one of the pioneers of computing with a primary focus on analog computers. His work on memex, a precursor of hypertext and the World Wide Web, influenced many generations of computer scientists (Zachary 1997). Besides research, he also founded Raytheon, a major US defense contractor. Later, he became dean of the school, vice president of MIT, and chairman of the National Defense Research Committee. During the war, he essentially laid the foundation for government funding of innovative research, principles that are still useful and practical today. In recognition of his scientific and policy contributions, he received the Edison Medal, Hoover Medal, and the National Medal of Science.

After the war, his essay pointed to the urgent need to use accumulated knowledge more effectively (Bush 1945a): "Professionally our methods of transmitting and reviewing the results of research are generations old and by now are totally inadequate for their purpose….truly significant attainments become lost in the mass of the inconsequential."

Recruitment and Retention of Research Faculty

Not with the intensity of recruiting football and basketball stars but universities vigorously compete for talented and well-funded researchers. Attracting star researchers from other universities with the simultaneous transfer of their funded projects is one of the best-known ways toward increased research productivity. Indeed, such approach compares favorably with the complementary effort of improving the performance of an existing pool of faculty researchers through continuous improvement initiatives.

Many people view *recruitment of researchers* as the hallmark of successful academic leadership at the level of department chairs or deans. Of course, hiring away productive researchers can generate some interinstitutional tensions, and therefore the process should be delicately managed. The process of hiring research faculty has two distinct steps, attracting qualified candidates and negotiating arrangements for success.

Attracting the Best Minds. Attempts to invite accomplished scientists should start with the development of the inventory of desirable qualifications. Subsequently, an extensive list of potential national or international candidates needs to be compiled based on influential publications and grant funding. Afterward, a brief invitation should be mailed to open the lines of discussion.

A simple mail merge can deliver the message that someone may like to receive. For example, Congratulations, your scholarly accomplishments generate a lot of interest and excitement. It is not just me but also my colleagues saying many appreciative comments about your research in X. Our University is seeking an experienced leader to lead X in the College of X. The College and the University have been steadily growing year after year. Research is multiplying impressively, thank our generous support practices. Our ambitious vision is to develop X at our University. I would like to discuss this leadership position and the many opportunities of our campus with you. We can provide very competitive packages. Of course, this is just an exploratory discussion, and no commitment of any sort would be expected.

Negotiating for Success. Faculty members considering a change of institution need to consider a variety of factors that may profoundly influence their chances for success. Among the important factors are the following: salary and multiyear startup support; laboratory, instrumentation, and physical facilities; intellectual resources and scholarly environment; support for grant application and post-award management; institutional policies; and support for entrepreneurship and innovation commercialization.

Based on the analysis of a large number of research faculty offer letters, a multitude of common challenges emerge, and some recommendations can be made. Such recommendations can not only protect institutional financial interests but also make the research enterprise sustainable, including hiring more and more accomplished researchers.

Occasionally but memorably, universities can be victimized by startup shopping. The aspiring researcher negotiates a generous startup package, launches successful research, and gets a million-dollar grant. When the original startup expires, the researcher, by now well funded, negotiates for a bigger startup in another institution and transfers to the highest bidder.

The list below identifies several challenges and illustrates remedies for the letters of offers to new research faculty members:

1) A typical need is clarity regarding the exact amounts and length of initial support provided by the hiring university. For example, in year one the following startup support will be provided in the categories of travel ($X), supply ($), personnel ($X), uncovered research faculty salary ($X), and designated X ft2 research space ($X); in year two...; and so forth. The typical length of startup support is X years with the possibility of a one-year extension based on approval.

2) Occasionally, newly hired research faculty members assume that the startup support is an absolute mandate, but the performance expectation is optional. In the case of multiyear research grants, NIH and NSF also make it clear that funding is contingent on congressional appropriations. Therefore, it is reasonable to include in offer letters prepared by universities as well: the startup funding is subject to availability of budgetary resources in the given year.

3) To address the need for performance thresholds, the university should set clear expectations. It may be beneficial to present the startup support and performance expectations in one table for side-by-side comparison: each year research performance thresholds have to be met to be eligible for further support in the subsequent year. The university expects X number of grant applications; X amount of grant applications; X percentage of salary coverage; and X amount of grant and contract funding in year one, in year two..., and so forth.

4) Not infrequently, newly hired researchers assume that startup support will come as a large, unrestricted cash donation ready for spending at any time throughout the startup period and beyond. Such perceptions fail to recognize the hard work and great budgetary effort of most institutions with limited resources. Therefore, it is appropriate to include the following: the university may use diverse funding sources at its discretion to cover startup-related expenses.

5) The great benefits and tremendous opportunity for startup funding often remain marginally appreciated by inexperienced researchers. Such cases of credit denial are not a healthy start for long-term productive employment experience. Therefore, it is appropriate for institutions to set expectations: the university expects a written acknowledgment of the startup support in all resulting papers and presentations.

6) In some cases, it may be prudent to include provisions that reduce the temptation for startup shopping. Applicable provisions may not be legally

enforceable but can indeed create a common sense of expectation signed by both the research faculty candidate and the institution. Similarly, to the usual requirements after sabbaticals, the institution may want to include the following: after receiving startup support, the faculty member agrees to continue full-time duties at the university for a period of not less than the preceding period of startup support.

Ultimately, it is not unreasonable to expect that research faculty hiring negotiations should be built on a simple business plan or at least back of the envelope calculation how the revenues and expenses will balance on the short and midterm. Among the items to be considered, it should be recognized that most federal research grants are transferable, but some can be retained by the original institution even after the principal investigator is hired away. Due diligence regarding the transferability of grants is highly recommended.

Drafting an agreement or offer letter with the newly hired faculty members can set clear directions and encourage the right kind of performance. As mentioned above, legal scholars may note that not all of the above-listed provisions are legally enforceable. However, goodwill and honest commitment to put the mission first can be the right start for successful research faculty hiring.

Susumu Tonegawa is a professor of biology and neuroscience at MIT. For his landmark discovery of the genetic principle for generation of antibody diversity, he was the recipient of the Nobel Prize in Physiology or Medicine in 1987. Essentially, his discoveries solved the puzzle on how less than 19 000 genes of the human body can form millions of antibodies by rearranging themselves (Tonegawa 1983).

Born and raised in Japan, Tonegawa received his PhD from the University of California, San Diego, followed by a postdoctoral fellowship at the Salk Institute in San Diego. Subsequently, he moved to the Basel Institute for Immunology in Switzerland where his most famous immunology experiments were performed.

After his landmark discoveries in immunology, his interest turned to an entirely different field of life sciences: brain functions and memory. Tonegawa became a pioneer again in the application of optogenetics and biotechnology in neuroscience. His studies focus on the molecular, cellular, and neural circuit mechanisms of learning and memory (Kitamura et al. 2017).

At MIT, he was the founding director of the MIT Center for Learning and Memory that later evolved into the Picower Institute for Learning and Memory, an impressive multidisciplinary research initiative. After stepping down from leading the Picower Institute, he was right away approached to lead the powerful RIKEN Brain Science Institute in Japan.

These centers collaborate and compete with another interdisciplinary initiative of MIT, the McGovern Institute for Brain Research that employs Nobel laureate Robert Horvitz, Ann Graybiel who received the US National Medal of Science, and five members of the US National Academy of Sciences. These and other powerful, cutting-edge research institutes illustrate the challenges of not only recruiting but also retaining eminent scientists in an intellectually stimulating environment.

Peer Comparisons of Research Productivity

Priority setting and measuring progress are among the essential responsibilities of leadership. This may not be a simple task given the widely known but not particularly effective measures of research productivity. It is well recognized that the general public expects research will improve their well-being and health through better practices, products, and services (Pollitt et al. 2016).

Many research universities and institutions rely on the established measures of success (i.e. number of publications, number of citations, number of articles in *Nature* or *Science*, or top 25% of journals, number and amount of federal grants received, and others). However, extensive reliance on these rigid, traditional indicators is often blamed for the current crisis of non-reproducible research (Alberts et al. 2014; Collins and Tabak 2014).

A logical extension of traditional measures and strategic directions is the promotion of engaged and team science to make research more closely aligned with recognized scientific opportunities and societal needs. Such priority settings can monitor the number of articles with external collaboration, international collaboration, and industry collaboration. Additional examples of tracking opportunities include interdisciplinary of publications, industry article citation, and publications cited in patents applications. The chapter on quality improvement lists a large number of indicators to measure performance in the direction of the three major outcomes of research in life sciences.

At research institutions that emphasize research supporting economic development, performance may be upgraded along the lines of patents filed, patents awarded, the number of publications cited in patents applications, co-patents with industry, startups initiated, R&D from industry, science and engineering staff, the ratio of R&D to institutional income, or licensing revenues. Apparently, in large multidisciplinary research institutions, some laboratories or researchers will be in a better position to perform well on the scientific dimension, while others will be better in the public health or economic dimensions.

At a time when concerns are growing about the quality and reproducibility of research results, institutional improvement programs and corresponding measures of quality are becoming necessary. Such measures may include the number of retracted articles, failure of reaching enrollment target in clinical trials, the ratio of clinical trials without reported outcomes, ratio of never cited articles, and others.

Researchers and research institutions are increasingly urged to demonstrate the societal benefits and potential practical value of their studies (e.g. statements of broader impact in NSF proposals or public health impact in NIH proposals). Correspondingly, research administrators and national research project coordinating centers have an increasing responsibility of measuring and ultimately improving the societal impact of scientific work in these dimensions. Corresponding tracking indicators may include completed clinical trials, contributions to FDA-approved products, and contributions to clinical practice guidelines.

Most improved performance in science can be defined not only by consistently increasing research productivity in comparison with past but also by improving relative ranking in comparison with peers. The top performers of research improvement are getting ahead of their group of similar universities (not to be confused with institutions that have always been high flyers). Identification of the group of most improved institutions is helpful in exploring their associated indicators of success.

Comparing research productivity among universities is necessary to identify best practices for strengthening university research programs. Based on these considerations, measurement of university research performance should help to identify what works. Ultimately, more effective strategic priority setting should advise universities in streamlining research activities and improving scientific output.

Improving research needs the good understanding of the research enterprise, priority setting, and consequent action. It is impossible to define an effective strategy for research expansion without the wide-ranging analysis of institutional performance in comparison with national/international trends and competing initiatives.

Various trends of research progress can be explored and compared by data trends of selected universities. For illustrative purposes, the annual grant funding statistics over a period of 25 years is examined based on the NIH Reporter data. Figure 21.1 shows the NIH funding expansion at three selected research universities.

The top part of the chart shows total NIH funding for these universities over a period of 25 years. At that starting point, University A had twice as much funding as University B, and University A had six times more funding than University C. At the end of the study period, the absolute difference between University A and C grew from about $30 to $270 million, a difference that is far above and beyond what can be explained by inflation.

A superficial look at the chart may suggest that University A has far the best strategic approach to developing its research agenda productivity. However, it

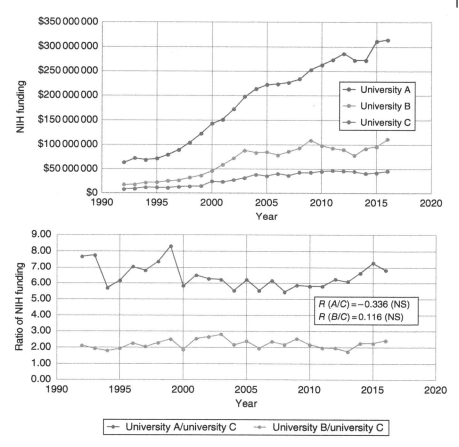

Figure 21.1 Funding trend for three universities (upper chart: growth over time; lower chart: relative to each other). *Source:* Data from NIH RePORT data (2017).

needs to be considered that University A was already funded at a higher level in the beginning and subsequently the average percent growth has been rather similar to the other charted universities. In other words, the soaring trend of University A reflects only the power of the same percentage growth.

Apparently, compounding growth is the most potent force in expanding research. In other words, institutions starting with a higher research portfolio become progressively bigger just by maintaining the same speed of expansion. One may say, compounding research expansion makes the rich richer, progressively. In other words, by exercising the same level of research readership and achieving the same rate of expansion as before, the high-performing research universities will get disproportionately larger and better funded over time.

As shown at the bottom part of the figure, it turns out that the relative ratios of the NIH funding remained nearly constant, i.e. University B makes twice as much and University A makes six times more NIH funding than university C. Apparently, the research growth rate was mostly the same at each of the three universities annually. None of the universities showed strategically superior leadership that would alter their relative position to the other universities in this comparison.

Smaller and beginner research institutions need to implement radical new actions if getting ahead of others is the stated goal of research performance improvement. Consequently, there is a huge need for an entirely new package of interventions that can radically expand research while remaining committed to respecting individual researchers and avoiding the type of centralized voluntarist dictatorship that can backfire in the field of scholarly creativity.

Catapults of Success: Education of the Youth

Educational institutions have a formidable opportunity and responsibility in raising talented scientists. There are many good educational programs, and some of them are exemplary. However, a few educational institutions demonstrated towering performance when it comes to preparing future scientists. Their production of scientific talents is far above and beyond what could be explained by random variation. The fate of the empire depends on the education of the youth – as stated by Aristotle (cited by American Annals 1837). These educational success factors deserve more study and active consideration.

In an interview, Nobel laureate Robert Lefkowitz (2012) proudly highlighted that he graduated in the *Bronx High School of Science*. In 1938, this specialized and preeminent school was established for academically talented students selected from among the best performers of high school test takers in New York. Beyond science and mathematics, the school also has strong programs in humanities and social sciences. Over the years, Bronx Science produced eight Nobel laureates, seven in physics and one in chemistry. Beyond the most famous award winners, this public high school also educated countless other outstanding scientists. In the US National Academy of Sciences, there are 29 graduates of Bronx Science.

The *"Fasori" Lutheran High School* was a uniquely accomplished secondary school in Hungary (Budapest-Fasori Evangélikus Gimnázium). Before World War II, the school educated several Nobel Prize winners and significant thought leaders. Among the students were Eugene Wigner (Nobel laureate physicist), John Harsányi (Nobel laureate economist), John von Neumann (mathematician), Edward Teller (theoretical physicist), and Kálmán Kandó (inventor engineer). Other accomplished students included Imre Kálmán (composer), Sándor Petőfi (popular poet of the nineteenth century), and Theodor Herzl (father of the modern state of Israel). The school's legendary faculty included László Rátz, teacher of mathematics, educational reformer, and later principal of the school. Interestingly, the selection of Fasori School students was not based on a wide net of hyper-competition. Instead, the formula for success was probably more reliant on a culturally inclusive environment, ambitious faculty, and outcome-oriented educational programs.

Between 1964 and 1972, many talented physicians came for postdoctoral training in the *National Institutes of Health (NIH)*. Apparently, the remarkable symbiosis of clinical background and bench research played a notable role in the evolving success. These Public Health Service slots offered military draft exemption and were highly competitive at the time of the Vietnam War. During this period, the NIH research training of physicians proved to be astoundingly efficient and produced nine Nobel laureates so far. Two of them, Mike Brown and Joe Goldstein, published an essay discussing the golden era of training in research (Goldstein and Brown 2012). Research fellows were not only trained by brilliant scientists but also benefited from the leadership of several NIH career scientists, themselves recipients of the Nobel Prize. With the latest tools of biochemistry and molecular biology, they all addressed fundamental problems that were relevant to medicine.

Selection of students, self or through competition, appears to be a crucial influencer of success in science later. Perhaps the right conversation with a student should start not with "did you learn anything today" but with "did you ask a good question today," as physics Nobel laureate Rabi learned from his mother (Cited by Sheff 1988). Curiosity, work ethic, and a sense of optimism are highlighted by Robert Lefkowitz as hallmarks of good students. Are you lucky? – this is one of his standard questions to applicants. "You can only be successful if you're happy with what you're doing," echoed Brian Kobilka, one of his former students and corecipient Nobel laureate (Kobilka 2017).

Robert Lefkowitz was born in New York and graduated from Columbia University College of Physicians and Surgeons. After completing further clinical and research training, he joined the faculty at Duke University Medical Center and was soon promoted to professor of medicine. Simultaneously, he also became one of the longest-serving investigators of the Howard Hughes Medical Institute.

Primarily, cells are informed about the external world through hormones that trigger intracellular messages with the help of receptors. For a long time, pharmacological data were consistent with the presence of receptors, but there was skepticism regarding their actual existence up to the 1970s. Jointly with Brian Kobilka, Lefkowitz's research answered the questions by identifying the G protein-coupled receptor family, essentially detectors of odors, hormones, neurotransmitters, and other ligands, demonstrating their existence and describing their mechanism of action. Three senses, taste, smell, and vision, are mediated by these receptors. The research of Lefkowitz sequenced proteins of the receptor molecule and explored its structure. Today, it is recognized that about half of existing drugs act as agonists or antagonists on these cell receptors. The better understanding of the receptors opened up opportunities for designing new drugs much faster.

"If you look at the history of science, important discoveries, changes often come from outsiders" as it was pointed out by Lefkowitz. After receiving the Nobel Prize, the chair of the Duke University chemistry department wrote a letter inviting him to join the department, which he accepted with the condition of not attending the faculty meetings. The American Chemical Society also invited him to become a member. According to his university president, the physician Lefkowitz was discovered as a chemist only after receiving the Nobel Prize in Chemistry.

Evidence-based Public Policies for the Knowledge Society

The surge of science and technology has repeatedly shown its immense power on charting the future of people and society. Policies based on sound scientific evidence can be particularly effective in improving public health and promoting economic development. With the growing public appreciation of research, scientists enjoy the high social prestige and greatly expanded opportunities but also increased responsibilities.

Looking at the 10 greatest public health achievements of the twentieth century can make the many benefits of sound public policies abundantly clear (Centers for Disease Control and Prevention (CDC) 1999). All listed landmark achievements have not only solid scientific foundation but, during the phase of implementation, also benefited from well-informed and evidence-based policymaking (e.g. widespread immunizations, motor vehicle safety, tobacco control, safety requirements at the workplace, fluoridation of drinking water, food safety, and family planning).

One of the most compelling examples is the remarkably successful tobacco control campaign. From the early 1950s, when the cigarette was recognized as a cancer-causing agent, the impact of evidence on public policies can be traced well. Examples of tobacco control actions of the US government include the following: in 1967, the Federal Communications Commission ruled that the fairness doctrine applies to cigarette advertising and stations advertising cigarettes must provide airtime to antismoking messages. In 1971, cigarette advertising became prohibited on radio and television. In 1975, cigarettes were discontinued in rations given to soldiers and sailors. In 1987, Department of Health and Human Services made its facilities smoke-free. In 1994, Occupational Safety and Health Administration introduced regulation to prohibit smoking in the workplace. In 2010, the law prohibited cigarette advertising by referring to "light," "low," "mild," or any similar descriptor.

Concepts of Evidence. In scholarly discussions, scientific evidence is reproducible, generalizable, and peer-reviewed information. It is commonly assumed that reproducibility and generalizability are supported by data, facts, or indisputable logical deduction (e.g. experimental results, proven mathematical formulas). In many ways, properly vetted scientific evidence is as close to reality as possible.

Mathematical proofs are known to last for thousands of years without losing their validity. The prevailing article of Sackett on evidence-based medicine (Sackett et al. 1996) urged avoiding nonexperimental approaches when asking questions about medical therapies. While such definition has a ring of obviousness, clearly expression and reception of evidence remain profoundly human functions.

In evaluating the health effects of some chemicals in the environment, there are a large variety of sources of evidence when the principle of engaged research is applied (e.g. epidemiology, chemical structural analysis, laboratory studies with rodents, *in vitro* cell line studies, primate research, and wildlife observations). Individual studies may provide some valuable but limited evidence. When the evidence from the diverse sources is converging to support a particular scientifically sound conclusion, there is an opportunity for evidence-based policy formulation (Krimsky 2005).

Legal proceedings apply a very different standard of evidence regarding admissibility of expert witnesses' testimony during US federal legal proceedings. A witness who is qualified as an expert by knowledge, skill, experience, training, or education may testify in the form of an opinion or otherwise if (i) the expert's scientific, technical, or other specialized knowledge will help the trier of fact to understand the evidence or to determine a fact in issue, (ii) the testimony is based on sufficient facts or data, (iii) the testimony is the product of reliable principles and methods, and (iv) the expert has reliably applied the principles and methods to the facts of the case (Rules 702 Testimony by Expert Witnesses; Federal, Federal Rules of Evidence 2017).

The role of stories can be the particularly challenging aspect of the political process. In conversations, scientists often make references to samples, populations, esoteric concepts, and jargon. On the other hand, individual stories and the difference certain policies can play a key role. Stories represent one of the most compelling forms of communication. Furthermore, stories are also useful to connect abstract scientific theories and observations to individual lives and public health. This is one of the reasons why the ring of evidence described in the earlier chapter includes not just scientific theory and sample-based experimental results but also illustrative stories.

Science has its fair share not just by improving understanding of major issues and generating thoughtful options for action but also by creating confusion and controversies. Some of the confusion can be beneficial when relevant scientific evidence is unavailable or major studies contradict each other. Discouraging politicians from enacting policies in the absence of supportive evidence is probably beneficial (e.g. mandating longer hospital insurance coverage when there are no demonstrated benefits of longer hospital stay).

At the same time, it does not build respect for science when experts present highly confident but opposite scientific conclusions on two sides of an issue. Studies indicate that people overestimate their actual performances on difficult tasks and that uncertainty can be covered up by overconfidence (Moore and Healy 2008). Occasionally, trial lawyers and legislators make the sarcastic point that "for every PhD, there is an equal but opposite PhD" (sometimes called Gibson's law cited by Singh 2001).

As mentioned earlier, Hannah Arendt, one of the most influential philosophers and political thinkers of the twentieth century, pointed out that, unlike opinion, the truth is "beyond agreement, dispute, opinion or consent" (Arendt 1967). Truth does not depend on how many people accept it. She also highlighted the worrisome tendency of "blurring of the dividing line between factual truth and opinion" and in that way transforming factual truth into a simple opinion that is subject to debate.

Evidence-based Policy Development. Our understanding of the world, the health of the public, and economic progress are tremendously influenced by research discoveries. As a result, there is a need to cultivate public discourse and foster policies to recognize the implications of new scientific discoveries and promote beneficial innovation.

Public discussions and the political process are famously complex, confusing, and unpredictable. In democracies, large numbers of policy proposals are generated all the time, but only a small fraction of them become actual policies. In the US Congress, the number of introduced bills is well over 10 000 in any given cycle, but fortunately, less than 5% of them become public laws. Considering a large number of policy proposals and bills, many legislators are not familiar with the issues before they vote. The lengthy process creates many opportunities not only for failures but also for public input, including advice from scientists with pertinent expertise.

The perennial classic case of data-driven public health advocacy comes from the cholera epidemic in the mid-nineteenth century in London. English physician John Snow (1855) mapped out the locations of cholera cases and demonstrated uneven distribution across the city. The evidence pointed to the public water pump on Broad Street as one of the sources of the outbreak. Subsequently, he convinced the local council to remove the handle from the Broad Street pump, leading to a decline in cholera cases of the surrounding area. Soon afterward, the message spread across London, and others also started to advocate the provision of clean water.

In the age globalization and pressure to innovate, policy transfer, lesson drawing, policy convergence, and policy diffusion are increasingly recognized opportunities. In cases of transfer, knowledge about policies, administrative arrangements, institutions, and ideas in one political setting are used to develop policies in another (Dolowitz and Marsh 2000). To take one example, the idea of the pension system, government-run financial support for older and disabled members of society, started in Bismarck's Germany in 1889 and spread throughout the industrialized world afterward. More recently, there have been innumerable examples of policy transfers (e.g. Silicon Valley-style innovation support policies from the United States to several other countries, patient data protection policies from Europe to the United States, or welfare-to-work programs from the United States to Britain).

Advocacy by Scientists. Complexities of the political process increase the stakes and also responsibilities of those who are familiar with various issues, including scientists, to speak up and share the benefits of their knowledge in the public discourse. The general public needs to be reminded of the value of well-substantiated scientific evidence in decision-making, including public policies and legislation.

Scientists need to develop a basic understanding on how public policies impact success or failure in the knowledge society and how policy development can be appropriately influenced. Such skills should include the ability to respond with clear and comprehensible information to questions of community leaders and the general public. It should also include the responsibility of alerting the public and legislators when the accumulating evidence urges societal action. In other words, scientists should be prepared to provide answers, sometimes even before questions are asked.

Recognizing the influence of targeted policies on public health, there have been numerous initiatives for the development of model laws to assist states and other jurisdictions in addressing major public health needs. A review of 107 model public health laws showed that the most common models were for tobacco control, school health, and injury prevention (Hartsfield et al. 2007). Unfortunately, only 6.5% of the model laws provided appropriate reference to scientific information (e.g. research-based guidelines).

The somewhat emotionally driven grassroots movements of the political process can represent a particular challenge to the witnessing scientists. For example, mastectomy length of stay health insurance coverage became one of the prominent discussions in the US 103rd Congress. The political pressure was enormous: a newspaper survey of 225 patients was published stating that 100% were outraged. In the US Congress, a bill requiring 48 hours of hospital coverage attracted 218 cosponsors. In the political debate about mastectomy length of stay, scientific studies and expert opinions showed something very different: the relevant NIH consensus statement did not mention the length of stay as an issue of health-care quality. A clinical research study by surgeons concluded that "discharge on the day following surgery safe and cost-effective" (Clark and Kent 1992). Ultimately, peer-reviewed research publications were mostly disregarded in the political debate.

A particular area of policy advocacy by scientists is the lobbying for better working conditions and more funding for research. In the age of increased bureaucratization of research and hyper-competition, it is quite understandable when scientists advocate for regulatory changes that lessen the burden on creativity and increase support for innovative research. In recent years, the American Association for the Advancement of Science (AAAS), the world's largest general scientific society, and Research!America in the life sciences field have been particularly effective coalition builders in the advocacy for science.

Advocacy for more funding competes with other industries and constituencies. Besides appropriate networking, the key to success is the demonstration of societal benefits that can be expected from increased funding. Often, the point of reference is an international comparison on supporting research or promoting science-based innovation. As recipients of public funds, scientists have the obligation to explain to taxpayers on how their resources are being spent. Again, broader understanding of the ultimate outcomes of science should be a distinct advantage: impact on future research, public health improvement, and economic development.

References

Alberts, B., Kirschner, M.W., Tilghman, S., and Varmus, H. (2014). Rescuing US biomedical research from its systemic flaws. *Proceedings of the National Academy of Sciences* 111 (16): 5773–5777.

Arendt, H. (1967). Reflections: truth and politics. *The New Yorker* 43 (25): 49–88.

Aristotle (1837). Cited by American Annals of Education and Instruction, vol. 7, p. 53.

Banfield, R., Eriksson, M., and Walkingshaw, N. (2017). *Product Leadership: How Top Product Managers Launch Awesome Products And Build Successful Teams.* Sebastopol, CA: O'Reilly Media.

Bush, V. (1945a). As we may think. *The Atlantic Monthly* 176 (1): 101–108.

Bush, V. (1945b). *Science, the Endless Frontier: A Report to the President.* Washington, DC: US Government Printing Office.

Centers for Disease Control and Prevention (CDC) (1999). Ten great public health achievements – United States, 1900–1999. *MMWR. Morbidity and Mortality Weekly Report* 48 (12): 241.

Clark, J.A. and Kent, R.B. 3rd (1992). One-day hospitalization following modified radical mastectomy. *The American Surgeon* 58 (4): 239–242.

Collins, F.S. and Tabak, L.A. (2014). NIH plans to enhance reproducibility. *Nature* 505 (7485): 612.

Colyvas, J.A. (2007). From divergent meanings to common practices: the early institutionalization of technology transfer in the life sciences at Stanford University. *Research Policy* 36 (4): 456–476.

Dolowitz, D.P. and Marsh, D. (2000). Learning from abroad: the role of policy transfer in contemporary policy-making. *Governance* 13 (1): 5–23.

Federal Rules of Evidence (2017). Michigan Legal Publishing Ltd.

Goldstein, J.L. and Brown, M.S. (2012). A golden era of nobel laureates. *Science* 338: 1033–1034.

Hartsfield, D., Moulton, A.D., and McKie, K.L. (2007). A review of model public health laws. *American Journal of Public Health* 97 (Suppl. 1): S56–S61.

Kitamura, T., Ogawa, S.K., Roy, D.S. et al. (2017). Engrams and circuits crucial for systems consolidation of memory. *Science* 356: 73–78.

Kobilka, B. (2017). You can only be successful if you're happy at what you're doing. Brian Kobilka, Nobel Laureate. https://www.youtube.com/watch?v=v5RLICHv CFE&feature=youtu.be (accessed 7 November 2017).

Krimsky, S. (2005). The weight of scientific evidence in policy and law. *American Journal of Public Health* 95 (S1): S129–S136.

Lefkowitz, R. (2012). The chemistry of discovery. *Duke Magazine* (5 November).

Lindelof, P. and Lofsten, H. (2004). Proximity as a resource base for competitive advantage: university-industry links for technology transfer. *Journal of Technology Transfer* 29 (3–4): 311–326.

Moore, D.A. and Healy, P.J. (2008). The trouble with overconfidence. *Psychological Review* 115 (2): 502.

NIH Research Portfolio Online Reporting Tools (RePORT) (2017). https://report. nih.gov/ (accessed 30 June 2018).

Pareto, V. (1896). *Cours d'économie Politique*, reprinted as a volume of Oeuvres Completes. Droz, Geneva, 1965.

Pollitt, A., Potoglou, D., Patil, S. et al. (2016). Understanding the relative valuation of research impact: a best–worst scaling experiment of the general public and biomedical and health researchers. *BMJ Open* 6 (8): e010916.

Sackett, D., Rosenberg, W., Gray, J.A.M. et al. (1996). Evidence-based medicine: what it is and what it isn't. *British Medical Journal* 312: 71–72.

Sheff, D. (1988). Izzy, did you ask a good question today?. *New York Times* (19 January).

Singh, R.S. (ed.) (2001). *Thinking About Evolution: Historical, Philosophical, and Political Perspectives*, vol. 2. Cambridge: Cambridge University Press.

Snow, J. (1855). *On the Mode of Communication of Cholera*. London: John Churchill.

Tonegawa, S. (1983). Somatic generation of antibody diversity. *Nature* 302 (5909): 575–581.

Zachary, G.P. (1997). *Endless Frontier: Vannevar Bush, Engineer of the American Century*, 261. New York: Free Press.

22

International Collaboration and Competition

[A] new chapter has been opened...not by the work of a single investigator, but by that of many...who have not been influenced in their work by political boundaries and, distributed over the whole surface of the earth, have devoted their powers to an ideal purpose, the advance of knowledge.
Willem Einthoven (1925)[1]

As summed up by the prophetic conclusions of Willem Einthoven's Nobel lecture in 1925, the pursuit of science has always been a profoundly universal and multicultural endeavor. From the *Royal Library of Alexandria* through the four great inventions of ancient China (compass, gunpowder, papermaking, and printing) to today's international scientific collaborations, conferences, and commerce, the open, universal, and inclusive nature of scientific research and technology development has been one of its greatest cool factors. As also expressed from the biotech side by Robert Wood Johnson I, the founder of Johnson & Johnson, "We are all fortunate, in that we are engaged in manufacturing products to be used throughout the world for the relief of pain and suffering" (1908).

The pursuit of science not only helps us understand nature and society but also connects with the global community. For the first time, Koch formulated his postulates about linking bacteria to the cause of an infectious disease in a presentation at the 10th International Medical Congress in 1890. The most prestigious scientific awards are also international (e.g. Nobel Prize, Lasker Award, Japan Prize, Gairdner Foundation International Award). In 2016, all six US-based Nobel Prize winner scientists were foreign born.

Today, collaboration and competition coexist in international science. The many transcontinental joint projects, the constant circulation of talents, ideas, and publications, and the numerous discoveries that have multinational roots

---

1 *Nobel Lectures, Physiology or Medicine 1922–1941* (1965). Amsterdam: Elsevier Publishing Company.

Innovative Research in Life Sciences: Pathways to Scientific Impact, Public Health Improvement, and Economic Progress, First Edition. E. Andrew Balas.
© 2019 John Wiley & Sons, Inc. Published 2019 by John Wiley & Sons, Inc.

show the power of formal and informal collaboration. The essence of international scientific competition is the relentless race to make the landmark discoveries first. Many leading scientists are not only good collaborators but also vigilant observers of their international competitors. As the time-honored wisdom states, competition makes us better and helps to bring out the best in people.

When Watson and Crick started working on the structure of the DNA in Cambridge, the dreaded competitor was Nobel laureate Linus Pauling at the California Institute of Technology (Caltech). Making matters more difficult, Pauling got a head start as he was already an expert in studying the nature of the chemical bond and the structure of complex substances. Later, Watson recalled the start of the DNA race in his book: "Within a few days after my arrival, we knew what to do: imitate Linus Pauling and beat him at his own game" (Watson 1998).

Scientific competition can also be illustrated by the seminal paper of Pyotr Ufimtsev, a Russian mathematician, demonstrating that the strength of a radar return is proportional to the edge configuration of an object, not its size (Ufimtsev 1971). In the 1970s, Lockheed analysts reviewing foreign literature found Ufimtsev's paper openly published in the otherwise secretive Soviet Union and realized that a large, stealthy airplane could be built based on this principle (i.e. the F-117A Nighthawk). Intriguingly, Ufimtsev later received for his discoveries both the USSR State Prize and the Leroy Randle Grumman Medal. The impact of Cold War also illustrated the power of competition as federal support for research substantially increased in the United States after the Soviet launch of Sputnik.

Scientific research is a dynamically expanding sector worldwide. Research is not just becoming a basic requirement of university faculty appointments, but the number of research institutions and their workforce is also rapidly growing. According to the World Bank and UNESCO statistical definition, "researchers are professionals engaged in the conception or creation of new knowledge, products, processes, methods and systems, as well as in the management of these projects" (Organization for Economic Co-operation and Development 2001).

It is increasingly recognized that a country's global leadership can be built on its scientific and technological superiority. Correspondingly, the number of people working in the research enterprise has been steadily growing worldwide. In 2013, there were 7.8 million full-time equivalent researchers. Between 2007 and 2014, the growth was 20% in the number of researchers. The worldwide distribution is quite uneven as 72% of researchers are in the Big Five countries (China, European Union, Japan, Russian Federation, and the United States). Researchers account for 0.1% of the global population (UNESCO 2015).

The Flagship Role of Research Universities

International competition and collaboration need to be studied in a larger context to recognize factors that are accelerating or impeding innovation. In academic knowledge production, entrepreneurial universities and their various collaborations with industries are in focus. The triple helix model of university–industry–government is an attempt to conceptualize essential relationships in studies of innovation in the knowledge-based economy at the national and international levels (Etzkowitz and Leydesdorff 2000). It captures the dynamics of both communication (knowledge base) and organization (knowledge infrastructure).

Key questions of the knowledge-based economy need to be addressed, including the dynamics of economic wealth generation, knowledge-based novelty production, and geographic variations (Leydesdorff and Meyer 2008). The triple helix model effectively highlights the intertwining relationships that define success or failure in the knowledge society. Studies of variation indicate that new knowledge tends to be produced in regions of higher population density, but only if there are no counteracting or limiting policies.

Research universities are increasingly recognized as sources of breakthrough ideas and technologies that can accelerate social and economic development. The traditional role of universities in educating professionals for the future is supplemented by a more direct and immediate contributory role in the knowledge society.

International assessments of research university performance are gradually turning toward innovation achievements of scholarly activities. One of the most noted assessment systems is the Reuters Top 100: The World's Most Innovative Universities. This annual ranking considers patents filed, the percentage of patents granted, and commercial impact of research as measured by publications cited in patent applications.

Country-by-country distribution of *most innovative universities* reflects international competition. Based on the Reuters top 100 ranking in 2017, Figure 22.1 shows the country distribution of the 100 innovative universities (Ewalt 2017). Obviously, the statistics need to be assessed not just regarding the number of universities but also the population size and economic strengths of various countries. There are two particularly striking aspects of the distribution of the most innovative universities. The United States is represented by nearly half of the top 100 most innovative universities.

The lack of representation of regions of developing countries including South America and most of Africa is particularly striking. This is not about the shortage of talents as accentuated by the many physics, chemistry, and physiology Nobel Prize winners from Egypt, Brazil, India, Morocco, South Africa, or Venezuela. To facilitate local response to local needs, create more equitable

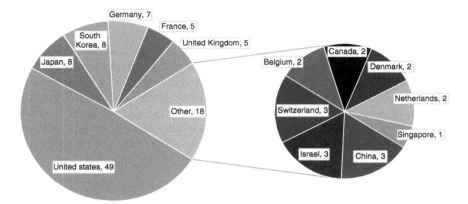

Figure 22.1 Country distribution of the 100 most innovative universities. *Source:* Data from Ewalt (2017).

Aziz Sancar is a Turkish-American molecular biologist specializing in DNA repair and regulation of the circadian clock. His parents had no education but strongly supported education for their children. Sancar studied at Istanbul University (MD) and the University of Texas, Dallas (PhD). Currently, he is a professor at the University of North Carolina, Chapel Hill.

In his studies of bacterial DNA damage, Aziz Sancar showed how certain protein molecules, repair enzymes, can correct errors in DNA damaged by ultraviolet (UV) light. When this essential repair system is defective, people exposed to sunlight develop skin cancer. Together with Paul Modrich of Duke University, Sancar demonstrated how other substances can also damage the nucleotide excision repair system and how cells recognize them. The discovery of nucleotide excision repair improved understanding of how the living cell works and how we can develop better treatments that protect against DNA damage.

Sancar's research has greatly enhanced our understanding of essential processes of basic biology, cancer, and aging. In 2015, he was awarded the Nobel Prize in Chemistry along with Tomas Lindahl and Paul L. Modrich for their studies of DNA repair. He is also an elected member of the American Academy of Arts and Sciences, the US National Academy of Sciences, and the Turkish Academy of Sciences. He is the first Turkish-born scientist to win a Nobel Prize.

In his telephone interview following the announcement of the 2015 Nobel Prize, Sancar made the comment "I am of course honored to get this recognition for all the work I've done over the years, but I'm also proud for my family and my native country and my adopted country. Especially for Turkey, it's quite important" (Smith 2015). He also noted in his autobiography: "I believe that there are three characteristics essential for a successful scientist: creativity based upon knowledge, hard work, and perseverance in the face of failure" (Sancar 2014).

opportunities, and engage more talents, better supporting research universities of the developing world should become one of the top priorities internationally.

The Next Silicon Valleys: What Works and How It Works

Major scientific breakthroughs and practical innovations often come from the intellectually fertile environment of major institutions and densely populated areas, rarely in isolation and seclusion. Characteristically, 18 of the top 20 technology companies are located in two centers of gravity: the US West Coast and the east coast of China (Candelon et al. 2018). They benefit from rich ecosystems of research universities, startups, suppliers, venture capital, and customers.

When it comes to centers of innovation, *Silicon Valley* in California is considered one of the most admired success stories, a veritable destination of international pilgrimage. Silicon Valley, the home of fast-paced actions and expectations of quick results, sometimes characterized as the high-velocity labor market with its own legal and economic framework. Certainly, there is a unique social climate behind the major technology innovations of Silicon Valley.

Fascination with the achievements of Silicon Valley has been truly global, and more and more countries recognize the importance of establishing and promoting similar clusters of innovation. Over the years, Indian PM Modi, French President François Hollande, and leaders from Israel, Ireland, New Zealand, Turkey, Russia, the Netherlands, Lithuania, and Malaysia visited Silicon Valley, among others.

Many countries are actively supporting the development of *clusters of technology* innovation like France's Station F in the heart of Paris; Silicon Fen (otherwise called the Cambridge Cluster) in the United Kingdom; Zhong Guan Cun technology hub in Beijing, China; Skolkovo Innovation Center near Moscow, Russia; and many others. There is certainly a sense of excitement and special opportunities in these clusters of innovation that are modeled after Silicon Valley.

There has been a profound shift in the appreciation of *entrepreneurship* over time. In the past before 1990, entrepreneurs were often viewed as anomalies and outsiders, just small businesses that are not terribly important. After witnessing the tremendous economic growth coming through technology firms in the United States, the prevailing view is that "more startups are better" and there is little that entrepreneurship cannot achieve (Braunerhjelm 2014). Indeed, there has been increasingly convincing empirical literature suggesting that small- and medium-sized enterprises are powerful job creation machines that also generate a disproportionately large share of innovations. In 2002, public funding to small firms already exceeded that given to universities in the United Kingdom. A more recent study of Sweden suggested a similar pattern of support (Lundström et al. 2014).

It is unmistakable to see that leading research universities that are known to be prime producers of cutting-edge science are also forming alliances with clusters of innovation and venture capital communities. The close collaboration between Silicon Valley and the academic life sciences community manifests partly by luring talented faculty members to industry and partly by actively developing projects with research universities (Hayden 2015).

In the development of local policies, clusters of entrepreneurship and innovation are gradually getting higher priority than attracting large enterprises. Spatial clustering can be observed in all major industries, and there are studies showing similar trends in the innovation enterprise. In such localities, expert workforce, intercompany exchanges of ideas, and labor mobility all play key roles in promoting robust growth. More and more politicians talk about the creation of next Silicon Valley as it is becoming the dream economy of innovation.

There is a definite relationship between city employment growth and initial entrepreneurship. According to the cross-sectional studies of Glaeser et al. (2012), cities with greater startup shares in 1982 were associated with greater subsequent employment growth over the 1982–2002 period. Kerr (2010) studied city patenting growth after breakthrough inventions. Analyses of the 10 largest cities per technology split by their breakthrough share in 1975–1984. Breakthrough patents were defined as the top 1% of each technology's patents during the same period as measured by citations subsequently received. Based on the observations, there has been significant city growth in US patenting following breakthrough technologies.

As Chatterji et al. (2014) pointed out, there are various levels and dimensions of policy actions that support innovation and startup success. In providing government support, the two key issues are the geographic location and specificity of recipient selection. Based on the level of targeting, policy actions can be general, industry specific, or firm specific. As governments and policymakers are not very good at picking winners, firm-specific interventions appear to be the most hazardous and least reliable. Industry-specific policy setting tends to be more effective but can also be challenging. Choice needs to be made between supporting old industries and new industries. Ultimately, industry-specific innovation policies have to support well-selected sectors to be effective (e.g. Boston innovation zone).

At the national and state levels, policy actions to promote innovation clusters can include changes in the overall tax code, improved patent policy, relaxed noncompete clauses, industry-specific tax provisions, research subsidies, loan guarantees for specific firms, and others. County and city policies that promote innovation may include citywide tax and business regulations, interurban transportation, tax breaks, contracts tied to local firms, empowerment zones, local infrastructure development, industry-specific local infrastructure, firm-specific infrastructure development, and many others.

The regulation of new entry and removal of noncompete clauses can greatly influence entrepreneurship and innovation. Enforceable noncompete clauses are considered obstacles to the labor movement and information sharing among companies, and therefore they are obstacles to innovation. Similarly making new entries to the market more difficult creates obstacles of launching startups, and well-established status quo companies will have major advantages.

Government, policymakers, entrepreneurs, and venture investors all play key roles in creating and growing clusters of innovation (Engel 2015). Strong research universities and pro-business policies are also essential assets in the development of innovation and entrepreneurship zones.

Recently, traditional Silicon Valley IT companies started aligning themselves with life sciences. Google multiplied spending on life sciences research. The promising side of today's partnership between Silicon Valley and life sciences is the venture capital-type expectation of developing workable, useful products on a fast-paced timeline. The companies explore a range of wearable devices and other health-care projects. The innovative technology projects include development of wireless glucose sensors in contact lenses, a promising research idea that was already licensed by Novartis. IBM Watson, the signature artificial intelligence project of the company, is now actively promoted and tested in the field of health-care applications. Apple introduced ResearchKit for researchers to write apps that collect patients' data from mobile phones.

Health care has always been an attractive area due to its enormous market size, public appeal, and revenue potential. However, no company should expect a simple and easy ride to success. Many high-tech adventures in health care will not be fully successful. Between the 1960s and 2000, numerous IT companies experimented with the presumably lucrative hospital information system development (e.g. DEC, IBM, Hewlett-Packard, Honeywell, and many others). However, most of these early ventures into the development of electronic health records failed to live up to expectations and ended in the abandonment of the field. Today, systems of electronic health records are indeed spreading but predominantly coming from dedicated and specialized health information system vendors.

Christiaan Barnard was a South African clinical researcher and cardiac surgeon who performed the first human-to-human heart transplant in 1967. Barnard came from a family of very modest means; his father was a missionary with minimal salary. The family had four sons, sending three through college. Christiaan Barnard got his medical degree from the University of Cape Town.

When Dr. Barnard designed the pioneering heart transplant surgery, a pivotal facilitating factor was his appointment as the head of the Department of Experimental Surgery at Groote Schuur Hospital while also holding a joint appointment at the University of Cape Town. He was an outstanding candidate for the post as he has been a hard-charging and determined pioneer of surgical innovation. His early work on intestinal atresia is a much-cited reference point in the surgical correction of this developmental malformation. He made major contributions to heart valve surgery, the treatment of complicated congenital heart defects, and cardiac hypothermic storage, among others (Toledo-Pereyra 2010).

In the early 1960s, several surgeons were exploring the possibility of heart transplant operation and demonstrated that the technique was feasible in animals. However, the ethical and legal dilemma of removing the beating heart of a still living donor was daunting (Hoffenberg 2001). The contrarian Barnard decided to proceed with the removal of the heart of a severely brain-injured patient. This decision led to the first successful heart transplantation, but the recipient died 18 days after the surgery due to rejection. Just 6 weeks later, he performed the second heart transplant surgery, and the patient survived for 18 months. Based on recent WHO statistics, more than 5000 heart transplantations are performed worldwide annually.

Mostly through the downstream effects, his work redefined the concept of life and death, the notion of brain death, and also the circumstances when surgeons can remove and replace the beating heart of a living person. Barnard was a highly talented experimenter with a big ego and a penchant for publicity. He famously summed up the dramatic turn of his life after the first heart transplantation: "On Saturday, I was a surgeon in South Africa, very little known. On Monday, I was world-renowned" (Cooley 2001).

Scientific Innovation and Economic Competition

In the worldwide statistics of the research enterprise, Japan, United States, Korean Republic, Sweden, Germany, Canada, Netherlands, United Kingdom, and France lead regarding the number of patents per 10 million inhabitants between 2009 and 2013. The same countries were also leading in the relative intensity of scientific publications per million population.

Various countries follow very different priority settings, and the quest for a growth strategy that works continues. What works and how it works are not trivial questions, but they are essential to understanding the dynamics of innovation and the role of entrepreneurial, innovative research universities.

Between 2009 and 2013, the economy of Israel grew at an impressive pace by 28% (Getz 2017). The strong performance was largely related to the growth of the high-tech sector and the tremendous investments in science, technology, and startups (Baskaran 2017). Progressively, Israel is called the startup nation. A particular challenge is an evolving gap between knowledge-and technology-intensive industries and traditional industries. The first one is providing much higher wages and greater growth rates, while the second one is lagging in productivity.

China is implementing an ambitious program to transition from the world's factory status to a technology- and innovation-based economy. Simultaneously, the country is addressing environmental concerns and labor regulation issues. In China, the gross domestic revenue spending on R&D is 2.08% exceeding the comparable ratio of the European Union (Baskaran 2017). The number of publications in medical sciences more than tripled between 2008 and 2014 (Cao 2017). The research and development sector is rapidly growing in China. The country aims to make scientific papers of Chinese scientists among the most cited and also becoming one of the top four source countries of patented innovations.

Based on all major indicators of output, the research and development sector has also been rapidly expanding in India. There has been significant growth in the number of patents granted nationally or abroad and in the share of high-tech exports in total exports. The number of scientific publications from India more than doubled between 2005 and 2014 (Mani 2017). India is particularly successful in developing high-tech industries, particularly space technology, pharmaceuticals, and computer and information technology services. While the IT companies tend to be foreign owned, the highly successful Indian biotech industry has been largely homegrown. Challenges include the uneven concentration of innovation in nine industries and six states.

There has been noteworthy growth in the research enterprise of Brazil. The number of PhD degrees obtained in Brazil grew from 8 982 in 2005 to 15 287 in 2013 (de Luna Pedrosa and Chaimovich 2017). Over the same period, the number of scientific publications has more than doubled. There was a jump in the number of Brazilian journals being tracked by the Thomson Reuters. The disparities of growth are reflected by the fact that a disproportionately large part of Brazilian PhD's scientific papers and new patents come from the state of São Paulo in Brazil. São Paulo has a well-funded system of state universities and a very active São Paulo research foundation; both have very large autonomy and are steadily funded from a fixed share of the state's sales tax revenues.

International Variation in Research Policies

The research enterprise is expanding, and many different views and cultural perspectives enrich the scientific discourse. Obviously, growing inclusiveness in the production of science is generating innumerable benefits globally. Meanwhile, *international variations* in ethics and regulatory standards of countries represent an enormously challenging aspect of international competition.

Overly relaxed ethics standards of certain countries give a seeming advantage to less scrupulous research while also raising some very troubling ethics questions. As a result of the globalization of clinical research, the number of countries serving as overseas trial sites more than doubled in 10 years between 1995 and 2005 (Glickman et al. 2009). The lower cost of clinical research studies is a powerful driver.

On the other hand, the ethnic and genetic makeup of the sample can be greatly influenced by choice of international clinical trial sites. There have been some unfortunate cases when developed country companies relocated clinical studies to lower standard countries, effectively bypassing well-established human subject protections.

According to a report by the China's Food and Drug Administration (CFDA), 80% of clinical trials data were fraudulent (Woodhead 2016). The report found widespread deception and a chaotic situation in the clinical trial industry. There were cases of discrepancies between original data in the records and submitted files, complete lack of raw data, selective reporting of results, fabricated data, and underreporting of side effects.

In some countries, less sophisticated and more relaxed standards may jeopardize the trustworthiness and humanistic mission of scientific investigations. Particularly in certain areas of stem cell research and genetic engineering, significant issues have been raised by the lower standards of certain countries. The number of stem cell-based clinical trials is rapidly growing, but analyzers of the field pointed out that therapeutic rhetoric must be moderated to reflect scientific reality and prevent high-profile failures (Li et al. 2014).

For example, it became too easy to conduct stem cell research in some countries where life sciences studies have the potential of generating defective embryos and harming human health (King and Perrin 2014). There are calls to harmonize standards for producing clinical therapies from pluripotent stem cells (Andrews et al. 2014). Such harmonization of country standards should cover issues of sourcing tissue from deceased individuals, criteria for the use of human cell lines, and transplants into animals. Other requirements include addressing issues of donor consent, immortal cells, shipping and tracking, privacy and confidentiality in cases of distribution, and also patentability of stem cell discoveries. There is a need to reduce the likelihood of the therapeutic misconception. The rush from bench to bedside should be avoided. Ultimately, harmonization of rules should strengthen the reproducibility and safety of various products coming from stem cell research.

While quite startling ethics issues can surface in various countries, it would be a mistake to stigmatize a few selected countries as ethics and human subject protection controversies come up periodically everywhere. The case of the celebrated South Korean stem cell researcher who committed major fraud or the numerous article retractions and debarments by NIH of US researchers and many others can illustrate the truly global distribution of research ethics concerns.

Global Networks to Accelerate Research

Engaged scientists are open up for broad and international collaboration. When an experiment fails, discussions with other researchers can effectively help in generating new ideas and overcome any setbacks.

To make a difference in science and society, the researcher needs to have a *global network* of colleagues around the world. In developing such networks, the scientist looking around can find those researchers who do similar but not necessarily identical work and have complementary expertise. These types of cross-border partnerships can also be used to develop competitive grant applications together.

Today's advanced science demands international collaborations in many areas. The expenses of many promising experiments are exorbitant, and only multi-country effort can effectively finance it. Especially in physics, there are many famous examples of experiments that need to be accomplished, but no single country has sufficient resources to realize them. Such collaborations include the International Space Station or the CERN Large Hadron Collider, representing the collaboration of thousands of scientists from over 100 countries, as well as hundreds of universities.

There are things that academia can do better, and there are things that the biotech industry can do better. Therefore, the search for potential international partners and collaborators should extend to industry as well.

In life sciences, there are many shining examples of international collaboration. Perhaps, the best-known example is the multinational collaboration effort that was needed to sequence the human genome. The discovery of CRISPR/Cas9 gene editing system also illustrates the power of international cooperation: Japanese, Spanish, French, and US scientists built on each other's work and collaborated in achieving this discovery.

Another contemporary example is the investigation of exosomes, small cell-derived vesicles. They can be found in many biological fluids, including blood, urine, and cell cultures. Historically, they were considered extracellular debris but now recognized as potential mediators of intercellular communication. Their role is attracting the attention of researchers all over the world. Cancer cells release a particularly large number of exosomes (e.g. exosomes released by aggressive subclones) (Schillaci et al. 2017). Correspondingly, their role in promoting tumor growth and potential use as diagnostic biomarkers are being investigated by academic researchers and biotech companies worldwide.

The international collaboration of scientists is in stark contrast with the huge variations in accepting and interpreting scientific evidence. One of the classic examples is tobacco control, an area where extensive research has been done, and a vast amount of evidence is available publicly. In fact, the body of evidence or the level of understanding is the same in every country. However, a comparative study of UK and Japan tobacco policies showed major variations and very divergent policy translations (Cairney and Yamazaki 2017). The mixture of political will and scientific evidence can be very different in various countries. All too often, the evidence does not speak for itself, and scientists have to accept the responsibility of advising community leaders.

Ahmed Zewail was an Egyptian-American Nobel laureate and director of the Physical Biology Center for Ultrafast Science and Technology at the California Institute of Technology. His childhood was spent in Rosetta where the famous stone was discovered and Alexandria, the home of the celebrated library in antiquity. After graduating from Alexandria University, he moved to the United States to complete his PhD at the University of Pennsylvania.

To study atoms within molecules rearranging themselves as they form new molecules, Zewail observed chemical reactions on extremely short, 10^{-15} seconds timescales (called femtosecond). In his elegant pump and probe experiments, he monitored reactions by using ultrashort light pulses at the timescale of chemical reactions at the molecular level. Eventually, he became recognized as the father of femtochemistry by studying atomic-scale dynamics of the chemical bond with the help of ultrafast lasers. An area of application is femtobiology; ultrafast infrared spectroscopy has been used in the study of photosynthesis to investigate the protein response signal to the energy or electron transfer (Di Donato and Groot 2015).

Ahmed Zewail never lost his connection with the great traditions of the Egyptian and Arabic culture. His autobiography points to the brilliant singer Umm Kulthum who had a major influence on his appreciation of music. Her unique music gave him a special happiness, and her voice was often in the background while he was studying mathematics and chemistry (Zewail 1999).

His international awards also include the King Faisal International Prize, the Wolf Prize in Chemistry, Carl Zeiss International Award, and much more. In speeches and articles, Zewail has been a vigorous advocate for modernizing science in the Arabic-speaking world. He reminded people of the historical greatness of their science and urged investment in education and fundamental research (Warren 2016). His drive led to the creation of the Zewail City of Science and Technology in Giza.

"I believed that behind every universal phenomenon there must be beauty and simplicity in its description. This belief remains true today" (Zewail 1999). His Nobel lecture highlighted the joys of multicultural laboratory: "For 20 years I have always looked forward to coming to work with them every day and to enjoying the science in the truly international family" (Zewail 2000).

Innovative Minds with a Sense of Social and Global Responsibility

International competition is emerging to make countries the most attractive setting in which to study and perform research. Such policies should help to develop, recruit, and retain the best and brightest students and scientists from within the country and throughout the world.

As mentioned in Chapter 9, the power of attracting innovative minds is also illustrated by the pattern of Nobel Prizes in the twentieth century. Between 1901 and 1932 the number of Nobel Prize winners for physics and chemistry in Germany was 16, and the comparable number was five in America. Between 1950 and 2000, seven people won Nobel Prizes in Germany, and the comparable number was 67 in America. The extraordinary turn started when the Nazi ideology came to power.

Policymakers of the *knowledge society* want to make sure their country is the premier place in the world to innovate. Such ambitious initiatives include investing in downstream activities like manufacturing and marketing, creating high-paying jobs based on innovation, realigning tax policies to encourage innovation, modernizing the patent system, and ensuring affordable broadband access.

In life sciences, an important milestone of global success is when an innovation becomes universally essential. As defined by the World Health Organization (WHO), essential medicines are "those drugs that satisfy the health care needs of the majority of the population; they should, therefore, be available at all times in adequate amounts and appropriate dosage forms, at a price the community can afford." This concept of the WHO has already been extended into other areas as well (e.g. medical devices, model regulatory frameworks, and others).

Considering the very large number of diseases that exist out there and the small number of diseases that can be effectively treated, most diseases can be rightly viewed as neglected diseases. We need fundamental biomedical research, not because someone knows the end from the beginning but because such research lays the broadest foundation for successful practical innovations. According to Emdin et al. (2015), there is only a weak association between global burden of disease and number of published randomized trials coming from clinical research.

Research in life sciences should pursue opportunities that will improve individual and public health. With the words of Avedis Donabedian, the father of medical outcomes research, "In all my work I have tried to embody the passionate conviction that the world of ideas and the world of action are not separate, as some would have us think, but inseparable parts of each other. Ideas, in particular, are the truly potent forces that shape the tangible world" (Best and Neuhauser 2004).

Sound science is becoming a fundamental human right in terms of pursuing research and also benefiting from discoveries. To advance the knowledge society that is innovative, competitive, and prosperous, there is a need to educate and train people who can go out and make the world a better place. To achieve that, many dispersed centers of intellectual leadership are needed in an interconnected world. Thought leader scientists reach out and make a difference with a great sense of social and global responsibility.

References

Andrews, P.W., Cavagnaro, J., Deans, R. et al. (2014). Harmonizing standards for producing clinical-grade therapies from pluripotent stem cells. *Nature Biotechnology* 32 (8): 724–726.

Baskaran, A. (2017). UNESCO Science Report: Towards 2030. *Institutions and Economies*, pp. 125–127.

Best, M. and Neuhauser, D. (2004). Avedis Donabedian: father of quality assurance and poet. *Quality and Safety in Health Care* 13 (6): 472–473.

Braunerhjelm, P. (2014). *Twenty Years of Entrepreneurship Research: From Small Business Dynamics to Entrepreneurial Growth and Societal Prosperity*. Entreprenörskapsforum. http://entreprenorskapsforum.se/wp-content/uploads/2014/03/20_years_of_e-ship_web.pdf (accessed 5 June 2018).

Cairney, P. and Yamazaki, M. (2017). A comparison of tobacco policy in the UK and Japan: if the scientific evidence is identical, why is there a major difference in policy? *Journal of Comparative Policy Analysis: Research and Practice* (16 May): 1–16.

Candelon, F., Reeves, M., and Wu, D. (2018). 18 of the top 20 tech companies are in the Western U.S. and Eastern China. *Harvard Business Review* (3 May).

Cao, C. (2017). China. In: *UNESCO Science Report: Towards 2030* (ed. A. Baskaran), 125–127. Institutions and Economies.

Chatterji, A., Glaeser, E., and Kerr, W. (2014). Clusters of entrepreneurship and innovation. *Innovation Policy and the Economy* 14 (1): 129–166.

Cooley, D.A. (2001). Memoriam: Christiaan Barnard 1922–2001. *Texas Heart Institute Journal* 28 (3): 165.

de Luna Pedrosa, R.H. and Chaimovich, H. (2017). Brazil. In: *UNESCO Science Report: Towards 2030* (ed. A. Baskaran), 125–127. Institutions and Economies.

Di Donato, M. and Groot, M.L. (2015). Ultrafast infrared spectroscopy in photosynthesis. *Biochimica et Biophysica Acta (BBA)-Bioenergetics* 1847 (1): 2–11.

Emdin, C.A., Odutayo, A., Hsiao, A.J. et al. (2015). Association between randomised trial evidence and global burden of disease: cross sectional study (Epidemiological Study of Randomized Trials – ESORT). *British Medical Journal* 350: h117.

Engel, J.S. (2015). Global clusters of innovation: lessons from Silicon Valley. *California Management Review* 57 (2): 36–65.

Etzkowitz, H. and Leydesdorff, L. (2000). The dynamics of innovation: from National Systems and "mode 2" to a triple helix of university–industry–government relations. *Research Policy* 29 (2): 109–123.

Ewalt, D. (2017). Reuters top 100: the World's most innovative universities – 2017. *Reuters* (27 September).

Getz, D. and Tadmor, Z. (2017). Israel. In: *UNESCO Science Report: Towards 2030* (ed. A. Baskaran), 125–127. Institutions and Economies.

Glaeser, E., Kerr, S., and Kerr, W. (2012). *Entrepreneurship and Urban Growth: An Empirical Assessment with Historical Mines*. NBER Working Paper no. 18333. Cambridge, MA: National Bureau of Economic Research.

Glickman, S.W., McHutchison, J.G., Peterson, E.D. et al. (2009). Ethical and scientific implications of the globalization of clinical research. *The New England Journal of Medicine* 360: 816–823.

Hayden, E.C. (2015). Tech titans lure life-sciences elite. *Nature* 526: 484–485.

Hoffenberg, R. (2001). Christiaan Barnard: his first transplants and their impact on concepts of death. *BMJ: British Medical Journal* 323 (7327): 1478.

Kerr, W.R. (2010). Breakthrough inventions and migrating clusters of innovation. *Journal of Urban Economics* 67 (1): 46–60.

King, N.M. and Perrin, J. (2014). Ethical issues in stem cell research and therapy. *Stem Cell Research & Therapy* 5 (4): 85.

Leydesdorff, L. and Meyer, M. (2008). The triple helix model and the knowledge-based economy. *Scientometrics* 58 (2): 191–203.

Li, M.D., Atkins, H., and Bubela, T. (2014). The global landscape of stem cell clinical trials. *Regenerative Medicine* 9 (1): 27–39.

Lundström, A., Vikström, P., Fink, M. et al. (2014). Measuring the costs and coverage of SME and entrepreneurship policy: a pioneering study. *Entrepreneurship Theory and Practice* 38 (4): 941–957.

Mani, S. (2017). India. In: *UNESCO Science Report: Towards 2030* (ed. A. Baskaran), 125–127. Institutions and Economies.

Organization for Economic Co-operation and Development (2001). *OECD Science, Technology and Industry Scoreboard 2001: Towards a Knowledge-based Economy*. Paris: Organization for Economic Co-operation and Development.

Robert Wood Johnson's letter (1908). www.kilmerhouse.com (accessed 31 October 2017).

Sancar, A. (2014). Biographical. Nobelprize.org. Nobel Media AB 2014. http://www.nobelprize.org/nobel_prizes/chemistry/laureates/2015/sancar-bio.html (accessed 29 October 2017).

Schillaci, O., Fontana, S., Monteleone, F. et al. (2017). Exosomes from metastatic cancer cells transfer amoeboid phenotype to non-metastatic cells and increase endothelial permeability: their emerging role in tumor heterogeneity. *Scientific Reports* 7: 4711.

Smith, A. (2015). Telephone interview with Aziz Sancar following the announcement of the 2015 Nobel Prize in Chemistry (7 October 2015). The interviewer is Adam Smith, Chief Scientific Officer of Nobel Media.

Toledo-Pereyra, L.H. (2010). Christiaan Barnard. *Journal of Investigative Surgery* 23 (2): 72–78.

Ufimtsev, P.Y. (1971). *Method of Edge Waves in the Physical Theory of Diffraction* (No. FTD-HC-23-259-71). Foreign Technology Div Wright-Patterson AFB, OH.

UNESCO (2015). *UNESCO Science Report: Towards 2030 by United Nations Educational, Scientific and Cultural Organization (UNESCO)*, 820p. Paris: UNESCO Publishing.

Warren, W.S. (2016). Ahmed Hassan Zewail (1946–2016). *Nature* 537 (8 September): 168.

Watson, J. (1998). *The Double Helix*. New York: Simon and Schuster.

Woodhead, M. (2016). 80% of China's clinical trial data are fraudulent, investigation finds. *British Medical Journal* 355: i5396.

Zewail, A. (1999). Biographical Les Prix Nobel. In: *The Nobel Prizes 1999* (ed. T. Frängsmyr). Stockholm: Nobel Foundation.

Zewail, A.H. (2000). Femtochemistry: atomic-scale dynamics of the chemical bond using ultrafast lasers (*Nobel Lecture*). *Angewandte Chemie International Edition* 39 (15): 2586–2631.

List of Award-winning Scientists and Serial Innovators

Innovative Research in Life Sciences: Pathways to Scientific Impact, Public Health Improvement, and Economic Progress, First Edition. E. Andrew Balas.
© 2019 John Wiley & Sons, Inc. Published 2019 by John Wiley & Sons, Inc.

Subject Index

Innovative Research in Life Sciences: Pathways to Scientific Impact, Public Health Improvement, and Economic Progress, First Edition. E. Andrew Balas.
© 2019 John Wiley & Sons, Inc. Published 2019 by John Wiley & Sons, Inc.